Signals and Communication Technology

More information about this series at http://www.springer.com/series/4748

Ayman Radwan · Jonathan Rodriguez
Editors

Energy Efficient Smart Phones for 5G Networks

Springer

Editors
Ayman Radwan
Jonathan Rodriguez
Instituto de Telecomunicações
Campus Universitário de Santiago
Aveiro
Portugal

ISSN 1860-4862 ISSN 1860-4870 (electronic)
ISBN 978-3-319-10313-6 ISBN 978-3-319-10314-3 (eBook)
DOI 10.1007/978-3-319-10314-3

Library of Congress Control Number: 2014951142

Springer Cham Heidelberg New York Dordrecht London

Printed on acid-free paper

Springer is part of Springer Science+Business Media (www.springer.com)

Preface

Every new release of iPhone or Android smart phone spurs new applications and services, with advanced display screens to deliver an exceptional quality of experience to the enduser. However, this comes at a cost placing stringent demands on mobile battery consumption and lifetime. Indeed, this trend is set to continue with the advent of 5G services, and the need to be always connected at any place at any time. If we do not adopt a proactive stance, the enduser will be constrained to the nearest available electricity socket which is a stark irony to the philosophy behind 5G communications.

Although 5G systems are still in their infancy, it is widely accepted among the research community and the key stake holders that the evolution toward 5G is believed to be the convergence of Internet services with legacy mobile networking standards leading to the commonly used term "mobile Internet" over heterogeneous networks (HetNets), with very high connectivity speeds. Moreover, green communications will probably play a pivotal role in this evolutionary path with key mobile stake holders driving momentum toward a greener society. Another aspect that can be deduced based on the current trends of emerging services is the continuous rise in the traffic demand.[1] Based on such vision, it is safe to conclude that energy and cost per bit reduction, service ubiquity, and high-speed connectivity are among the main design drivers for next generation networks.

Toward this end, this book aims to address technology trends and challenges leading toward the next era in mobile communications on the terminal side, namely 5G-ready handsets. The book adopts a multidisciplinary and interdisciplinary stance toward handset design, a necessary ingredient if such handsets are to really takeoff in next generation mobile systems. In this context, the scope of this book covers a wide range of research areas spanning from RF design, mobile cooperation, context awareness, to seamless roaming among multiple technologies, all of which working in synergy toward providing energy efficient seamless ubiquitous mobility and

[1] Demestichas, P., Georgakopoulos, A., et al.: 5G on the horizon: key challenges for the radio-access network. IEEE Veh. Technol. Mag. 8(3), 47–53 (2013).

high-speed connectivity within a HetNet environment. The book also addresses potential business models to secure the adoption of the proposed technologies by the key mobile stake holders.

In more detail, the book is organized in a well-defined structure, starting with Chap. 1 which defines prospective green and smart scenarios that will potentially exist in future generations of networking. It is worth mentioning here that the 5G research roadmap is still in its infancy, with commercial deployment only foreseen around 2020. Despite this fact, the majority of the research community agrees on the key design requirements that will drive technology solutions under the 5G umbrella. On one hand, there will be a heterogeneity of communication systems where stakeholders aim to capitalize on existing deployments for providing voice and low data rate services. However, it is expected that this will be complemented by new overlay tiers based on small cells. By reducing the communication distance between the base station and the mobile handsets, small cells are able to provide significant energy saving and high-speed data comparable to the indoor femto cell paradigm available in the market. On the other hand, device-to-device communications are also expected to be a prominent feature of 5G to support the demand for proximity based services while also acting as a vehicle for off-loading traffic from the core network(see Footnote 1). These widely accepted design requirements and technology paradigms provide the inspiration for Chap. 1, that suggests not only candidate networking scenarios for next generation mobile systems based on current roadmaps, but tailor-made to provide high-speed and energy-efficient connectivity to the mobile handset based on the notion of cooperation. These scenarios provide the framework for the technology solutions developed throughout this book and aim to paint a gradual picture of a 5G handset, in terms of requirements, technology features, and capabilities.

Exploiting cooperation and intelligent networking is one piece of the energy saving jigsaw, since a large chunk of the power consumption on the mobile handset is attributable to the radio frequency (RF) circuitry. In fact, if there are no energy saving strategies in place, future mobile handsets could become too hot to handle requiring fan cooling systems; therefore, next generation handsets will unavoidably need to be energy conscious, smart, and multimode in nature to support seamless operation over legacy/future emerging technologies (LTE, LTE+, HSDPA, 3G) in HetNet environments. This vision gives way to stringent design requirements on the RF system design. Addressing the RF perspective to handset design, Chaps. 2 and 3 focus on studying new hardware design concepts that can provide flexible and energy efficient multistandard transceivers. Chapter 2 addresses the global transceiver architecture design and the antenna frontend unit, which is bridged to the RF frontend via a tunable matching network; the latter alleviating the effect of perturbations caused by the user's head or hand. Moreover, Chap. 3 tackles the critical issue of power amplifier design in future mobile handsets. Power amplifiers are considered a key consumer of power in handheld devices. The effect is expected to multiply when ubiquitous broadband connectivity becomes the standard. Chapter 3 hence presents three innovative energy efficient power amplifier designs that aim to reduce the energy consumption in next generation handsets, leading to longer battery lifetime and stronger market penetration.

Chapter 4 shifts the perspective to handset mobility given a rich heterogeneous networking environment. Although this was partially achieved in 4G systems, the handover use-cases are limited and lack coordination in terms of energy efficient mobility. In an attempt to push further the boundaries on network mobility to provide tighter integration between diverse networking standards, we investigate energy efficient handovers that include macro-femto use-cases. In the first instance, this chapter studies specific handover algorithms between cellular networks and WiFi, in addition to handover between macro and femto cells in LTE-A, complemented by emphasizing the network functionalities of both IEEE 802.11 and LTE-A that enable mobility between these networks. Based on this platform, we then go beyond by taking energy efficiency to the forefront and propose energy efficient handover algorithms for integrated WiFi-LTE Femto networks using context information to reduce the overall energy consumption of mobile devices, while maintaining the required QoS.

In addition to heterogeneity, cooperation is extensively used in legacy mobile networks for improving link reliability and coverage through approaches such as relaying. These are proven technologies that are showing promising potential for energy saving through cooperation with proxy mobile devices. In Chap. 5, we tackle the topic of cooperative communications by investigating how cooperation may be used for energy saving by exploiting short-range cooperation in synergy with long-range. The chapter proposes a new cooperative approach based on coalitional game theory that uses this as a tool to increase the overall energy saving gain of the network, rather than solely of the mobile user. The game is also used to fairly distribute the gains among players hence motivating players to cooperate; additionally, the game identifies selfish users and free riders and forces them to be excluded from the cooperation process making this technique attractive for use in real operating scenarios.

Next generation mobile systems are expected to be intelligent in nature, enabling attractive business models such as targeted advertisement, as well as providing a platform for operators to effectively exploit their network resources in an era where spectral resources are at a premium. In this book, we exploit the notion of cognition in the sense of smart networks to gather useful context both in the network and the terminal side in order to promote effective use of battery resources. Toward this end, Chap. 6 elaborates on context information and architecture based on legacy mobile networking standards to deliver the notion of a smart 5G mobile phone, and indeed provides the framework for many of the innovative smart building blocks that are elaborated throughout this book. The chapter then concludes by proposing two discovery algorithms which exploit context information to discover available networks and mobile devices in the vicinity, again the key driver being that the discovery approaches in legacy devices are key consumers of power since they are blind in nature.

Following the footsteps of the previous chapter, Chap. 7 reinforces the notion of smartphones, and how this can be exploited for enhancing network operation.

The chapter elaborates on how mobile devices and networks can use positioning context for enhancing network performance in terms of community-based sequential paging and location-aided scheduling for fractional frequency reuse in LTE-A relay network.

Having proposed multiple energy efficient solutions for mobile handsets, Chap. 8 represents the venue for showcasing the proofs of concept on exploiting cooperation and cognition for energy saving in next generation handsets. Chapter 8 presents three showcases to demonstrate the gains achieved by the proposed solutions in a practical networking environment. The showcases constitute an energy efficient short-range cooperation testbed, vertical handover platform for quantifying energy savings, and a horizontal showcase demonstrating the benefits of the RF hardware building blocks proposed (including envelope tracking power amplifier, Doherty power amplifier, tunable matching network integrated circuit, and a miniature multiband antenna). The details of each showcase are described, along with the scenario to be demonstrated.

Having discussed numerous energy saving solutions at the technical level, it would be difficult to observe the take-up of such technologies in the marketplace without the support of viable business models that highlight the potential benefits in terms of revenue. Therefore, the final chapter of this book discusses the business potential for driving the technology solutions based on cooperation and femto cell integration, as part of the 5G paradigm. We present a number of existing business models, which can be adopted and tailored to the proposed cooperation solutions, alongside new business use-cases for the cooperation scenario as well as the Femto cell scenario. The different players in each business model are defined and the interactions among them are clearly shown through elaborative diagrams.

The motivation for this book is to address energy efficient handsets for next generation networks, which is looming toward the so-called 5G paradigm. It is widely accepted that next generation networks will not only evolve from legacy architectures to preserve market acceptance, but will also include disruptive approaches to ensure innovation and the high-speed and energy-efficient targets that are sought after. In this respect, the scenarios suggested in this book are based on cooperation and cognition that indeed drive the proposed technology solutions and may be considered as pieces of the 5G jigsaw. However, it is true to say that 5G is still in its infancy and the final 5G scenario could somewhat take a different path. However, the solutions we propose are technology agnostic and in fact could coexist with a plethora of underlying technologies.

The editors believe that this book represents a significant step forward on the way toward 5G handsets, and built based on commonly agreed design drivers that are still valid for future generations of mobile networks.

Ayman Radwan
Jonathan Rodriguez

Acknowledgments

This work adopts a visionary stance toward the evolution of 5G handsets placed in motion by key design requirements that include energy efficiency, multimode operation, cooperation, and smart enabling features. These proliferate to all aspects of handset design leading a new generation of mobile devices that are able to provide high-speed data services at any time, place, and anywhere. The inspiration for this timely work emanated from the editors at the Instituto de Telecomunicações (IT) with vast experience in green communications design, that have coordinated major international research efforts on next generation handsets. To provide a complete and energy efficient design from holistic perspective requires the coming together of experts from different scientific disciplines ranging from the RF domain to networking experts, and business modeling. In particular, the editors would like to thank the FP7 ICT-C2POWER project, that has contributed countless hours of collaborative work to make this highly interdisciplinary scientific contribution possible.

The RF architecture plays a pivotal aspect in future handsets, where the power amplifier, antenna unit, and filter design must work in synergy to provide a holistic energy gain. We would like to thank the QREN CATRENE (CA505) BENEFIC (PO Centre 38887) that also provided some of the inspiration for the RF technology challenges tackled in the scope of this book.

Finally, the editors would like to thank the 4TELL research group (IT) who provided technical support throughout the compilation of this book, and last but not least Miss Claudia Barbosa, who has spent endless hours assisting with the final editing.

Ayman Radwan
Jonathan Rodriguez

Contents

About the Editors

Ayman Radwan received his Ph.D. from Queen's University (Canada) in 2009, his Master of Applied Science (M.A.Sc.) from Carleton University (Canada) in 2003, and his B.Sc. from Ain Shams University, in Cairo, Egypt in 1999. In January 2010, he joined Instituto de Telecomunicações (Aveiro, Portugal) as a Senior Research Engineer. Since then, he has specialized in coordinating and managing European projects, within the FP7 and CELTIC Plus programs. His main research interests include radio resource management, quality of service and quality of experience provision, and green communications. He is the author of several articles including journals, conference publications, and patents. He is an active IEEE member, acting as TPC member and reviewer for a number of respected journals, magazines and conferences.

Jonathan Rodriguez received his Master's degree in Electronic and Electrical Engineering and Ph.D from the University of Surrey (UK), in 1998 and 2004 respectively. In 2002, he became a Research Fellow at the Centre for Communication Systems Research and was responsible for coordinating Surrey involvement in European research projects under framework 5 and 6. Since 2005, he is a Senior Researcher at the Instituto de Telecomunicaçoes (Portugal), and founded the 4TELL Wireless Communication Research Group in 2008. The 4TELL group currently constitutes 30 researchers with a Project portfolio that includes 10 ongoing European collaborative research projects. He is author of more than 160 scientific publications, served as general chair for several prestigious conferences and workshops, and has carried out consultancy for major manufacturers participating in DVB-T/H and HS-UPA standardization.

Chapter 1
Toward 5G: Smart and Green Scenarios

Álvaro Gomes, Tiago Moreira, Jacek Kibilda, Marcin Filo, Radoslaw Piesiewicz and Ayman Radwan

Abstract The first wave of 4th Generation systems is finally being deployed over Europe, providing a vehicle for broadband mobile services at anytime, anyplace and anywhere. However, mobile traffic is still growing (Cisco in Visual Networking Index: Global Mobile Data Traffic Forecast Update 2012–2017, [1]) and the need for more sophisticated broadband services will further push the limit on current standards to provide even tighter integration between wireless technologies and higher speeds, requiring a new generation of mobile communications: the so-called 5G. The evolution towards 5G is considered to be the convergence of internet services with legacy mobile networking standards leading to the commonly used term "mobile internet" over heterogeneous networks (HetNets), with very high connectivity speeds. In addition, green communications seem to play a pivotal role in this evolutionary path with key mobile stake holders driving momentum towards a greener society through cost-effective design approaches. In fact it is becoming increasingly clear from new emerging services and technological trends that energy and cost per bit reduction, service ubiquity, and high speed connectivity are becoming desirable traits for next generation networks. This chapter provides an introduction to the solutions presented in the book, by investigating potential smart and energy efficient scenarios for the 5G paradigm, which explores how cooperation working in synergy with seamless handovers using context information can provide the platform for significant energy saving at the mobile device. Given these scenarios, we define the system requirements and propose a functional architecture on which solutions proposed throughout this book are built.

Á. Gomes (✉) · T. Moreira
Portugal Telecom Inovação, Aveiro, Portugal
e-mail: agomes@telecom.pt

J. Kibilda · M. Filo · R. Piesiewicz
Information and Communications Technology Research Center,
Wroclawskie Centrum Badan EIT+, Wrocław, Poland

A. Radwan
Instituto de Telecomunicações, Campus Universitário de Santiago,
3810-193 Aveiro, Portugal
e-mail: aradwan@av.it.pt

A. Radwan and J. Rodriguez (eds.), *Energy Efficient Smart Phones for 5G Networks*, Signals and Communication Technology,
DOI 10.1007/978-3-319-10314-3_1

1.1 Introduction

Cognitive radio and cooperative networks are key disruptive technologies in the field of wireless communications and are seen as strong candidate for many future wireless applications [2]. Since their inception in [3] and [4], Cognitive radio has been mainly used to improve spectrum efficiency and the majority of cooperative concepts have been developed with the goal of enhancing the individual and/or group wireless link capacity; following a different path, this book investigates how these paradigms (cognition and cooperation) can be applied for reducing energy consumption of wireless mobile devices. As a first step and in order to define the book framework, three reference scenarios are defined:

- Scenario1: Homogeneous RATs-Cooperative cluster. This scenario addresses a hybrid combination of a unique RAT access and short-range cooperative cluster.
- Scenario2: Heterogeneous RATs cooperation. This scenario considers that several RATs are available in the location of a multi-standard mobile phone.
- Scenario3: Heterogeneous RATs-Cooperative cluster. The scenario envisages a more efficient use of the RATs available on a heterogeneous cluster which will be considered globally as a pool of opportunities rather than as independent technologies. In this way, Scenario 3 is a merge of features from Scenario 1 and Scenario 2, allowing cooperation among heterogeneous RATs and can be seen as an advance or final evolution of the previous scenarios.

Given the envisioned scenarios, a set of requirements are defined for the design of a new architecture that promotes the cooperation between radio access systems and mobile terminals. Central to the proposed architecture is a cognitive framework that determines how new sources of information are obtained, processed and exploited towards reducing energy consumption. These new sources of information can be referred to as context information, where mobile devices can learn and adapt their behavior in order to save power: for example, by using the appropriate wireless interface for the current application workload and keeping the others effectively turned off, the system can save energy. Additionally, based on context extraction, the mobile device is capable of being aware of the neighboring devices to allow the possibility for cooperation in order to achieve higher data rates, energy savings, and ultimately to extend battery lifetime. The proposed decision making process is a matter of choosing the most economic connection with respect to energy consumption that provides the required quality of service and will result in three key events: to keep the mobile device as it is, establish short range cooperation with neighboring mobile devices and/or perform handover to a different wireless system. The handover process is a basic mechanism for the intra/inter RAT cooperation, which has been engineered towards energy optimization during the different handover phases:

- System Discovery: mobile device must know which wireless networks are reachable at its current location.
- Handover Decision: mobile device then evaluates the reachable wireless networks to make a decision based on context awareness and energy saving strategy.

- Execution: if the mobile device decides to perform a vertical handover, it executes the vertical handover procedure required to be associated with the new wireless network.

Small cells or femto cells are one of the most promising solutions to increase the mobile terminal autonomy and QoS, due to the short distance between the user devices and the access point [5, 6]. Nevertheless there are several constraints to its deployment like inefficient handover process and intercellular interference management. The foreseen solution overcomes this situation with a cognitive approach supported by contextual information, that improves the handover decision making to determine whenever it is advantageous to perform handover and when to avoid unnecessary user cell switching. This knowledge will help to avoid "false handovers", for example, when a user is moving at high speed and tries to handover to a femto instead of a macro cell.

1.2 Reference Scenarios

1.2.1 Scenario1: Homogeneous RATs-Cooperative Cluster

Today's mobile terminals (MT) have the capability to access different radio systems and technologies, from the cellular systems GSM, UMTS and LTE (WAN) to the short range communication paradigms like WiFi (WLAN), Bluetooth or UWB (WPANs). Each technology was designed to address a specific environment and service requirements, and does not take into consideration the feature of close cooperation with other systems. Although some synergies between systems were already investigated, and some basic mechanisms to be implemented are already available (e.g. Vertical handover [7]), only the generalized availability of multi-mode mobile terminals with high processing capabilities will allow the implementation of efficient solutions. In the first envisioned reference scenario: short range cooperative cluster on homogeneous networks, the mobile device cooperate with other mobile devices using a short-range radio technology in order to increase the availability of an infrastructure based system (e.g. cellular system), getting better QoS or increasing its autonomy (low battery case).

Figure 1.1 depicts the most significant use case of scenario 1, where the mobile handset is entering a shadowed region and line of sight communications is challenging resulting in reduced data rate, and beyond that increasing the drain time of the battery. Nevertheless with a cooperative solution based on short range communication, the user terminals in the near vicinity can act as "helpers", or in other words cooperative users, that will simply create a short-range network and relay the information from the source till the terminal that has the best available connection with the core network. The final hop is a long range point-to-point connection that will transmit data packets at high speed. Hence by taking advantage of the cooperative users, the source user is able to extend its battery lifetime significantly since

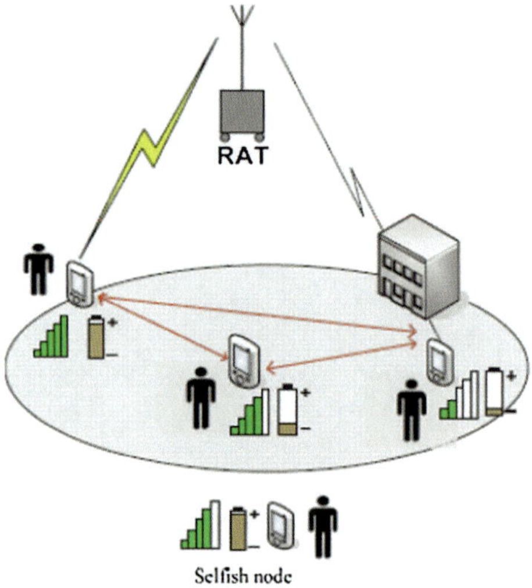

Fig. 1.1 Cooperative cluster in a homogeneous network [8]

it can transmit all its information in one shot at high speed, effectively as if the mobile device was within a small cell coverage zone. Setting up a cooperative cluster raises some questions:

- When to setup a cluster?
- Which terminals should be included or excluded from a cluster?
- How to maintain the cluster when a terminal is moving out of coverage?
- How to guarantee the fairness in the allocation of resources avoiding selfish behavior that could jeopardize the cooperation success?

The implementation of mechanisms to support a feasible and well performing cooperation behavior requires more information and measurements than the ones currently collected in present systems. Network and mobile devices should gather and process new information that allows the characterization of the terminal context at each moment; this issue is addressed in more detail in Sect. 1.4.3 of the current chapter.

The communication range, user mobility type and particularly the user speed should be analyzed in order to find the suitability of a terminal to join a cluster. The lifetime in the cluster coverage area and the service provided by the terminal coupled with the signaling cost will influence whether the end-user device should join and associate to the cluster.

The radio connections that carry the service to and from (Uplink and downlink) the core of the system are provided on different spectrum bands possessing variant propagation environments, which can place some constraints on the type of services that can be supported. For instance, the Internet, being best effort, can tolerate brief interruption, jitter and delays, hence represents the most adequate service for cooperative communications solution.

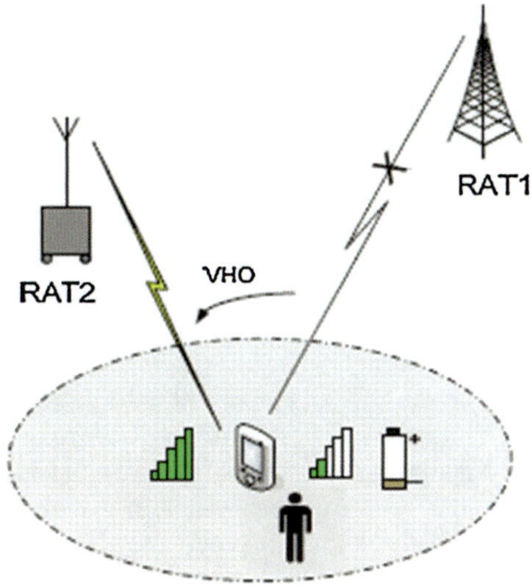

Fig. 1.2 Cooperation in a heterogeneous environment [8]

1.2.2 Scenario2: Heterogeneous RATs Cooperation

The evolution of mobile networking technologies is a constantly challenging domain, where new solutions aim to support broadband applications in a seamless and cost effective manner. A key driver behind the 4G vision is the possibility for a mobile user to access any network at anytime, anywhere. The enabler for this solution requires the network to evolve to allow the exchange of information and facilitate vertical handovers between different RATs. By using the full capabilities of mobile terminals operating on heterogeneous radio propagation environments and technology modes, high quality services can be provided to the mobile consumers. This cooperation needs to be coherent in order to achieve "effective" decisions. Figure 1.2 represents the second scenario, where mobile terminals are able to exploit heterogeneous Radio Access Technologies.

As can be seen from this figure, a multimode mobile terminal is currently connected to RAT1 (e.g. LTE Macrocell); however over time and with the increase of distance from RAT1, the received signal strength becomes weaker and eventually the mobile terminal needs to increase the uplink transmission power, which causes its battery to drain more rapidly. In order to save energy and increase the battery duration time, the mobile terminal, after scanning the surrounding available cells using context information as a guide (provided to the mobile handset from the centralized context architecture), finds that a new cell RAT2 (e.g. WiFi Access Point) has better signal. So after evaluating the availability of resources in RAT 2 to provide service continuity, a decision is made to execute a vertical handover from RAT1 to RAT2. With an innovative mechanism to roam between heterogeneous RATs, mobile terminals are able to achieve higher throughput, better network stability and simultaneously reduce their energy consumption.

The presented scenario, to become a successful approach, needs to take into account a variety of parameters, such as latency or bandwidth that can vary depending on the application in use.

The high level mechanism that is needed to make this scenario feasible is described by three sequential phases.

1. The first phase handles the network discovery, where the mobile terminal performs network scanning to know which wireless networks are available. The context architecture described later in the chapter should provide assistance in discovering available networks, so as to decrease energy consumption within this phase.
2. The second phase handles the handover decision, where the mobile terminal evaluates the best wireless RAT to access based on its preferences.
3. Finally the third phase handles the handover execution, where the mobile terminal performs the vertical handover to the new RAT. This point is where the vertical handover procedure is performed and finalized, and the point of attachment is moved from one RAT to another.

The integration of multiple RATs brings some challenges that will be addressed by the proposed architecture that includes changes on the network side, and also on the terminal side. In short, the network side needs to have some innovative features that can handle the mobile terminal handover messages coming from any RAT. On the other hand, the mobile terminal needs to integrate several network interfaces to be able to connect to several RATs.

As a part of this scenario, the envisioned scenario considers handover between macro and femto-cells. This use case contemplates the features of Femto-cells (small cells [9]) and handover to and from these cells, as well as the improvements that can be achieved by introducing context information (can be multiple variables such as user location, routine habits, expected indoor movement, cell placing, etc.). This can provide better and more efficient algorithms, but also provide more realistic and predictive handover decisions.

1.2.3 Scenario3: Heterogeneous RATs-Cooperative Cluster

Scenario 3 envisages the optimum utilization of spectrum and energy, exploiting all available nodes (BSs/APs and terminals) in the vicinity. In fact, this scenario integrates scenario 1 (cooperative clusters) and scenario 2 (heterogeneous RATs cooperation); hence providing more efficient solution than the individual approaches per se. The challenge to design and implement a system that addresses such complex scenario is more easily addressed if scenarios 1 and 2 are analyzed in detail first.

Figure 1.3 shows an example of scenario 3. In this case, the service is initially provided to Terminal B by RAT2, but because the terminal is running out of battery, or the possible maximum transmitted energy or data rate is not sufficient for

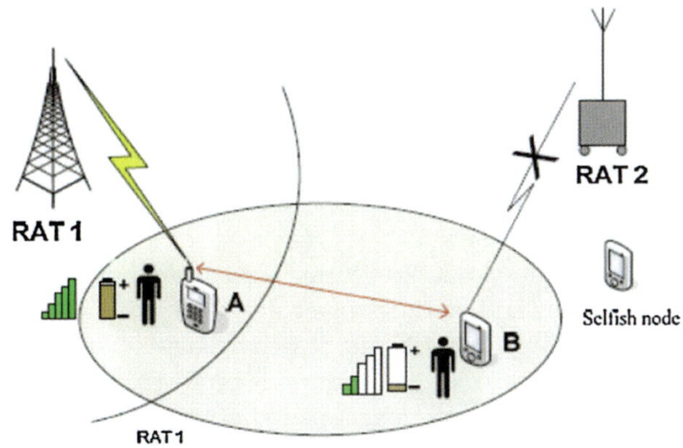

Fig. 1.3 Cooperative cluster in a heterogeneous network [8]

the specific service accessed, the service will thence interrupted. However, Terminal A has a short range communication interface and is in the range of Terminal Band furthermore is willing to cooperate; the service can then be provided by RAT1 using terminal A as a relay. The handover between RAT2 and RAT1and the maintenance of more than one radio connection in a mobile environment is a more complex process than the presented approaches. These require the improvement of today's wireless systems, particularly with the incorporation of new information sources, and beyond that a context-aware framework that is able to collect, process, and maintain information both on the user and network side. This will support more effective decisions making for communication algorithms/protocols, and provide a platform for seamless integration of new business models that rely on context information.

1.3 System Requirements

Based on the previous scenarios, legacy wireless systems and on the 3GPP System Architecture Evolution (SAE) [10], a set of system requirements were defined for the design of a new architecture that intends to promote the cooperation between radio access technology systems and mobile handsets. The most significant requirements are listed here:

- Several radio systems (including cellular and non-cellular) should be considered at least: 3GPP (LTE and HSPA) [10], WiFi [11] and WPAN (Wireless Personal Access Network).
- New system is based on the 3GPP SAE in order to take advantage of a widely approved and adopted workframe that promotes the integration of different wireless systems, 3GPP and Non 3GPP.
- Inter RAT mobility will be supported by exploiting optimized vertical handover capabilities, particularly the network discovery process, in order to speed up the process and save energy on the mobile handset side.

- New information, associated with a specific context should be gathered and provided to the system on both the terminal and network sides in order to support more elaborate decisions that will implement the optimization process and the new features provided by the system. The several functional components responsible for the flow and treatment of information will be incorporated in the 3GPPSAE in order to guarantee the evolution of the present system towards the proposed concept.
- A multimode and multi technology mobile terminal, according to the selected radio technologies, is mandatory.
- The new system should be designed to have a good performance in the different environments (urban, sub-urban and rural) and provide a high range of services concerning its QoS requirements like:
  - Voice over IP
  - Video Tele Conferencing
  - Audio Streaming
  - Video Streaming
  - Interactive Games
  - HTTP Services
  - Email Download
- The proposed solution should present good performance and significantly enhanced energy efficiency on the handset side, when compared to the current system without a significant increase in the computational efforts in order to allow the use of existing mobile handsets and other networking equipments.

1.4 System Architecture

The previous sections described the foreseen 5G Scenarios showing several approaches where the mobile terminal is able to improve the energy efficiency while keeping or enhancing the achieved quality of service. In order to enable this concept, a detailed system functional architecture needs to be defined for enabling the additional information exchange between the network and the mobile terminal. In this context, there are several applications that need to be considered such as energy efficient Vertical and Horizontal Handovers and Inter-User equipment cooperation for power saving. Taking this into consideration a whole new approach needs to be envisaged; however the general idea is based on the capabilities of the EPS (Evolved Packet System) [10, 12]. Figure 1.4 depicts the high level system architecture that is envisaged to be deployed under the umbrella of the EPS, highlighting the information flow within the network and the mobile terminal logical entities. This figure also shows the functionalities divided into two different parts: the network and handset functionalities, that in unison realize the main target objective of increasing the performance and the energy efficiency of future mobile terminals.

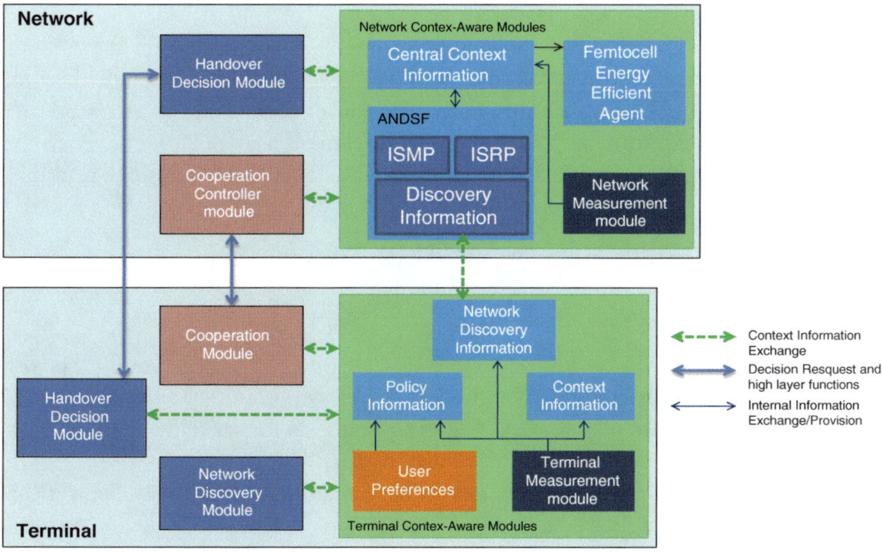

Fig. 1.4 System architecture

1.4.1 General Description

The system architecture shown in Fig. 1.4 can be split into two main parts that are logically connected in order to exchange specific information to fulfill the requirements of the module: Network Functionalities and Mobile Terminal Functionalities.

1.4.1.1 Network Functionalities

Cooperation Controller Module (CCM)

The *Cooperation Controller Module* acting as negotiation mediator is responsible for any decision related to the cooperation among mobile terminals. In other words, this module is the core of Scenario 1, where any negotiation information between two mobile terminals (the relaying MT and the source MT) is validated. Additionally, it is also responsible for triggering the authentication (in the 3GPP Authentication, Authorization and Accounting [AAA] Server [13]) of the requesting MT and determining the IP-Connectivity Access Network (IP-CAN) type, which is used to determine the proper charging rate for the relaying services. In the case of the ePDG oriented variant for session modification, the CCM is responsible for handling the IP-CAN session modification on behalf of the relaying MT. This module is logically connected with the mobile terminal Cooperation Module needed for Scenario 1.

Handover Decision Module (HDM)
This module is responsible for any handover decision performed on the network side. In addition, any handover request from the mobile terminal is handled by this module; the final decision takes into consideration every network and mobile terminal real time conditions, supported by the information exchange between other modules. This module is logically connected with the mobile terminal Handover Decision Module (HDM) to improve the handover decision in Scenarios 2 and 3.

Network Context-Aware Modules (NCAM)
The NCAM is a collection of different context-aware modules that individually handle different types of information:

- *Central Context Information* Module is responsible for gathering network specific information requested by the HDM or Cooperation Module (CM). This module has the capability to handle internal requests for other modules and take decisions whether the information is useful or not.
- *Access Network Discovery and Selection Function* (ANDSF) collects ISMP (Inter-System Mobility Policy), ISRP (Inter-System Routing Policy) and AND (Access Network Discovery) information, and delivers the information to the MTs. This module is logically connected with the Network Discovery Information module on the UE side due to the fact that some information can be directly provided to the mobile terminal, upon request, in order to decrease the information lag of real time decisions. ISMP, ISRP and AND have been specified by 3GPP [10]. The AND information should be extended to include information about UEs, which are willing to participate in cooperation and information of BSs/APs power class/transmission power. In order to support Scenario 1, where node discovery process is a requirement, some functionalities need to be improved or included in this module. Positioning methods, such as GPS, network based solutions like measurement of Time-Of-Arrival, or other localization techniques need to be integrated for this purpose.
- *Femtocell Energy Efficient Agent* module is an additional entity implemented as part of the Home NodeB (HNB). The main purpose of this entity is to optimize the context information targeted to small cells in order to provide improvement in energy consumption of handsets. In short, this module is used as input for the optimization process.
- *Network Measurement Module* is responsible for providing context information about the network status such as cell capacity, cell load, latency, number of connected users. In terms of deployment, this module should be located in the access network, which means in 3GPP it can be either in the Base Station (BS) or in the Radio Network Controller (RNC); while in trusted/untrusted non-3GPP access, it shall be located in the Access Point (AP). Specifically in the case of untrusted non-3GPP, the measurements obtained in this module need to be adjusted to the common standard and evaluated in terms of reliability (role taken by the *Context Aware Information* Module).

1.4.1.2 Mobile Terminal Functionalities

Cooperation Module (CM)
The *Cooperation module* on the handset side has two different roles, depending on the scenario. When the mobile terminal acts as a relay, the CM takes the role of *Cooperation server*, on the other hand when the mobile terminal is a source and requests to be relayed through another terminal, the CM takes the role of *Cooperation client*. The main differences are further elaborated hereafter.
The *Cooperation server* is responsible for provisioning the service to the requesting UE. By handling the decision making based on context information provided from the terminal context information modules such as power consumption, the estimated battery drain rate is taken into consideration. In addition to that, the information about current running services is also taken into consideration to estimate the required throughput that can be used by the handover policies. This module also features a monitoring process that continuously determines whether the cooperation is still advantageous, aborting it when it becomes disadvantageous.

The *Cooperation client* is responsible for different procedures such as the establishment of the connection, detachment from the relaying MT node. The cooperation client is responsible for initiating the formation of cooperative cluster, triggered when the *Cooperation Module* is notified about the existence of new nodes in the vicinity.

Handover Decision Module (HDM)
The *Handover Decision Module* is responsible for taking the decision for performing energy efficient vertical and horizontal handovers based on information provided by the *Terminal Context-Aware Modules* such as current battery state, throughput or processed parameters such as battery consumption rate. These decisions are controlled by certain constraints given by the policy information module and must be validated by the HDM on the terminal side before being executed. Regarding the cooperation scenario, the HDM needs to have information about the network state via the relaying MT. Using context information; decisions regarding cooperation are shared with the CM. The CM is then responsible for the application of the cooperation and cooperative strategy whereas HDM is responsible for making the HO decisions in the active and idle state (i.e. selection of paging/location update when the terminal is in the idle mode).

Network Discovery Module (NDM)
Specifically located on the terminal side, the *Network Discovery Module*s responsible for the optimization of the network/node detection and scanning by controlling the internal handset radio interfaces based on context information gathered from the Terminal Context Aware Module (TCAM).

Terminal Context-Aware Module (TCAM)
Using the same approach on the network side, the *Terminal Context-Aware Module* is a collection of different context-aware modules that individually handle different types of information.

- *Network Discovery Information* module collects information about the networks/mobile handsets in the neighbourhood, in terms of area coverage, physical proximity to other handsets, and topologies of cooperative clusters. The information is derived from the data provided by the Context Information module and/or is pulled from/pushed by the ANDSF on the network side which is logically connected.
- *Policy Information* module stores information about the policies applied to the handset, set by the network operator such as handover policies, requested and available levels of security for available networks. Those policies limit the number of degrees of freedom for HO decision-making and Node Cooperation. In addition to that, this module also receives information from the *User Preferences* module such as preferable RAT, or even with information of specific prohibited mobile access network.
- *Context Information* module(or context aggregator) gathers and stores context information from other modules inside the TCAM related to the radio environment such as received signal strength, used bandwidth, channel state information, power needed for transmission using current interface. Simultaneously, information about the capabilities of the device are also gathered and stored in this module. One key feature of this module is the capability to store history information about the application in use, including battery consumption rate and other energy related information.
- *User Preferences* module gathers the user defined settings regarding preferred and/or restricted networks, technologies, strategies (e.g. cost saving versus extending the battery lifetime), etc. In particular, this module contains a set of profiles that can be selected by the user, and information about which profile is currently active.
- *Terminal Measurement* module extracts low level information from sensors and radio interfaces. After processing the raw information such as battery power level, transmission power, RSS, the module provides processed information to the Context Information module. This module also receives requests from Context Information module that need to be collected by means of subscriptions. For instance, when setting up the scanning procedure for a radio interface, the NDM provides the PHY/MAC parameters of a certain technology to optimize the scanning procedure.

1.4.2 Deployment of Proposed Architecture on Evolved Packet System (EPS)

In the first quarter of 2013 the 3GPP family of systems (GSM, WCDMA-HSPA, LTE) provided mobile services to 90.4 % of the global subscribers (Informa Telecoms and media [14]). Taking into consideration the worldwide importance of these systems and in order to facilitate the integration, this section presents some guidelines about the deployment of the proposed functional architecture on top of the 3GPP EPS [10, 12].

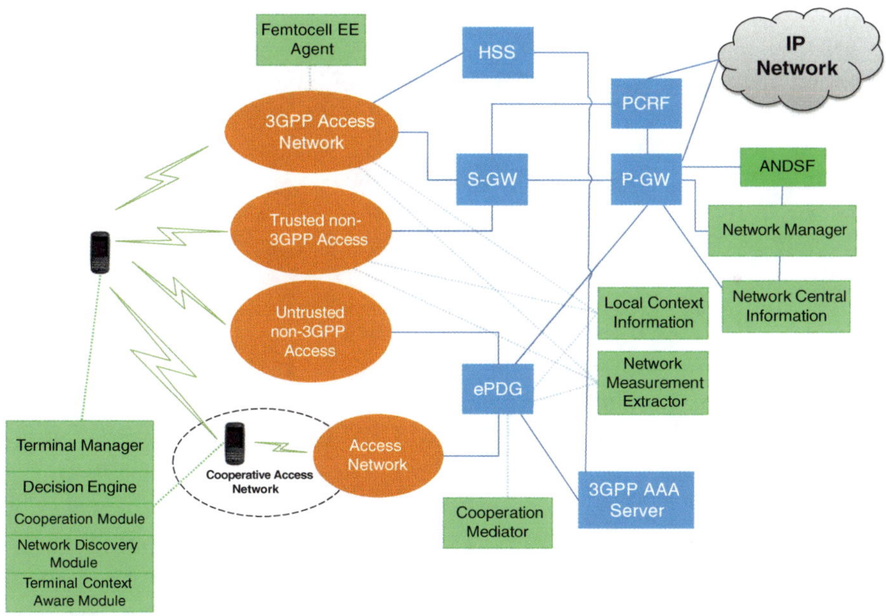

Fig. 1.5 Potential futuristic deployment of proposed architecture in EPS

As can be seen in Fig. 1.5, a simplified EPS functional architecture is depicted. The UE can reach the core network using E-UTRAN (LTE and LTE-Advanced); however this is not the only access technology supported. 3GPP specified support of multiple radio access technologies and also the mechanisms to guarantee the mobility of the service between them through "Handover". The idea was to bring convergence using a unique core network providing various IP-based services over multiple access technologies. The 3GPP specifications define how the interworking is achieved between an E-UTRAN, GERAN (GSM/GPRS) and UTRAN (UMTS/HSPA) [15].

The EPS also allows non-3GPP technologies to interconnect the mobile handset and the EPC. Non-3GPP means that these accesses were not specified in the 3GPP. These technologies include WiMAX, cdma2000, WLAN or fixed networks. Non-3GPP accesses can be split into two categories:

- Trusted non-3GPP accesses can interact directly with the EPC;
- Untrusted non-3GPP accesses interwork with the EPC using a network entity called the ePDG (Evolved Packet Data Gateway). The main role of the ePDG is to provide security mechanisms such as IPsec tunneling of connections with the MT over an untrusted non-3GPP access.

3GPP does not specify which non-3GPP technologies should be considered trusted or untrusted. The decision is left to be determined by the operator. The mobility is a fundamental requirement for the cooperation between different systems provided by the handover mechanisms. There are three types of mobility defined by EPS:

- **Inter 3GPP Networks**
  The handset handovers from e-UTRAN or GERAN to either another UTRAN or GERAN. To guarantee a seamless mobility the resources at the target network are reserved before the link is switched from the source s-GW to the target s-GW maintaining the p-GW.
- **Inter 3GPP and Trusted Non-3GPP Networks**
  To ensure the session mobility, a PMIPv6 (Proxy Mobile IPv6) or MIPv4 (Mobile IPv4) tunnel establishes the connection of the EPC to the non-3GPP network. When the handset or the network identifies a need to switch from one technology to another, the P-GW is used to coordinate the redirection of packets from the home RAT to the target RAT.
- **Inter 3GPP and Un-trusted Non-3GPP Networks**
  The handset handovers from/to a 3GPP network to/from an Un-trusted Non-3GPP network. The link is transitioned by PMIPv6 or MIPv4 tunnel from the s-GW to the e-PDG, where the Un-trusted Non-3GPP network is attached. From the e-PDG to the non-3GPP network, an IP-Sec tunnel directly to the handset is maintained.

Figure 1.5 depicts the deployment of the proposed functional blocks on top of the Evolved Packet System (EPS) architecture. The green colored blocks represent the new functionalities that need to be implemented at each different element of the general architecture of the EPS, in order to achieve the energy saving and performance gains provided by the proposed architecture

Some of the core elements of the architecture are the Terminal Manager (TM) and the Network Manager (NM), which are respectively located on each enabled UE and network. There is one NM in each Public Land Mobile Network (PLMN).

- TM is responsible for energy efficient HO decision making, inter-node cooperation negotiation, energy efficient HO trigger, policy maintenance, and context information collection on the handset side.
- NM is responsible for context information collection and energy efficient HO decision making at the network side. NM also includes ANDSF module, which maintains and distributes information and Inter-System Mobility Policies.

Previously, the chapter had defined two main scenarios (in addition to a third scenario that is an integration of the two main scenarios) to present the capabilities of the propped functional architecture. In the first scenario, the UE connects to the network via relaying nodes. The second scenario presents a HO between heterogeneous access networks, where the UE handovers to a different RAT in the vicinity. In both cases, the information about relaying UE or new access network can be pulled/pushed from the ANDSF (Access Network Discovery and Selection Function). When the UE discovers or receives information about new networks and/or nodes in the vicinity, it estimates the required transmission power. Based on the estimated transmission power and additional context aware information such as user mobility, data rates, network/cell load or battery consumption rate, the UE decides whether the new connection is more energy efficient.

If the UE decides that the new connection can be more energy efficient without compromising the required QoS, the NM checks if the new network configuration is feasible (or more efficient) from the network operator point of view, based on network context aware information gathered from different entities. If the NM accepts the handover decision, it reports back to the handset (two-way handshake) and an energy-efficient HO process is executed. If either the TM or the NM decides that the new access network is not more efficient or not feasible, the TM starts searching for other available networks/devices in the neighborhood.

In addition to adding new functional blocks to the EPS architecture, some updates to already existing functional entities are required. Each Serving Gateway, ePDG (secure access gateway for untrusted non-3GPP networks) and access gateways in trusted non-3GPP accesses need to support additional functionalities:

- Collection of local network measurement information (e.g. network load, cell load, and cell capacities);
- Information extraction and storage of new measurement data;
- Mediation in case of inter-UE cooperation. The mediator in this case is the chosen ePDG during the negotiation between the nodes. In rare cases, the Serving GW can be the optimal choice;
- The Node B and evolved Node B have also to host the Femtocell EE Agent that is responsible for the optimization of HNB parameters, particularly to manage the inter-cellular interference. This new cognitive approach, in order to facilitate the overall radio network optimization allows the wide spread of femtocell (small cell) deployment. The small cells can achieve significant power saving gains in the radiated energy when compared to traditional macrocell only deployment.

1.4.3 Context Awareness

Performance enhancement and energy saving mechanisms targeted to mobile handsets greatly rely on information provided by different sources. Those information combined aim at understanding the mobile handset surrounding environment. This context awareness provides the mobile handset with the ability to take efficient decisions and configure itself in order to minimize the energy consumption, i.e. to connect to the best available cell or even switch off certain interfaces that are currently not in use.

1.4.3.1 Context Aware Information

The basis of the cognitive radio is the ability to gather information from different sources (mobile equipment or network) and take decisions based on such information. However one major issue is to implement a module capable of extracting context information from the surroundings to enable an optimization and learning process, that provides support to energy saving strategies for UEs. To accomplish this, it is necessary to establish a common understanding of the definition of context within the proposed architecture.

Context, in general, can be seen as higher level knowledge, derived from an aggregation of information or modes of parameters describing the operational environment, including information such as time, geographical location etc. This higher level knowledge is derived in a "context engine". Operational context is closely linked to the actual application domain and the range of applicable situation/ environment information is dependent upon the actual use case. Context may include information about the UEs, network and applications. Information that form the context can be put into two basic categories: Information about the device itself (radio interfaces, battery life, maximum transmission power, processing power, etc.) and information about the network (cell load, cell capacity, QoS guarantees, etc.). Context itself is an aggregate of such individual information fragments, which are collected from various types of sources, including physical sensing of the radio environment, equipment data sheets, optimization policies, etc.

Specifically for the proposed architecture and solutions in this book, the context information are mainly targeting information fragments related to energy, i.e. required information are highly related to energy metrics, such as:

- Transmission power;
- Energy consumption per bit;
- Interference level (SNIR);
- Battery life;
- Current state of battery consumption;
- Power emission requirements based on regulations (block edge mask);
- Spectrum pollution.

The information gathered using the metrics above expresses the energy individually used by each mobile terminal; however the energy consumption must also be considered in a wider context. A context engine, aiming to facilitate decisions that lead to higher energy efficiency on the level of the entire network, needs to also consider the status and availability of neighboring UEs and BSs. There are a number of information fragments to be considered in such cases, including:

- Availability of neighboring handsets;
- Energy status (remaining power levels);
- RATs availability

There are several collected context information sets that can be used; however it has to be noted that context extraction and processing are time and resource consuming. There is always a trade-off between the amount and quality of the gathered information fragments, and the actual gains that can be achieved from having more accurate context descriptions. Finding this balance is one of the challenges of designing context aware architecture. Applying policy constraints at early stages to reduce resource wastage in context extraction is considered a useful strategy.

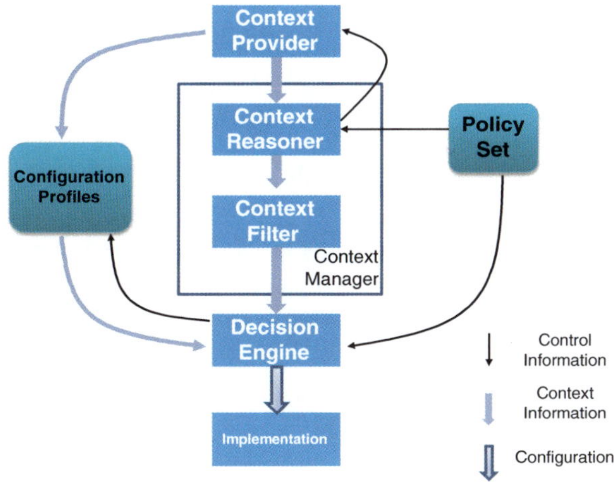

Fig. 1.6 Context awareness based on interactive decision making

Context awareness needs a system that interprets the available pieces of information. There are several logical entities that are involved in obtaining, processing and managing context awareness in such a way to create a learning and adaption process. A functional architecture of context awareness based on interactive decision-making is described in Figure, which shows the different functional entities involved in the process of context extraction.

In the functional architecture, the TCAM entity is responsible to collect and provide the required context information fragments. These information fragments serve as input for the context awareness and decision making process of the overall system as depicted in Fig. 1.6. The context information flows from the context provider through derivation of higher level knowledge and filtering, until it reaches the implementation procedure. During this flow, a range of processing steps and operations are required. The individual functions and roles of the different modules are described below.

1.4.3.2 Context Aware Modules

Context Provider

The *Context Provider* is the source of the raw context information. The Context provider attains its inputs directly from sensors or radio interfaces. It provides the information, without processing, to the *Context Manager* for higher level processing.

Context Manager

The *Context Manager* is responsible for information gathering, knowledge deduction and context filtering. Its functionalities are split into the *Context Reasoner* and the *Context Filter*. Before providing the context information into the Cognition Process, the context information needs to be gathered, higher knowledge has to be inferred, and then filtered to achieve the so-called operational context. Operational context describes all the possible allowable options that may be implemented, when a particular energy saving or performance enhancement strategy is applied. Operational context can limit the range of possible strategies; thus reducing the search space of the cognition process. From one side, the *Context Reasoner* has the role to generate rules for the *Context Filter* based on the constrains from the policy set that informs the *Context Filter* about which context information should be ignored, or which range of solutions may be excluded, from the other side. The *Context Filter* processes and filters context information based on the rules from the *Context Reasoner*.

Decision Engine

The core of the cognition process is the *Decision Engine that* optimally chooses a set of applicable strategies based on the available context. The *Decision Engine* is capable of considering all possible adjustments or combinations for saving energy and/or enhancing performance. One possibility is looking into a cross layer approach, whereby identifying all the important parameters and employing a dynamic combination of techniques to best adapt the radio in real time. Since this optimization problem needs a dynamic real time solution, reasonably fast efficient algorithms are required, which can be an equilibrium that is difficult to achieve. Given a hard optimization problem, it is often possible to find an optimum solution facing minimum space and time complexity. For small search spaces, classical exhaustive search algorithms can be applied, but for larger search spaces, special Artificial Intelligence (AI) techniques must be employed. AI techniques such as game theory, genetic algorithm and fuzzy logic are considered valid options for the definition of the *Decision Engine*.

In a network environment, the decision of a single node may have impact on the performance of other nodes and the entire network. For example, within the defined scenarios, nodes can form cooperative clusters to gain total energy saving with multiple nodes. There can also be cases where individual nodes will not have energy saving by joining the cooperative cluster. Clearly, conflicts between social benefits and self-interests may discourage the cooperation; however a defined business solution rewarding the most cooperative mobile terminals could overcome this difficulty, which is the topic addressed in Chap. 9. Another example of the usage of context is the network selection. When multiple nodes choose the same network for energy saving, the nodes may compete to access the same network resources, which may offset the expected gains. Thus, a MAC protocol which implements interactive decision making processes is needed. Game theory provides a mathematical basis for the analysis of interactive decision making processes.

Nodes are considered to be interconnected through a network. A node collects the context information from the network and the terminal itself. The context information from the network conveys the information of other nodes' impact on the entire network. The context information will be further processed and filtered, being the most relevant context information used for decision-making. Based on the context provided by the *Context Manager*, the *Decision Engine* calculates the utility and selects a strategy that maximizes the utility.

Implementation
The strategies selected by the *Decision Engine* are implemented in the *Implementation* block. The resulting configuration impacts both the node itself and the entire network. The impacts on the entire network can be observed by other nodes as context information from the network. Other nodes with the same context awareness entities may adapt their strategies according to the observation. The new strategies may again have consequences on other nodes' behavior. This adaptation may hence result in an iterative process. Ideally, an optimal equilibrium would eventually be reached and all nodes can have balanced energy savings.

Policy Set
In general, a policy is seen as a collection of rules. The rules define various operational aspects, including the allowable spectrum assignments, power levels, RATs, etc. for different UEs. The *Decision Engine* also takes inputs from the *Policy Set*, which act as constraints to limit the search space.

Configuration Profiles
Configuration Profiles can be seen as a database for building-up knowledge from successful energy saving strategies. Based on relevant operational context, a UE may use this build-up knowledge to generate performance enhancement strategies. The *Configuration Profiles* can also be personalized by the owner of the UE, enabling the efficient decisions to be manually controlled. This means that for example, if a certain consumer is not willing to cooperate with a specific mobile terminal, he can configure a profile that translates this into a decision.

1.4.4 Mobile Terminal Design Requirements

The chapter investigated technologies that will enable the increase energy efficiency of mobile handsets resulting in longer battery lifetime and standby time, as well as to enhance the overall performance in terms of higher data rates, lower delays, etc. This was achieved by various means that included:

- Use of short-range cooperative technologies to enable terminals to communicate via low pathloss channels when only one RAT is available.
- Selection of the most energy efficient RAT amongst those which can fulfill the requested QoS and provide the highest performance in terms of energy consumption.
- Use of heterogeneous cooperative techniques that will use low energy short range communication as a medium to relay information to terminals experiencing bad channel conditions with point-to-point long range systems. The terminal may be initially connected to one specific wireless technology, but eventually receive data through a distinct technology by using multi-hop short range connectivity in synergy with long range connectivity.

To address all the aforementioned use-cases requires a HetNet (Heterogeneous network) environment, which is aligned with the technology roadmap for 4G systems and beyond. In addition, the architecture proposed to support cooperation and energy efficient vertical handover, also requires re-engineering not only the network side, but new design requirements on the terminal handsets. In other words, future handsets will need to be multimode (e.g. LTE, HSPA, WiFi, Bluetooth), and multiband in nature providing the impetus for new approaches on the RF architecture, including energy efficient multi-standard transceivers, power amplifiers, RF front-end/antenna impedance matching, and miniature antennas. These topics are the subject of Chaps. 2 and 3.

1.5 Conclusion

Although research has already started discussion with regards to 5G mobile systems, deployment of those systems are not expected till around 2020. Despite the fact that the system requirements are still in its infancy, the motivation for 5G, and high level design requirements are clear: mobile traffic will continue to increase, higher data rates is of paramount importance, and smart systems that can exploit spectrum effectively and promote energy saving is a "must" [16]. These drivers create stringent design requirements for future emerging handsets. This book sheds light on innovative solutions to reduce the energy consumption of wireless mobile handsets, paving the road towards energy compliant 5G handsets. The book investigates possible energy efficient solutions for reducing energy consumption in the handset based on using widely adopted engineering paradigms that include cognition, and cooperative communications. In order to define a framework for the solutions of the book and promote the market uptake of this technology, potential scenarios need to be defined, that include a system functional architecture on which the proposed solutions have to be built.

This chapter defined two main scenarios which are envisioned to be part of the overall 5G system overview. The first scenario constitutes homogenous perspective where we use short range networking in synergy with long range connectivity to promote energy saving. We assume that the consumer has a typical point-to-point connection with the RAT, that degrades in quality when the mobile enters a heavily shadowed region resulting in a slow and power hungry connection. Exploiting the cooperation paradigm, the use is able to rely on mobile handsets in the near vicinity to establish an energy efficient connection back the same serving network. In the second scenario, the mobile handsets exploit a plethora of different RATs available, in addition to macro and femto cells, to always stay connected to the most energy efficient available RAT through smart energy aware handovers. Additionally, a third scenario is defined which integrate the properties of the first two scenarios, in order to exploit the advantages of short range cooperation in addition to energy efficient vertical handovers to provide an energy efficient connection back to the core network, which may not be the same service network but in fact the best available radio network available in terms of energy consumption. Based on the envisioned scenarios, a system functional architecture is defined that is carefully engineered in order to evolve legacy communication infrastructures, to overcome the inertia that operators may have when investing in new business models and solutions.

References

1. Cisco Visual networking index: global mobile data traffic forecast update 2012–2017. Available Online: www.cisco.com/en/US/solutions/collateral/ns341/ns525/ns537/ns705/ns827/white_paper_c11-520862.html (2012)
2. Rossetto, F., Zorzi, M.: Mixing network coding and cooperation for reliable wireless communications. IEEE Wirel. Commun. Maga. **18**(1), 15–21 (2011)
3. Laneman, J.N., Tse, D., Wornell, G.: Cooperative diversity in wireless networks: efficient protocols and outage behavior. IEEE Trans. Inf. Theory **50**(12), 3062–3080 (2004)
4. Ahlswede, R., Cai, N., Li, S.-Y.R., Yeung, R.W.: Network information flow. IEEE Trans. Inf. Theory **46**(4), 1204–1216 (2000)
5. Chandrasekhar, V., Andrews, J., Gatherer, A.: Femtocell networks: a survey. IEEE Commun. Mag. **46**(9), 59–67 (2008)
6. Golaup, A., Mustapha, M., Patanapongpibul, L.B.: Femtocell access control strategy in UMTS and LTE. IEEE Commun. Mag. **47**(9), 117–123 (2009)
7. Yan, X., Şekercioğlu, Y.A., Narayanan, S.: A survey of vertical handover decision algorithms in fourth generation heterogeneous wireless networks. Comput. Netw. **54**(11), 1848–1863 (2010)
8. ICT-248577 C2POWER consortium. D2.2: Scenarios, system architecture definition and performance metrics, deliverable 2.2 (D2.2) of C2POWER project (2010)
9. http://www.smallcellforum.org/
10. 3GPP TS 23.402: Architecture enhancements for non-3GPP accesses, v9.3.0
11. IEEE 802.11: Wireless LAN medium access control (MAC) and physical layer (PHY) specifications. (2007 Revision). IEEE-SA. 12 Jun 2007. doi:10.1109/IEEESTD.2007.373646
12. 3GPP GAN specification 43.318
13. 3GPP TS 23.203: Policy and charging control architecture, v9.3.0
14. Informa telecoms and media. Online: http://www.informatandm.com/ section/home-page/
15. 3GPP, E-UTRA and E-UTRAN overall description. TS 36.300 V10.1.0 (2010-10)
16. Demestichas, P., Georgakopoulos, A., Karvounas, D., Tsagkaris, K., Stavroulaki, V., Lu, J., Xiong, C., Yao, J.: 5G on the horizon: key challenges for the radio-access network. IEEE Veh. Technol. Mag. **8**(3), 47–53 (2013)

Chapter 2
Green Multi-homing RF Architectures

Abubakar Sadiq Hussaini, Issa Elfergani, Ayman Radwan, Jonathan Rodriguez, Laurent Dussopt, Alexandre Giry, Michael Pelissier, Sami Aissa, Frederic Fraysse and Dany Lenox

Abstract Next generation handsets will need to be energy aware so as to support 5G services, that are likely to be intelligent and bandwidth hungry, as well as support multi-mode operation (LTE, LTE+, HSDPA, 3G, WiFi among others) in Heterogeneous Networking (HetNet) environment. This vision gives way to stringent design requirements on the RF system design that in today's handset is a key consumer of power. This vision provides the impetus for new research lines that will encompass techniques and the implementation of functional entities so as minimize the carbon footprint in mobile 5G handsets. The performance of the future handset transceiver depends primarily on the performance of antennae and RF circuit designs. The future handset requires the transceiver to operate efficiently and to be reconfigurable. The current chapter and the sequel present a comprehensive study of new hardware components that can provide a flexible and energy efficient multi-standard transceivers architecture, with proof-of-concept validation for specific use-cases including LTE, TETRA and TETRAPOL. This chapter addresses the global transceiver architecture design in next generation handsets and the antenna front end unit, that is bridged to the RF front end with tunable matching network to provide an adaptive response for maximum power transfer.

A.S. Hussaini (✉) · I. Elfergani
Instituto de Telecomunicações, Aveiro, Portugal
e-mail: ash@av.it.pt

L. Dussopt · A. Giry · M. Pelissier
CEA-LETI, MINATEC Campus, Grenoble, France

S. Aissa · F. Fraysse · D. Lenox
CASSIDIAN, Elancourt, France

A. Radwan · J. Rodriguez
Instituto de Telecomunicações, Campus Universitário de Santiago, 3810-193 Aveiro, Portugal

A. Radwan and J. Rodriguez (eds.), *Energy Efficient Smart Phones for 5G Networks*, Signals and Communication Technology,
DOI 10.1007/978-3-319-10314-3_2

2.1 Introduction

This book addresses innovative solutions based on cognitive radio and cooperative strategies for future wireless multi-standard mobile devices towards energy saving whilst achieving high data-rates and better quality-of-service. These new technologies will rely on the multi-standard capabilities of handheld terminals to enable the use of the best available radio-access technology (RAT) or cooperative strategy depending on the available access-points and handsets, required data-rates, battery levels, and service costs among other metrics.

The objective of Chaps. 2 and 3 is to investigate flexible multi-standard transceivers at the hardware level and propose innovative technology blocks targeting energy saving as a new design metric, in addition to the traditional criteria of re-configurability, flexibility that are taken as standard in future emerging handsets. In this chapter, we study the design of multi-standard flexible transceiver and wideband miniature antenna along with a matching network to overcome perturbations caused by the hand or other material close to antenna. Whereas in Chap. 3, we tackle the crucial issue of Power Amplifier (PA) design to provide a complete solution for a full radio front end.

The next-generation of multi-mode transceivers needs to handle many standards with different modulation schemes and signal bandwidths. For example, multi-mode Tx performance metrics such as noise figure, linearity, power consumption and among others should be optimum for each bandwidth. This implies severe constraints on the architecture during the design process. The main goal of the first section in this chapter is to define a multi-standard architecture, which is compliant with the targeted standards which include TETRAPOL, TETRA/TEDS and LTE. The architecture is organized in two parts depending on whether the RF signal envelope is varying or constant: for a variable-envelope modulation (TETRA and LTE for instance), a Cartesian (I/Q) transmitter can be used; whilst for constant-envelope modulation (TETRAPOL for instance), a fractional PLL direct modulator with better noise and power consumption is preferred.

The second part of the chapter considers the optimized antenna units and matching circuitry. Modern handheld terminals put severe miniaturization constraints on the antenna in order to manage more space for additional functions, for example the screen, and battery to name a few. We outline the motivation for the miniature notch antenna, which we designed, implemented and characterized performance in the lab. It covers the GSM-900 band, as well as a wide band from 1.7 to 3.6 GHz. The sensitivity of its impedance to the environment was investigated experimentally and theoretically in the case of a metallic plane, user's hand and user's head, showing significant mismatch losses. The large variations in impedance provided the impetus for an adaptive tuning network that can secure maximum power transfer, even under extreme operating conditions.

A tunable matching network was designed for the miniature antenna based on a CLC pi-network topology using switched capacitors integrated in CMOS-SOI technology. The circuit operates over the 1.7–2.7 GHz band. The measurements show good agreement with the simulations. The tunable network was designed together with a sensing unit based on power detectors measuring the incident and reflected power levels at the input of the matching network. A reasonable accuracy of ±2 dB for the reflection coefficients above −8 dB is expected.

Finally, the chapter concludes by analyzing the performance of the TMN-Antenna module. The results show promising performances with significant efficiency improvement (above 25 %) under difficult operating conditions.

2.2 Multi-standard RF Architecture

Over the last decades, there has been an evolution in wireless communications towards multifunction and multi-standard mobile terminals. Reducing the number of external components to a minimum is the key when the same mobile terminal has to process several different standards. The growing economic and social impact of mobile telecommunication devices, together with the evolution of protocols and interoperability requirements among different standards for voice and data, is currently driving worldwide research towards the implementation of fully-integrated multi-standard transceivers.

Recently, numerous low-power and low-cost transceivers have been widely applied in our daily life along with the rapid growth of RF IC technology, such as building automation, remote keyless entry, consumer electronics, home automation, voice communication, sensor networks, health monitoring, and medical diagnosis [1–3]. Most of these wireless transceivers work in 433, 868 or 915 MHz license-free industrial, scientific and medical (ISM) frequency bands, but with different modulation schemes and different signal bandwidths [4, 5]. It is apparent that reliance on a single mode transceiver will not do the job. Therefore, to meet different applications using just one transceiver chip, it is necessary to implement a multiband multi-mode transceiver. Considering the stringent cost and power targets, it is necessary to select the lowest-cost CMOS technology and minimize the number of discrete external components.

2.2.1 Rx Multi-standard RF Architecture

Private Mobile Radio (PMR) is a field in radio communications systems widely used by, for example, security forces, and fire brigades. New technologies for high data rates are emerging, with very different usage and constraints, so that architectures need to be analyzed in two different parts, namely narrow-band and wide-band systems.

2.2.1.1 Narrow Band PMR

The legacy analogue PMR was used for voice in a walky-talky mode (call direct mode) or trunked mode as in civil cell networks. NBPMR have been standardized, as APCO P25 [6, 7] in the US, TETRA [8] in the EU and TETRAPOL [9] in France. These standards use digital modulation with the main motivation to secure the transmission. Some low rate digital features are used like short text messages.

These networks are mainly operated in the 400 and 800 MHz bands. Globally speaking, frequency bands are less standardized than for civil networks. Since customer frequency bands are widely spread, different versions of the same products may be needed to address all customers.

For instance, state-of-the-art 400 MHz products need around 50 MHz of bandwidth, grouping the uplink, downlink and direct mode channels. Different terminal bands are defined (380–430, 440–470 MHz) to gather costumer needs. It can be said that NBPMR terminals are able to manage a tremendous number of channels to deal with customer frequency versatility.

It is important to note that terminal usage is very different from regular public cell phone networks. For instance, instead of single user to single user communications, group calls are mainly used; many users listen to the same channel, while only one user transmits pressing a push to talk button. This makes terminals operate mainly in RX mode. The official Schengen scenario is 60 % standby, 35 % RX and 5 % TX, but the actual RX/TX ratio is usually high. This is why the optimization of power consumption in RX mode is very important. Direct mode is mainly used when a trunked network is not available.

Narrow Band PMR (NBPMRs) are designed to be compliant with channels used for analogue PMR networks. The typical channel step is 12.5 or 25 kHz and the modulation bandwidth from 6 to 20 kHz. Furthermore, channels can be offset by ±6.25 kHz to ease the coexistence of different customer networks.

These narrow channels drive all the PMR architecture, as it requires:

i. the synthesizer phase noise to be low in adjacent channels with a very low offset;
ii. high selectivity at adjacent channel, which is very close to centre frequency;
iii. very small synthesizer steps (at 400 MHz, the minimum step is 64,000 times smaller than signal frequency).

The legacy of analogue networks make interferer levels very high: the typical selectivity to bi-adjacent interferers is 70 dB. The above points i and ii mean that very high quality coefficient components (as Δf is very small) are required for local oscillators and filters.

Local Oscillator Architecture

To achieve a low phase noise, the Phase Lock Loop (PLL) filter cut-off frequency has to be low compared to the channel step, in order to filter out the noise contribution of the reference (TCXO) and the phase detector at all interferer frequency offsets; hence, the Voltage Control Oscillator (VCO) is a unique contribution. An integer PLL block diagram is shown in Fig. 2.1.

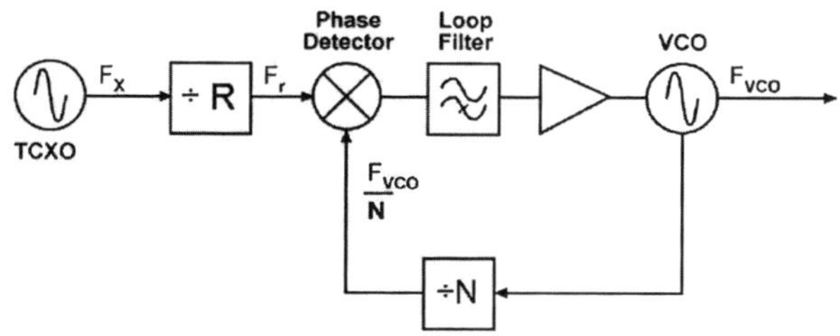

Fig. 2.1 Integer PLL block diagram

The drawback of the low cut-off frequency is the long settling time of the local oscillator; however nowadays PLL provides speed-up techniques and no rapid change is required for NBPMR due to the narrow channel bandwidth. So settling time is usually not a major issue.

For any architecture, the local oscillator relies on a high-end VCO. Moreover, low phase noise VCO design is a key knowledge to design NBPMR terminals. The required quality factor is so high that, until now, on chip VCO are not able to reach NBPMR requirements.

A common trick to improve the local oscillator phase noise is to design it at four times the required frequency: when the signal frequency is divided by four, the phase noise is improved by 12 dB (20*log10(4)). This provides good improvements, while high Q can be maintained around 1.5 GHz for the 400 MHz band. For the 800 MHz band, the benefit is less obvious and a balance between the phase noise improvement and the Q value drop has to be studied carefully. A division by two may also be a valid option.

Integer or fractional PLLs can be used: a fractional PLL gives fractional spurious, which may be an issue considering that the typical NBPMR rejection specification is better than 70 dB, whereas an integer PLL needs a comparison frequency equal to the minimum frequency step; this means a high level of phase detector noise, which may be difficult to filter by the PLL loop filter. Eventually, the choice is made according to the PLL phase detector noise and the spurious level.

The typical share of the local oscillator in the RX total power consumption is around 25 %. Considering the harsh phase noise requirements, the design has to be done with discreet components. Only a fine partitioning without over killing specification and an optimized electronic design may reduce the power consumption.

Analog Versus Digital Filtering Partition

One trend of civil market receivers is to use digital filters. It sounds very attractive, as digital filters can have very high Q, their response is not sensitive to the process or temperature, and their response can be adjusted with filter coefficients without hardware redesign. However, moving filtering functions to the digital domain increases the requirements on Analog to Digital Converter (ADC):

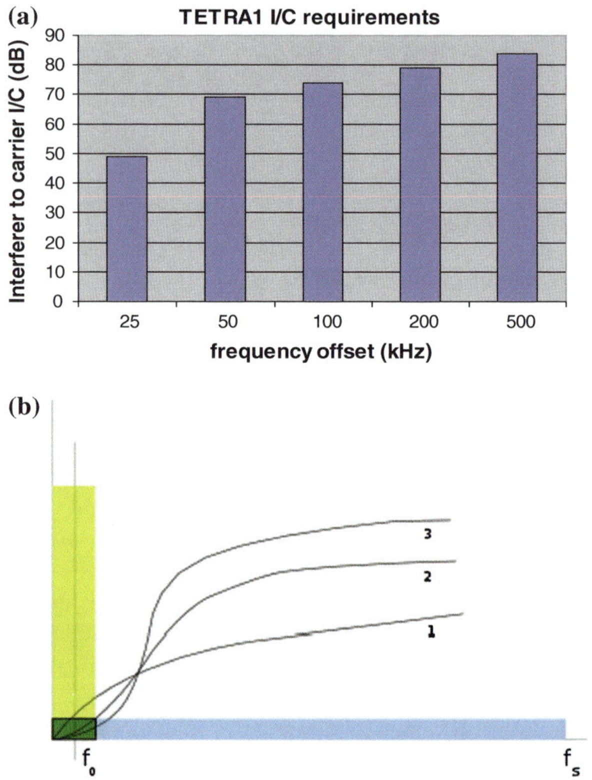

Fig. 2.2 **a** TETRA1 I/C from ETSI EN 300-392-2; **b** ΣΔ noise shaping

i. sampling rate has to be increased to sample interferers;
ii. interferer level increases with the frequency offset. Figure 2.2 shows the Interferer to Carrier ratio (I/C) for adjacent and blocking channels in a TETRA1 system.

Sigma-delta ADCs are a popular architecture for NBPMR: the idea is to oversample the signal and build the output signal using the sum of input signal variations (ΣΔ). Sigma delta ADCs can be optimized to shape the output noise out of the useful signal band (green part in Fig. 2.2).

It is clear that a NBPMR leaves room to increase the sampling rate to improve the dynamic range. Furthermore, it eases anti-aliasing filtering. However, it needs down-sampling and corresponding anti-aliasing filtering before delivering the signal to the DSP.

The required ADC dynamic range is the sum in dB of I/C, required C/N for requested BER, headroom for AGC management and a 10 dB margin (as ADC noise is partitioned to be much <RF part). The typical dynamic range for adjacent channel digital filtering only is more than 90 dB in the useful signal bandwidth. But ADC power consumption for this kind of dynamic is very high for handheld devices. Knowing that the ADC power consumption increases linearly with ΣΔ, oversampling or even increases by 4 each time 1 hardware bit (i.e. 6 dB) is added,

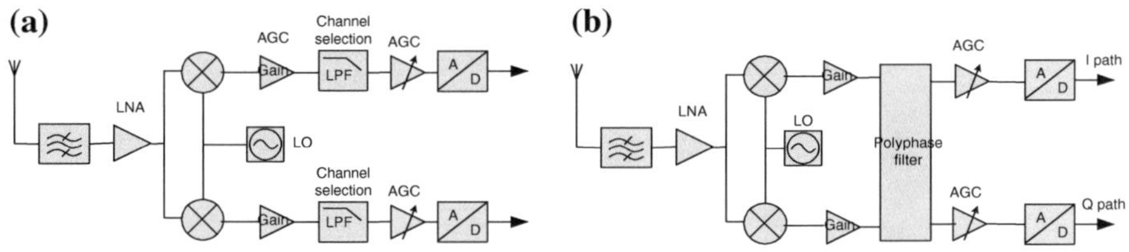

Fig. 2.3 **a** TETRA1 I/C from ETSI EN 300-392-2, **b** "low IF" receiver

bi-adjacent and higher channels cannot be filtered in digital domain. Actually, even a part of the adjacent channels is filtered in the analogue domain to reduce the ADC power consumption in most receivers.

The ADC power consumption, when high dynamic ranges are needed, explains the relatively rare usage of digital filtering in NBPMR terminals.

Architecture Choice

In both Low-IF and direct-conversion architectures, all receiver filtering functions are performed before the ADC. This means that high dynamic, to handle all interferers, has to be maintained in almost all receiver paths. Also, all constraints are set to base-band filters: high linearity, high attenuation for far jammers, high stability (to maintain adjacent channel filtering without filtering useful signal). This is so challenging that, according to our knowledge, theses architectures are not used for NBPMR (Fig. 2.3).

The traditional super-heterodyne architecture is popular for NBPMR. Filtering is usually split between IF using crystal filters technology, baseband and digital domain. The more signal gets through the RX path, the more interferers are filtered and gain may be applied to signal.

Crystal filters are able to reach sufficient quality factors with a centre frequency between 50 and 100 MHz. This makes the first mixer image frequency far enough to be filtered (terminal typical bandwidth is 50 MHz), so that no image rejection mixer is needed. This is a good point since the power consumption increases with the frequency. The crystal filter frequency is too high for sampling, so the signal is converted to a second IF, which may be 0 Hz. Additional filtering may be added at IF2 to reduce the ADC dynamic. Image rejection is often used for the second mixer as the consumption penalty is low and the image frequency close to the useful band (Fig. 2.4).

2.2.1.2 Wide-Band PMR

Giving details about wide band PMR (WBPMR) is difficult as it is still under development. It is designed using civil technologies like LTE for high-speed data transmission. In a wide-band system, many of the NBPMR constraints vanish; for instance, channels are more than 50 times wider so that filtering requirements are lower. Moreover, there is no reason for much more RX operating time than TX.

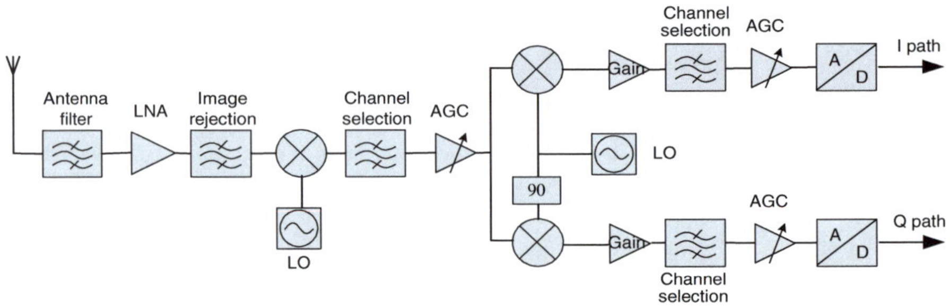

Fig. 2.4 Super-heterodyne receiver

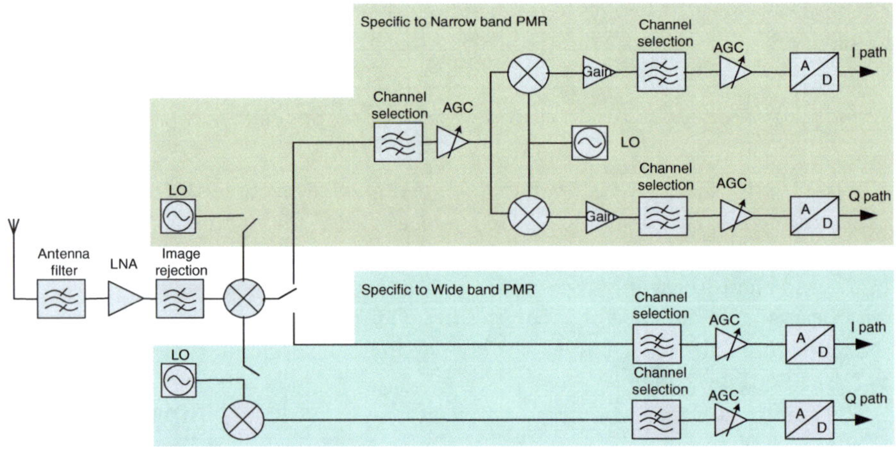

Fig. 2.5 The foreseen architecture of multi standard receiver

The foreseen architecture is direct conversion as it is popular in civil network devices. As less filtering is required, it can be implemented at baseband only. Modulation width is more than 1 MHz so the impact of 1/f noise is much less than for narrowband PMR. The base band design at IF = 0 Hz is not so challenging.

Faster local oscillator settling time and reasonable phase noise specification make wider PLL loop possible, relaxing VCO specification. It is worth to use several VCOs for wideband operation as narrow-band VCO power consumption is very high (Fig. 2.5).

2.2.2 TX Multi-standard RF Architecture

The next-generation of multi-mode transceivers needs to handle many standards with different modulation schemes and signal bandwidths. Multi-mode TX performances such as noise, linearity, power consumption and others should be optimum for each bandwidth. This implies severe constraints on the architecture to deal with during the design. The main goal of this section is to define a multi-standard TX architecture, which is compliant with the targeted standards which are TETRAPOL, TETRA/TEDS and LTE.

The architecture is organized in two parts depending on whether the RF signal envelope is varying or constant:

- For a variable-envelope modulation (TETRA and LTE for instance), a Cartesian (I/Q) transmitter can be used;
- For a constant-envelope modulation (TETRAPOL for instance), a fractional PLL direct modulator with better noise and power consumption is preferred.

2.2.2.1 Cartesian (I/Q) Direct-Up Transmitter

The direct up-converter, also commonly known as homodyne or Zero Intermediate Frequency, is composed of only one frequency conversion stage as shown in Fig. 2.6.

The base-band signals (I and Q), issued from a DSP, are converted to analogue via Digital to Analogue Converters. These signals are then filtered and directly modulated to RF frequency by two mixers driven in quadrature by an RF synthesizer.

The main advantages of this architecture are its simplicity and low number of components making it suitable for a high level of integration. Indeed, as the channel rejection is done dynamically by summing the mixers outputs, there is no need for an external image filter.

Such architecture is attractive for a complex I/Q modulation, but implies hard constraints on the IQ modulator. The IQ modulator is a multiplier providing the transposition to RF frequency. It needs the baseband information components (I and Q), and a carrier RF frequency provided by the RF synthesizer. The design of each component is critical as the noise added by the multipliers will be impossible to filter and can violate the out-of-band spurious emission. A major difficulty in the modulator design is the matching between the I and Q paths. Indeed, gain and phase imbalances create unwanted images (Fig. 2.7).

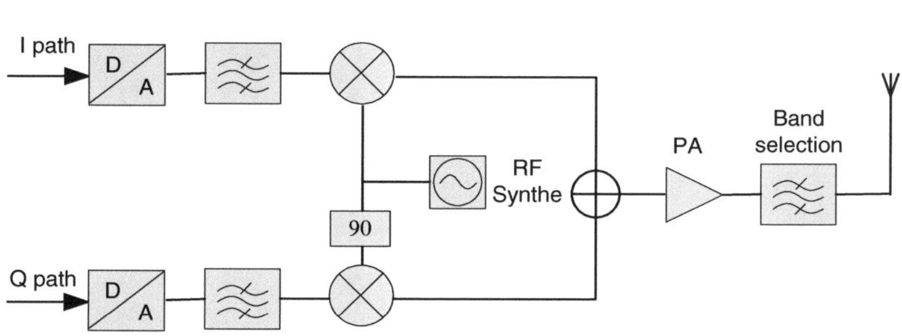

Fig. 2.6 Direct-up transmitter

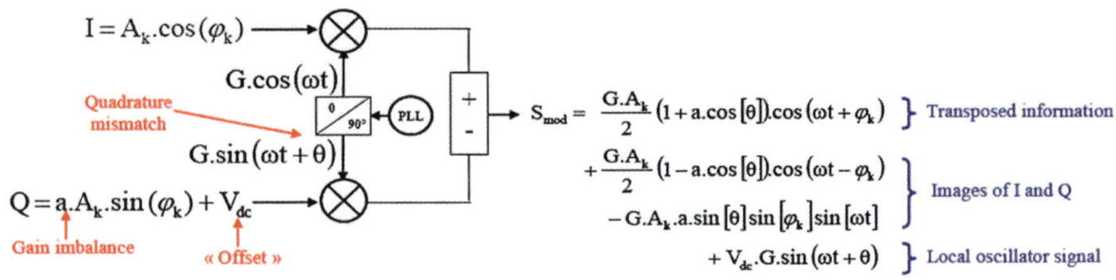

Fig. 2.7 Unwanted generated signals

The frequency synthesizer generates the RF signal required for transposition. This RF signal has to be stable with a high resolution to address all channels (the distance between two channels may be as low as 6.25 kHz). Moreover, the local oscillator phase noise should be compliant with the targeted standard performances especially ACLR1 and ACLR2. As a reminder, Adjacent Channel Leakage Ratio (ACLR) is defined as the ratio of the transmitted power to the power in the adjacent radio channel. This figure of merit represents the noise added by the transmitter in the adjacent channels. This noise can disturb the other users present in the same band and the receiver of the base station. The allowed limits are defined by the ACLR specifications. This is the reason that the noise added by local oscillator must not degrade the ACLR level.

After analyzing the targeted standards, it was found that only LTE could use directly this architecture by carefully designing each block. TETRAPOL is too restrictive with respect to wide-band spurious while TETRA would need linearization.

2.2.2.2 Cartesian (I/Q) Direct-Up Transmitter with Cartesian Feedback Loop

To be able to address the TETRA standard, a Cartesian Feedback Loop was introduced, as shown in Fig. 2.8.

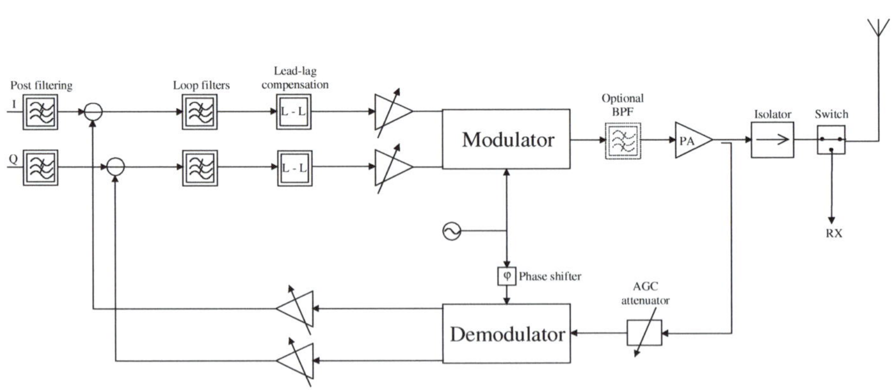

Fig. 2.8 Cartesian feedback Loop implementation

The information signal is split in I (In phase) and Q (Quadrature) and converted to analogue signal. These signals are compared with the feedback signals, then up-converted to RF frequency and finally summed into a single modulated signal. This modulated signal is then amplified and sent to the antenna. A part of the output signal is down-converted in order to recover I-feedback and Q-feedback signals.

This Cartesian loop helps improving the linearity by reducing the distortion of the circuits in the forward path such as the modulator and the power amplifier.

2.2.2.3 PLL for Phase Modulation

The two previous architectures are not suitable for TETRAPOL standard because of the noise generated by the multipliers that makes these solutions non-compliant with the out-of-band spurious emission specification. As TETRAPOL uses a constant signal envelope, another approach is used which consists of indirect modulation of the VCO through the control of the VCO frequency and yields to the simplest transmitter solution of those presented, as shown in Fig. 2.9.

Compared to a direct modulation of the VCO (Fig. 2.10), the D/A converter can be removed and the frequency drift defect during the modulation is eliminated.

The indirect modulator structure depicted in Fig. 2.9 is considered as the optimal topology for frequency or phase modulation. The architecture is composed of two main blocks:

- frequency synthesizer that generates the RF carrier. The frequency resolution must be lower than 1 Hz;
- transmit filter which is necessary to produce an efficient modulation signal.

The digital input data stream, which is first filtered by the transmit filter (FIR), has a Gaussian response. The data are then summed with a nominal value which corresponds to the carrier frequency and fed into a ΣΔ converter, the output of which controls the instantaneous divide value of the PLL. So, the nominal value set the carrier frequency and the variations correspond to the modulation of the output value by the input data stream.

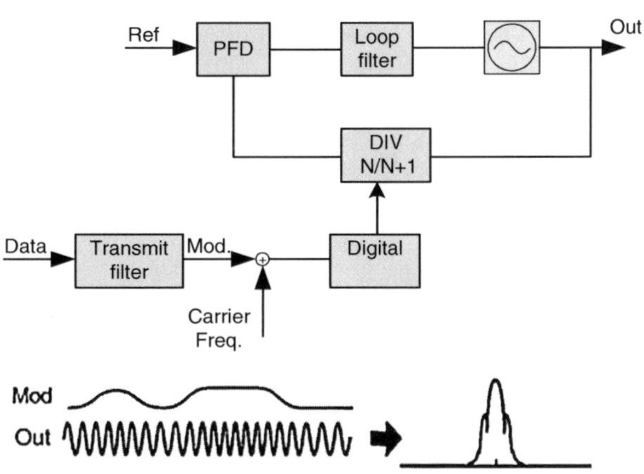

Fig. 2.9 Fractional-N modulator (indirect modulation of VCO)

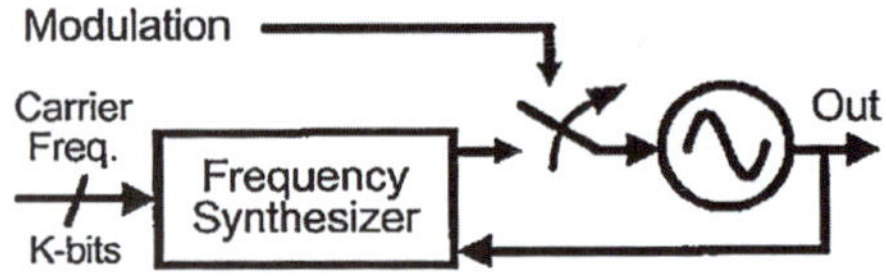

Fig. 2.10 Direct modulation of VCO

Figure 2.11 illustrates the transient and frequency behavior of such architecture.

The main important aspect is the noise behavior of the loop which has to be optimized in order to reach all the TETRAPOL performances. Figure 2.12 illustrates the results of a noise study in TETRAPOL environment which proves the feasibility of this architecture for TETRAPOL:

- PN_R4, PN_R3 and PN_R2 correspond to the phase noise contribution of the loop filter element;
- PN_VCO corresponds to the phase noise of the VCO at 400 MHz;
- PN_SD5 corresponds to the phase noise of the sigma-delta;
- PN_PFD corresponds to the phase noise of the phase frequency detector;
- PN_REF corresponds to the phase noise of the crystal reference;
- PN_total is the sum of all the noise contributors.

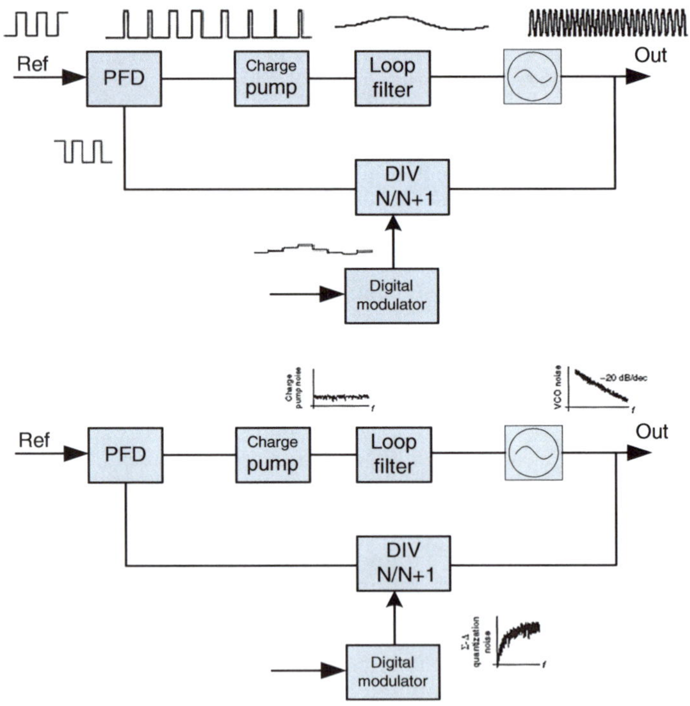

Fig. 2.11 Transient and frequency behavior

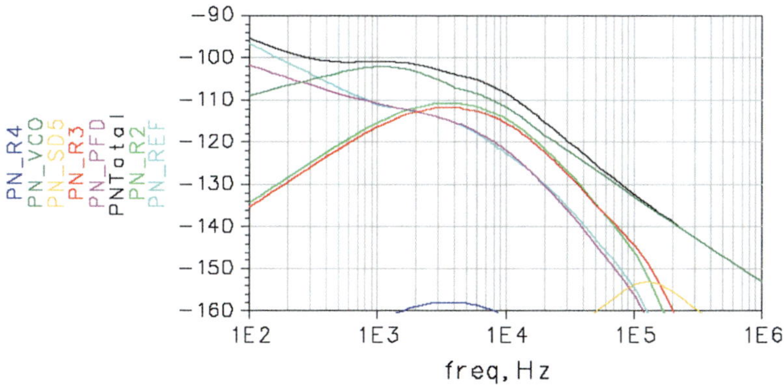

Fig. 2.12 Noise analysis

Generally, the indirect method, which involves the use of a phase locked loop (PLL) has the advantages of low power, feasibility of monolithic implementation, and phase coherence during frequency transitions and has a low phase noise, fine frequency resolution and fast dynamics.

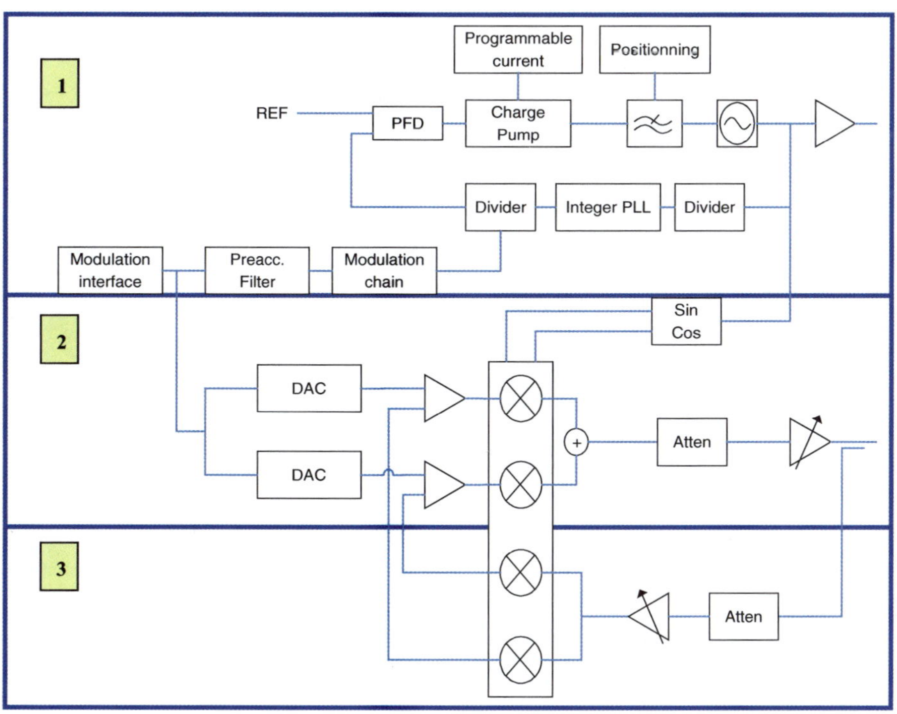

Fig. 2.13 TETRAPOL, TETRA and LTE multi-standard architecture

2.2.2.4 Global TX Multi-standard Architecture

Figure 2.13 illustrates the global TX architecture. In order to save silicon area and power consumption, the TETRAPOL fractional modulator is used as Local Oscillator when TETRA or LTE mode is activated.

In summary, the operation of this architecture is:

- When TETRAPOL mode is activated, only block 1 is ON while others are OFF;
- When LTE mode is activated, block 1 and 2 are ON;
- When TETRA mode is activated, block 1, 2 and 3 are ON.

2.3 Tunable Matching Network and Miniature Antenna for Multi-standard Handsets

2.3.1 Miniature Dual-Band Antenna

The rapid expansion of wireless communications demands the use of compact transceivers with multiband operation. This requirement presents a considerable challenge to the antenna designer, who has to deliver a small volume multiband antenna embedded into the transceiver's case. The main electrical features of these antennas are small size and multi-band operation. Cell phones have evolved from former dual-band designs (ex: GSM900 and 1,800 MHz) to more sophisticated designs such as those having GSM (global system for mobile communications, 860–980 MHz), DCS (digital communication system, 1,710–1,880 MHz), PCS (personal communication services, 1,880–1,990 MHz), universal mobile telecommunications system (UMTS, 1.9–2.17 GHz), WiBro (2,300–2,390 MHz), Bluetooth (2,400–2,480 MHz), satellite-digital multimedia broadcasting (S-DMB, 2,630–2,655 MHz), and Worldwide Interoperability for Microwave Access (WI-MAX) for IEEE802.16 bands (3.3–3.7 MHz). Many mobile handsets have design constraints on size and weight, partly driven by customer expectations, and partly owing to the increasingly ubiquitous use of location-aware applications. From an antenna design perspective, these constraints may be summarised thus to reduce the antenna size with unchanged, or improved, performance characteristics.

Several antenna size reduction techniques have been proposed over recent decades, including the use of high permittivity substrate, shorting pins, shorting walls, and modification of the geometry of the internal antenna [10–12]. Recently, another size reduction technique has been proposed, using the magnetic wall concept [13–15], the use of fractal geometry [16], employment of ground slots [17–20] and meta-materials technique [21–23].

2.3.1.1 Design and Nominal Performances

A miniature notch slot antenna was designed. The antenna is etched on a ground plane of size 110 × 50 mm^2, which is representative of typical current smartphones. A top and side view of the antenna is presented in Fig. 2.14a. The thickness of the PCB is 0.8 mm. The notch antenna is fed through a coupled microstrip line, near its centre, and loaded by a parasitic capacitive load near its short end. The antenna impedance matching is realized through an open stub and an inductive line section at the input, as shown in Fig. 2.14b. More details of the antenna design and optimization can be found in [24].

The antenna exhibits a dual-band response with a lower band at 815–1,010 MHz (VSWR < 3:1) covering the GSM standard (Fig. 2.15). The upper band at 1,840–3,365 MHz (VSWR < 2:1) covers the DCS-1800, IMT-2000, WiFi 802.11b/g/n, and LTE (band 7 and 38) standards. The WiMAX band at 3.4–3.6 GHz is also covered with a VSWR < 3:1.

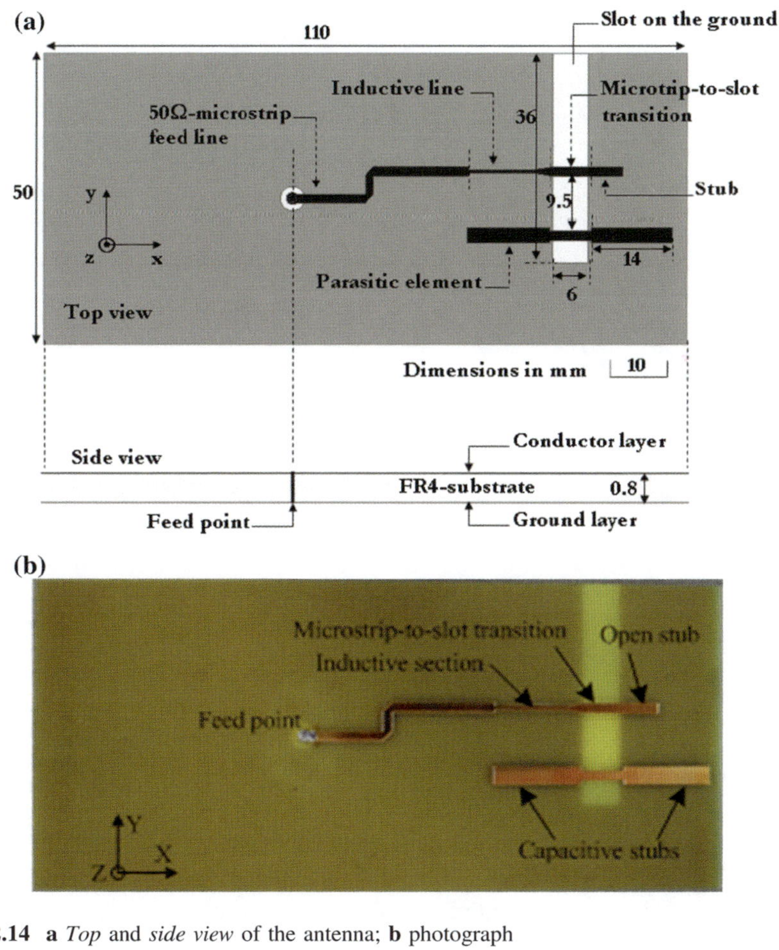

Fig. 2.14 a *Top* and *side view* of the antenna; b photograph

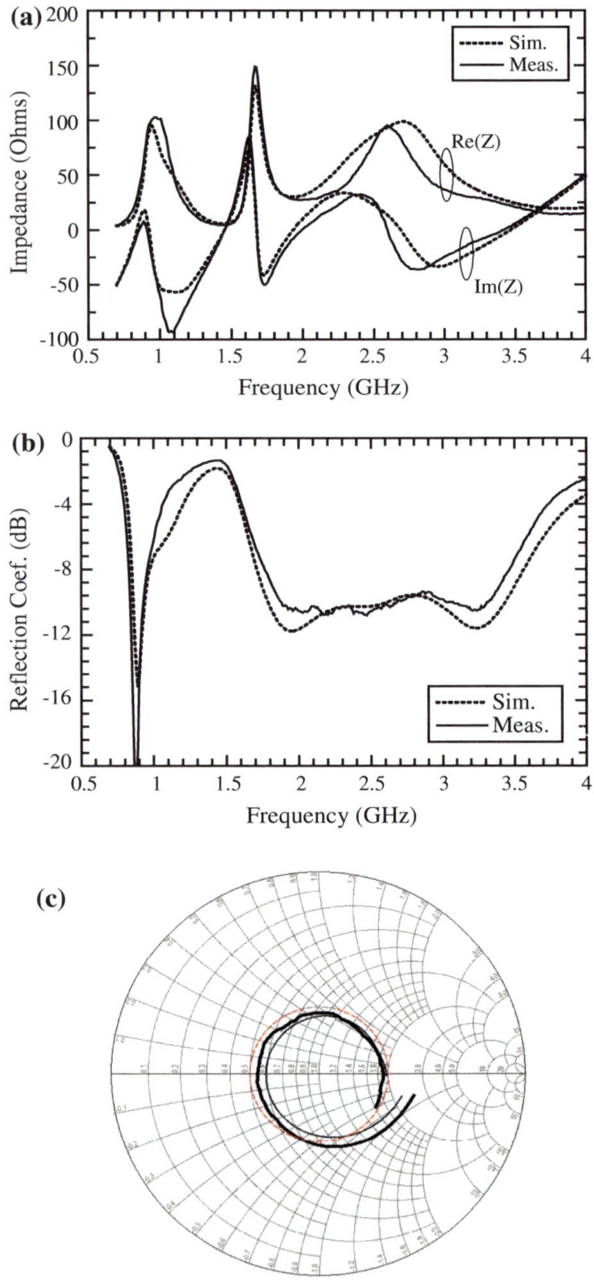

Fig. 2.15 **a** Measured and simulated impedance of the antenna; real and imaginary parts, **b** reflection coefficient and **c** impedance on the Smith chart at 1.7–2.7 GHz

The radiation patterns as well as the antenna gain were measured in an anechoic chamber (Fig. 2.16a, b) and are in good agreement with the simulations. The experimental efficiency of the antenna is above 70 % in each frequency band.

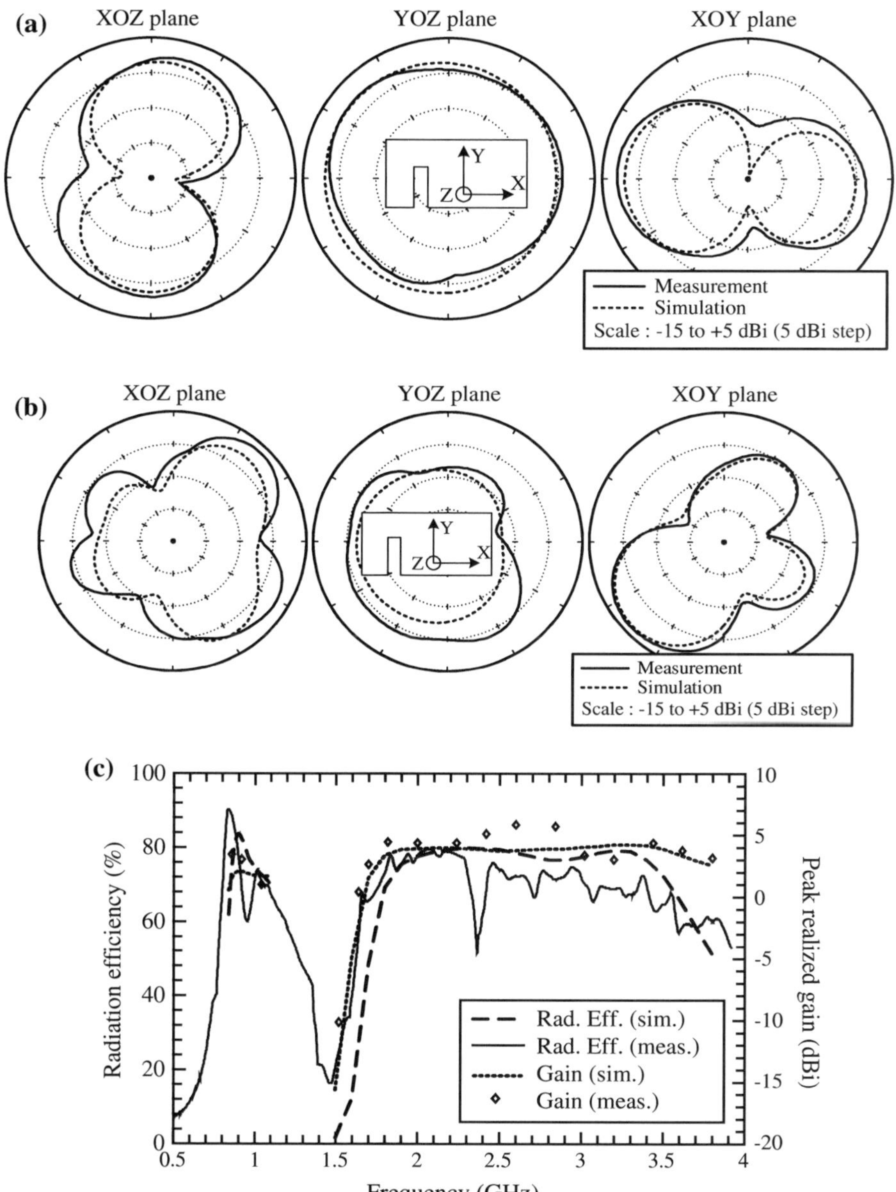

Fig. 2.16 **a** Measured and simulated radiation performances of the antenna; radiation patterns at 920 MHz, **b** radiation patterns at 1,800 MHz, and **c** gain and radiation efficiency

Overall, the antenna has been fully characterized experimentally and its performance is in very good agreement with the simulations.

2.3.1.2 Impedance Sensitivity to the Environment

The most significant perturbation of the antenna occurs in presence of a metallic objects in its close vicinity. We investigated the behavior of the antenna at a varying distance above a metal plane parallel to its ground plane (Fig. 2.17). The simulations and measurements exhibit a fairly good agreement. More specifically, the GSM-900 band is very significantly affected at distances closer than 50 mm. The 1.7–2.1 GHz band is also significantly mismatched at distances lower than 23 mm. In the case of the ISM-2400, LTE and WiMAX bands, the mismatch become critical for distances below 10 mm.

In the normal operation of a hand-held terminal, the user's hand usually impacts the antenna performances. We tested the antenna reflection coefficient in the case of different grips (Fig. 2.18). The exact hand grips are not the topic of discussion here, but rather the perturbation effect they have on the antenna. Most of hand grips do not result in a strong mismatch, i.e. the reflection coefficient remain below −6 dB. In some cases, where a finger is very close to the antenna's slot (Hand grip #2 and #6 in Fig. 2.18), the impact is stronger with reflection coefficients in the range of −3 to −5 dB. The corresponding reflection loss is in the range of 2–3 dB.

This assessment of the hand effect was done experimentally. Although some simulations can be done to get a more complete view of the antenna's performances (radiation pattern, efficiency), it is very difficult to implement a hand EM model close enough to a real hand grip to be able to perform simulation/measurement comparisons.

2.3.2 Tunable Matching Network

Matching impedance networks have become ubiquitous in all radio-frequency (RF) transmitters and receivers, especially in wireless mobile devices such as handheld computers (PDs) and cellular phones. Fixed matching networks are inserted between the power amplifier (PA) module and the antenna. The amplifier output impedance has to be matched to the low antenna impedance through an efficient matching network. A well matched RF system is required to increase the reliability and power efficiency of the system. However, antenna input impedance is affected by the presence of surrounding objects [25–28], and can fluctuate significantly with the antenna close to the human body or with the position of the hand on a handset that may bring a mismatch in which can cause more than half of the transmitted power to be reflected [29]. This leads to reduce the radiated power efficiency which in turns increases the demand on the battery.

To deal with this matter, tunable matching networks have been recently proposed [30–33]. Considering the size; matching range; and insertion loss constraints

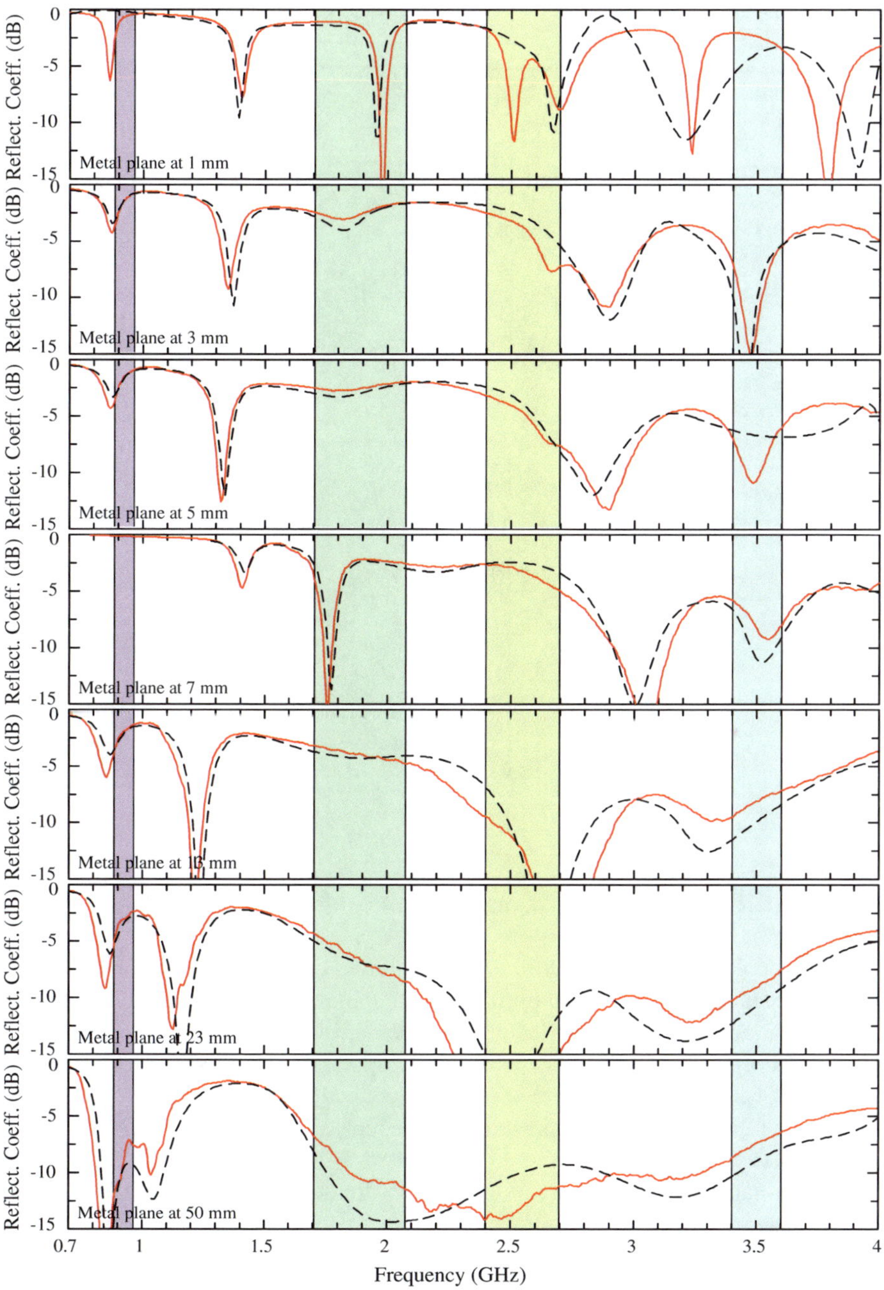

Fig. 2.17 Measured (*red*) and simulated (*black*) reflection coefficient of the antenna in presence of a metallic ground plane. The main telecom bands are highlighted: GSM-900 (*purple*), DCS-1800 and IMT-2000 (*green*), ISM2400 and LTE (*yellow*), WiMAX (*blue*)

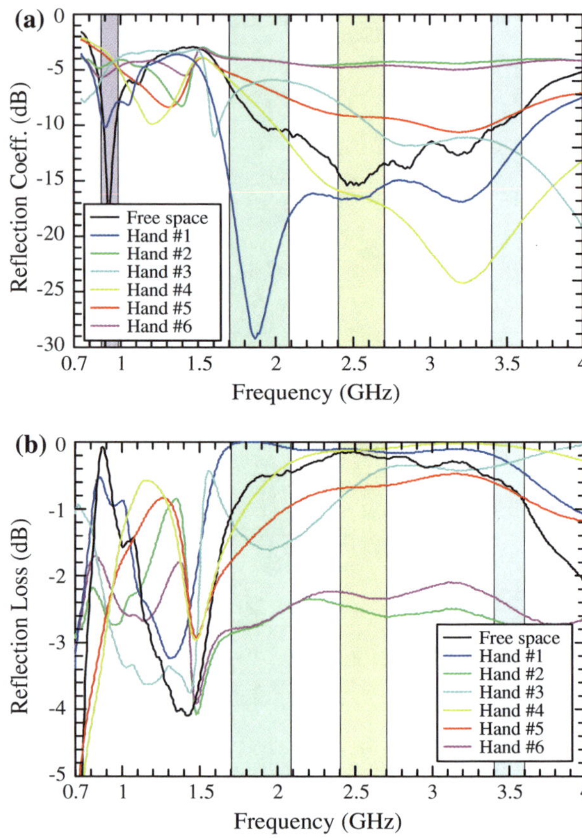

Fig. 2.18 Measured reflection coefficient (**a**) and corresponding reflection loss (**b**) of the antenna in presence of a hand with different grips. The main telecom bands are highlighted: GSM-900 (*purple*), DCS-1800 and IMT-2000 (*green*), ISM2400 and LTE (*yellow*), WiMAX (*blue*)

in practice, Π- and T-structured matching networks are commonly used in the tunable RF front-end.

The Tunable Matching Network (TMN) is based on a Pi-network composed of a fixed external inductor and two high power tunable capacitors integrated in a SOI CMOS 130 nm process. SOI CMOS provides an attractive trade-off among performance, cost and integration capability. Compared to standard bulk CMOS process, SOI CMOS features higher speed and reduced power consumption since the drain/source capacitances are substantially reduced. Moreover, the use of a high resistivity substrate enables high-Q integrated inductors, as well as excellent crosstalk isolation. Besides, by exploiting transistor stacking, SOI CMOS RF switch can be designed to handle arbitrarily high off-state voltages, which is a crucial feature for the implementation of a high power tunable matching network. Despite a higher wafer cost compared to standard bulk CMOS, the net impact of substrate cost on a fully packaged integrated circuit is quite small, and it is believed to further decrease in the near future, eventually reaching parity with standard bulk CMOS.

Implementation of the tunable capacitors is based on banks of binary weighted switched capacitors, where floating body NMOS transistors are used as low loss

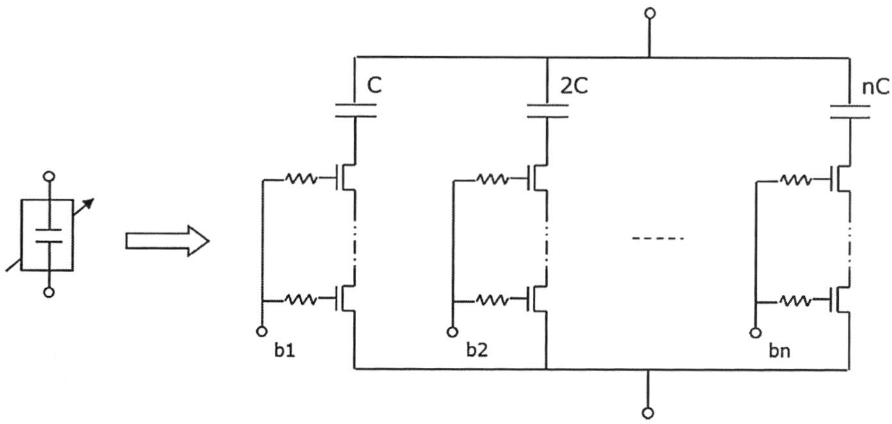

Fig. 2.19 Implementation of a tuneable capacitor in SOI CMOS

switches to select the appropriate capacitance value. As shown in Fig. 2.19, multiple transistors have been stacked in series to prevent breakdown in OFF state by providing voltage division of the high power RF signal.

Each switched-capacitor exhibits 32 states and has been designed to cover the range 0.7–2.8 pF with a minimum quality factor of 40 at 2.7 GHz and a maximum power rating of 36 dBm. The size of the chip including two tunable capacitors, a negative bias generator and a Serial Peripheral Interface (SPI) is 1 mm^2, and the power consumption is <1 mA under 2.5 V.

The chip has been directly attached on a dedicated FR4 test board as shown in Fig. 2.20. The external inductor is a surface mount device (SMD) from Johansson Technology and it has been soldered on the bottom layer of the test board.

A VHDCI connector is used to control the TMN which requires two analog inputs (VDD, PDN) for bias and three digital inputs (DCLK, DIN, DLOAD) for capacitors settings. Figure 2.21 shows the test setup used for small-signal characterization of the TMN, where a PXI module (not shown) is used to drive the SPI interface. A graphical user interface (GUI) has been developed under Labview to control the different instruments and allow acquisitions of S parameters of the TMN for different frequencies and capacitor states.

The TMN has been characterized experimentally in the frequency range of interest (1.7–2.7 GHz). The following figures show the measured smith chart coverage obtained for different inductance values, where the Smith chart coverage corresponds to the antenna impedance locus that can be matched to 50 Ω at TMN input (Fig. 2.22).

The TMN provides good impedance coverage at 2.2 and 2.7 GHz and is well centered on 50 Ohm at 2.2 GHz. At 1.7 GHz, impedance coverage is a bit limited and needs to be improved by further tuning of the inductor value or eventually by modifying the TMN configuration. Besides, the input impedance of the miniature dual-band antenna has to be checked to lie in the Smith chart coverage of the TMN.

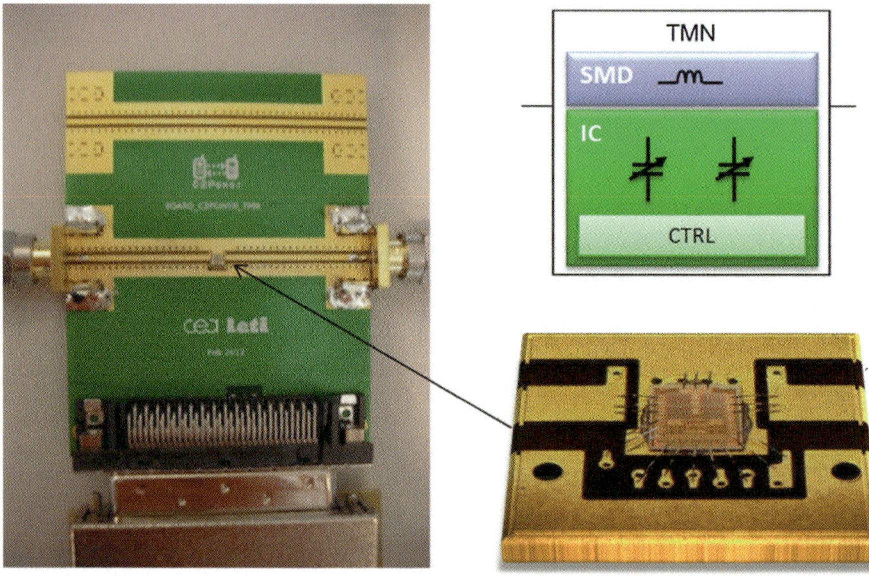

Fig. 2.20 Pictograph of the test board and tunable capacitors IC

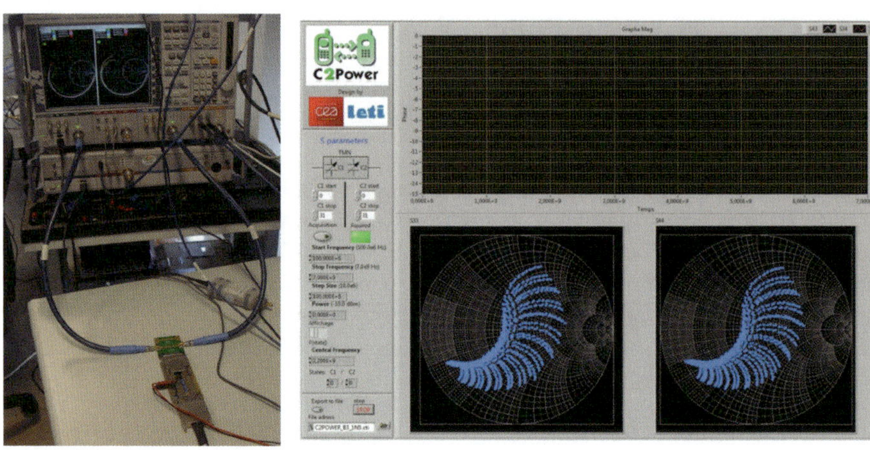

Fig. 2.21 Pictograph of the test setup (*left*) and TMN control user interface (*right*)

As shown in Figs. 2.23 and 2.24, a relatively good agreement is obtained between measurement and simulation for the 1,024 states of the TMN and in a large frequency range, including the band of interest.

At 2.2 and 2.7 GHz, a small shift between measurement and simulation can be observed and is partly due to higher losses than expected.

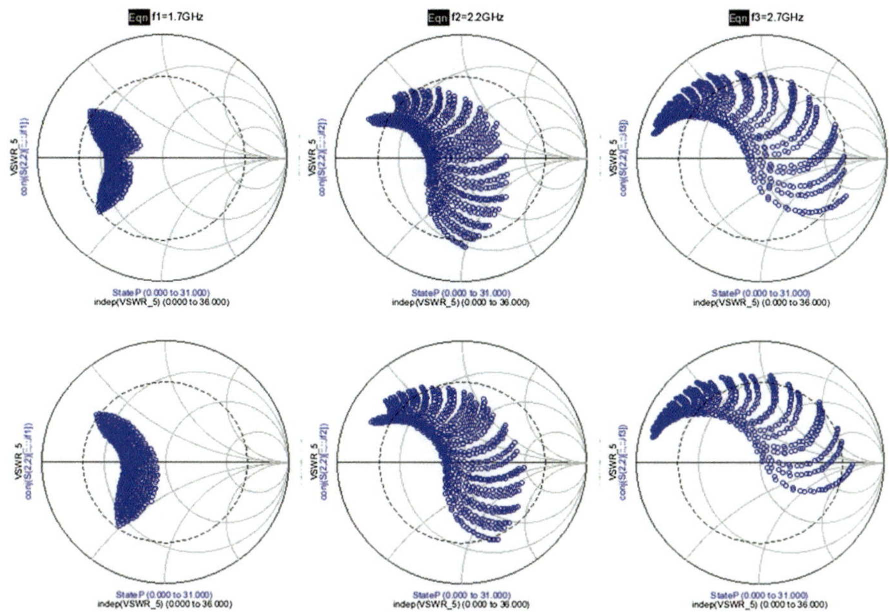

Fig. 2.22 Measured Smith chart coverage for L = 1 nH (*top*) and 1.5 nH (*bottom*) at 1.7, 2.2, and 2.7 GHz

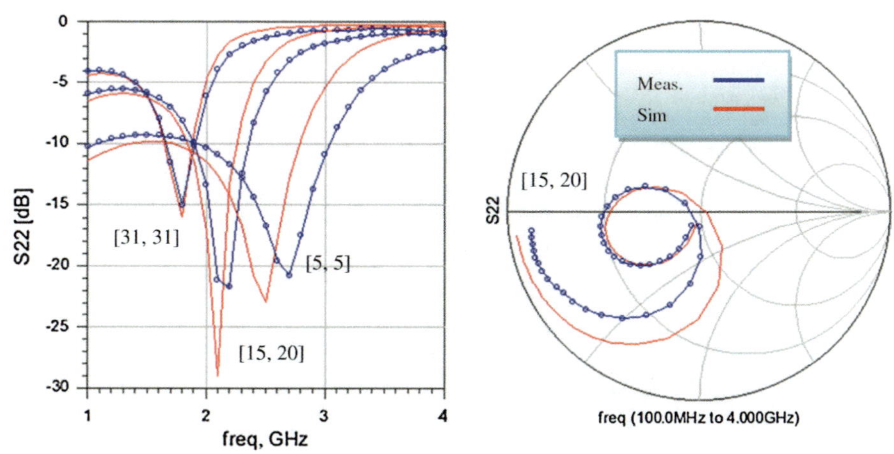

Fig. 2.23 Measured and simulated S22 for different TMN states (*left*) and frequencies (*right*)

2.3.3 Mismatch Sensing Unit

The schematic and a photograph of the TMN-Antenna module are presented in Fig. 2.25. The main transmission branch is composed of the input selection switch, a couple, the TMN and the antenna. Two detectors measure the RF power in the

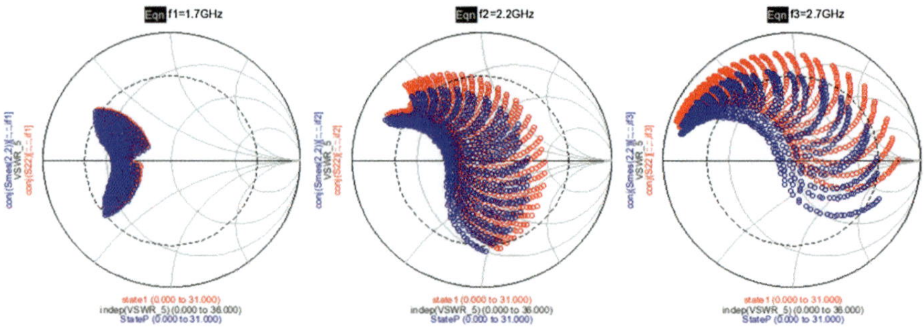

Fig. 2.24 Measured and simulated Smith chart coverage

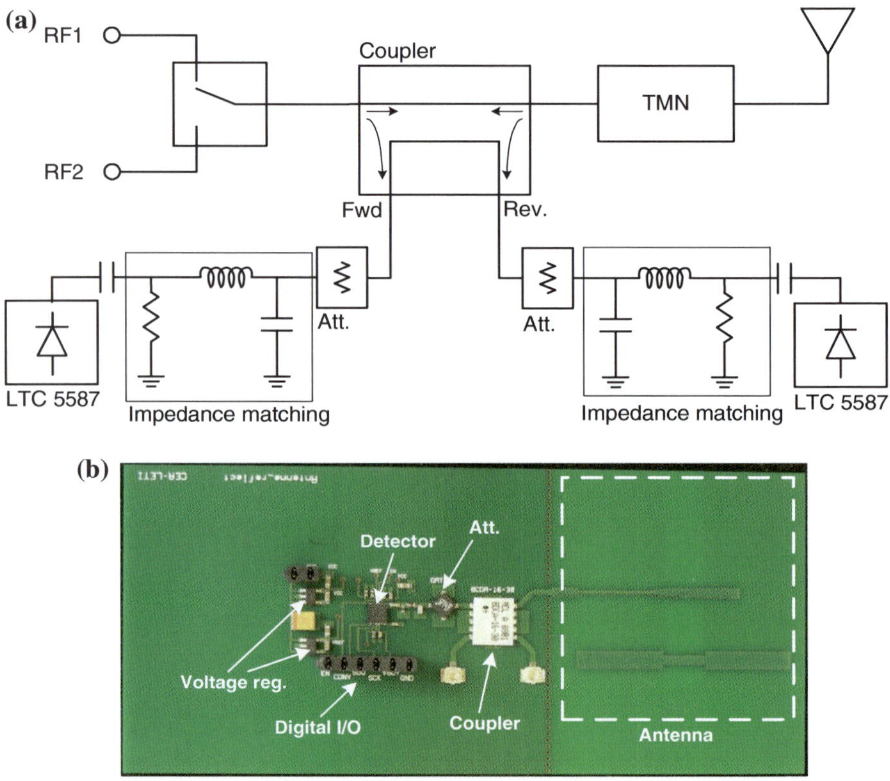

Fig. 2.25 **a** Schematic diagram of the TMN-Antenna module; **b** photograph of a preliminary prototype with a single detector and without the TMN

forward and reverse direction in order to compute the reflection coefficient. Each of these detectors is associated with an impedance matching network optimized for the 1.8–3.6 GHz band, as well as attenuators to adjust the input power to their dynamic range. The main components performances are described hereafter.

Table 2.1 Main performances of the input selection switch Hittite HMC544

Insertion loss	0.4 dB @ 2 GHz 0.6 dB @ 3.5 GHz	Input power for 1 dB compression point	36 dBm with 3 V control 39 dBm for 5 V control
Isolation	>13 dB	Control	0/+ 3 V or 0/+ 5 V < 2 μA

Switch Hittite HMC544 was chosen for its low insertion losses, high compression-point (it has to handle at least 1 W), and low power consumption. The main characteristics are summarized in Table 2.1.

The coupler is a Mini-Circuit BCDA-16−30+, which covers the 1,800–4,200 MHz band with about 16 dB coupling. Its insertion losses are typically 0.5 dB at 1,800–2,500 MHz and 0.7 dB at 3.5 GHz. Its directivity is typically 22 dB (Fig. 2.26).

The attenuator is from the GAT-xx+ series from Mini-Circuit and covers the DC-8 GHz band.

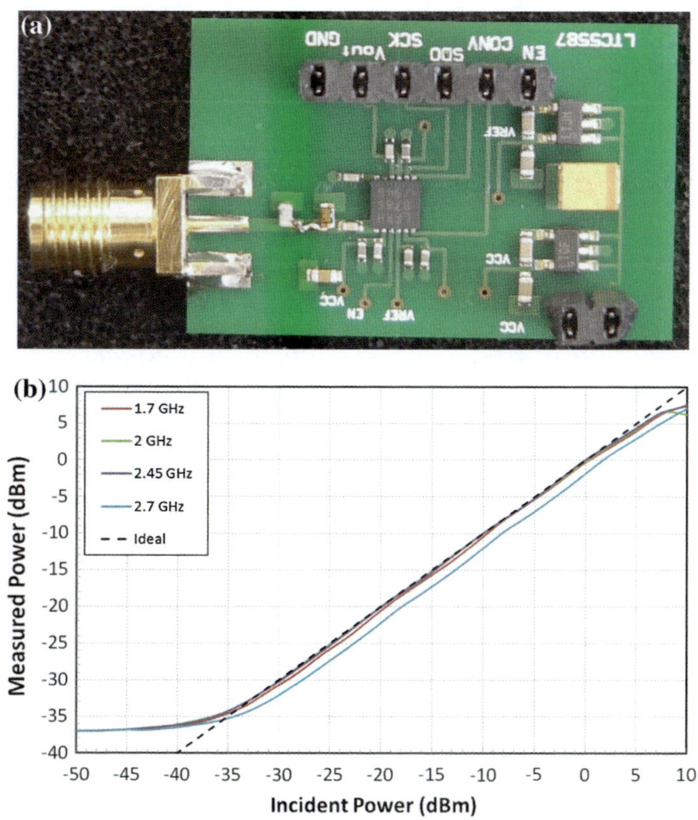

Fig. 2.26 **a** Photo of the detector test board; **b** measured response of the detector

Detector LTC5587 from Linear Technology is a 10 MHz to 6 GHz low-power precision RMS power detector with an integrated 12-bit serial analog-to-digital converter. Its detection range is −34 dBm to +6 dBm. Its control requires three digital inputs (EN, CONV, SCK) and the output is delivered through a single digital output (SDO). It is a low-power component (3 mA at 3.3 V and 500 ksps). The maximum sampling frequency is 500 ksps. The detector was tested with an optimized matching network for the frequency band 1.7–2.7 GHz.

In order to validate the capabilities of the coupler and detectors to measure accurately the reflection coefficient, we simulated the antenna associated to the coupler and the detectors, and calculated the reflection coefficient from the detectors outputs. The coupler and detectors S-parameters are given by the manufacturer; the S parameter of the antenna was measured in different cases: antenna in free space, antenna perturbed by a ground plane at different distances, antenna perturbed by a hand. In each antenna case, the actual measured reflection coefficient of the antenna (Γ_{meas}) was compared to the reflection coefficient calculated from the detectors outputs (Γdetect) over the frequency band 1.7–2.7 GHz. Figure 2.27b shows the error (Γdetect$-\Gamma_{meas}$) as a function of the actual reflection coefficient (Γ_{meas}). We can observe that the error is limited to ±2 dB, when the reflection coefficient is above −8 dB, which is the domain where the TMN and therefore accurate measurements of the reflected power are needed. At low reflection coefficient levels ($\Gamma_{meas} < -8$ dB), the errors are larger, which is expected since the power levels become close to the sensitivity of the detectors, but these errors are of no consequence since our goal is to know that the reflection level is below a certain level ($\Gamma_{meas} < -9.5$ dB (VSWR < 2:1) for instance) and not to have an accurate measurement.

2.3.4 Evaluation of the TMN-Antenna Performance

The TMN circuit simulations of the system have been performed to evaluate the expected performances. The simulation model takes into account the switched capacitors model, the series off-chip inductor, the interconnection parasitics (wirebonds) and the antenna.

The experimental antenna performances (impedance, efficiency) were introduced in the simulation in order to compute the actual radiated power taking into account the impedance mismatch loss and the losses of the antenna. To this purpose, the antenna is modeled as a 2-port component with the impedance of the antenna in port 1 and the effective transmission coefficient between port 1 and port 2. This transmission coefficient takes into account the reflection losses and an average efficiency of 70 % of the antenna.

Three typical cases are considered with respect to the antenna performances, corresponding to a favorable case and two critical cases:

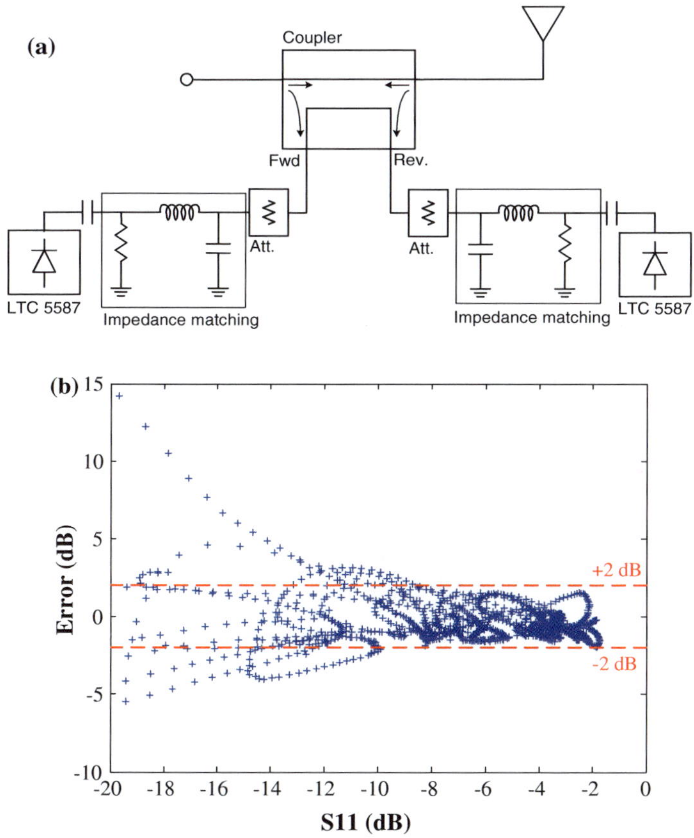

Fig. 2.27 **a** Schematic of the antenna associated to the coupler and detectors; **b** simulation of the error between the reflection coefficient calculated from the detectors output and the actual reflection coefficient

- Case 1: Antenna in free space;
- Case 2: Antenna at 3 mm of a metal plane;
- Case 3: Antenna in hand (A heavily perturbed position).

The simulations reported here are performed at three frequencies (1.7, 2.2 and 2.7 GHz) representative of the targeted frequency band. A series inductance value of 2.2 nH was used here, although this value is subject to future optimization. For each antenna case, the optimal TMN state (out of 1,024 possible states) was selected for highest radiated power. The results are reported in the following subsections. A schematic representation of the simulation setup of the TMN-Antenna is represented by Fig. 2.28.

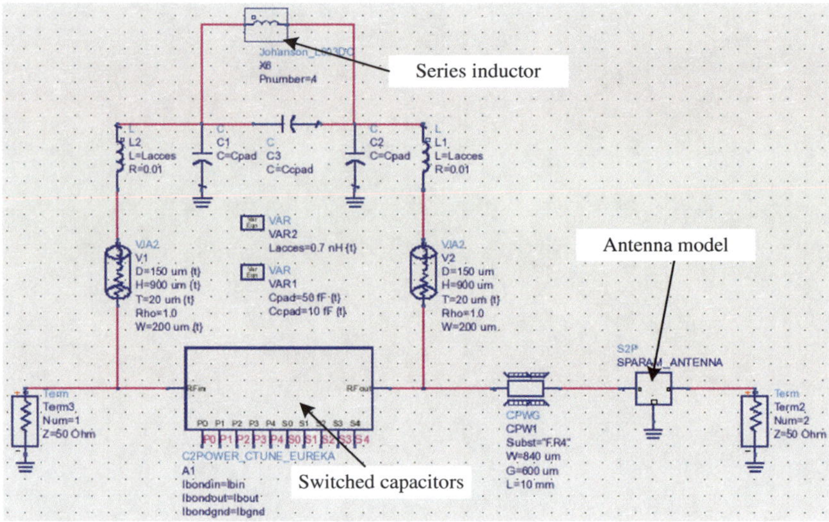

Fig. 2.28 Simulation schematic of the TMN-Antenna

2.3.4.1 Case 1: Antenna in Free Space

This is a favorable case since the antenna is unperturbed and therefore reasonably matched without TMN (S11 < −6.3 dB). Figure 2.29 and Table 2.2 report the simulation results. At 1.7 and 2.2 GHz, the reflection coefficient is strongly improved to −32 and −25.8 dB respectively. However, since there was no significant reflection loss to compensate, the final radiated power level is slightly degraded by 0.2–0.3 dB due to the insertion losses of the TMN. It is seen from the figure that TMN states are available with a VSWR < 1.5:1 at these two frequencies.

At 2.7 GHz, the TMN states are actually concentrated outside of the circle VSWR = 5:1, and the performances are here severely degraded with a reduction of the radiated power level by 3 dB. In this latter case, it is seen that the TMN is

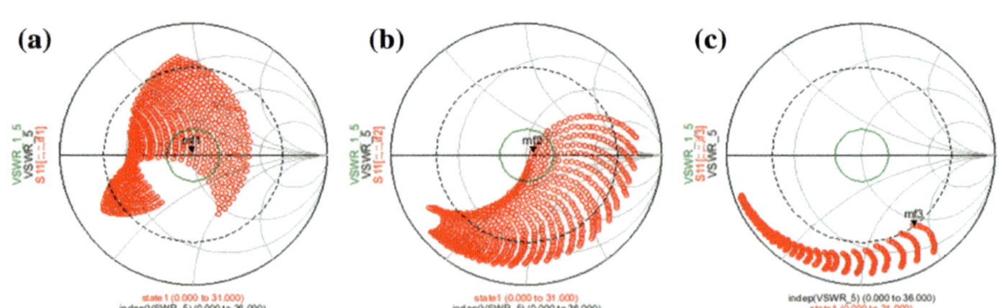

Fig. 2.29 Simulation of the 1,024 impedance states of the TMN-Antenna module in case 1 at **a** 1.7 GHz, **b** 2.2 GHz, and **c** 2.7 GHz

Table 2.2 Performance summary of the TMN-Antenna module in case 1

	1.7 GHz				2.2 GHz				2.7 GHz			
	State #		Performance (dB)		State #		Performance (dB)		State #		Performance (dB)	
	C1	C2	S_{11}	S_{21}	C1	C2	S_{11}	S_{21}	C1	C2	S_{11}	S_{21}
No TMN	–	–	−6.3	−2.7	–	–	−11.1	−1.9	–	–	−12.4	−1.8
TMN	28	21	−32.1	−3.0	0	4	−25.8	−2.1	0	0	−3.6	−4.8
Gain	–	–	–	−0.3	–	–	–	−0.2	–	–	–	−3.0

detrimental to the performances of the system and therefore needs to be optimized. The main optimization parameters are the series inductor and the transmission line length between the TMN and the antenna.

2.3.4.2 Case 2: Antenna at 3 mm of a Metal Plane

This case is one of the most critical cases considered with reflection coefficients as high as −1.6 dB.

Figure 2.30 and Table 2.3 report the simulation results. At the three frequencies, the reflection coefficient is strongly improved to below −11 dB. In turn, the radiated power is improved by 1 and 2.4 dB respectively at 1.7 and 2.2 GHz. At 2.7 GHz, the insertion losses of the TMN compensate the reflection loss reduction so that there is no net benefit.

2.3.4.3 Case 3: Antenna in Hand

We evaluated several hand positions to assess the influence of the user's hand on the antenna. One of the most critical was the hand position, where the user's finger touches the antenna slot, resulting in a reflection coefficient in the range −3 to −4 dB. Figure 2.31a shows a CAD view of the specific hand position.

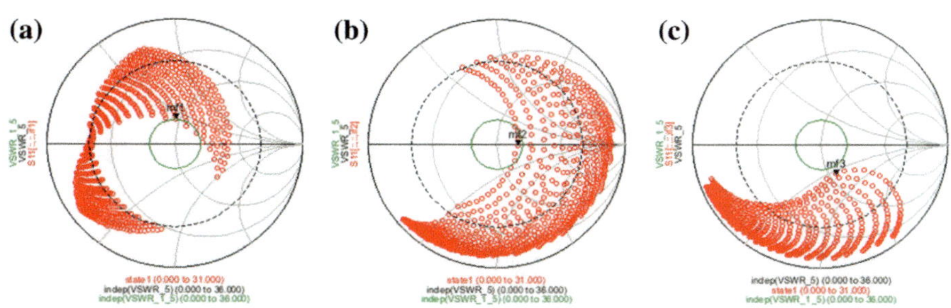

Fig. 2.30 Simulation of the 1,024 impedance states of the TMN-Antenna module in case 2 at **a** 1.7 GHz, **b** 2.2 GHz, and **c** 2.7 GHz

Table 2.3 Performance summary of the TMN-Antenna module in case 2

	1.7 GHz				2.2 GHz				2.7 GHz			
	State #		Performance (dB)		State #		Performance (dB)		State #		Performance (dB)	
	C1	C2	S_{11}	S_{21}	C1	C2	S_{11}	S_{21}	C1	C2	S_{11}	S_{21}
No TMN	–	–	−2.4	5.3	–	–	−1.6	−6.6	–	–	−7.2	−2.5
TMN	31	24	−14.1	4.3	18	0	15.7	−4.2	0	1	−11.4	−2.5
Gain	–	–	–	+1.0	–	–	–	+2.4	–	–	–	0.0

Figure 2.31 and Table 2.4 report the simulation results. At 1.7 and 2.7 GHz, the reflection coefficient is strongly improved to below −16 dB. In turn, the radiated power is improved by 1 and 1.5 dB respectively. Figure 2.31 shows that several

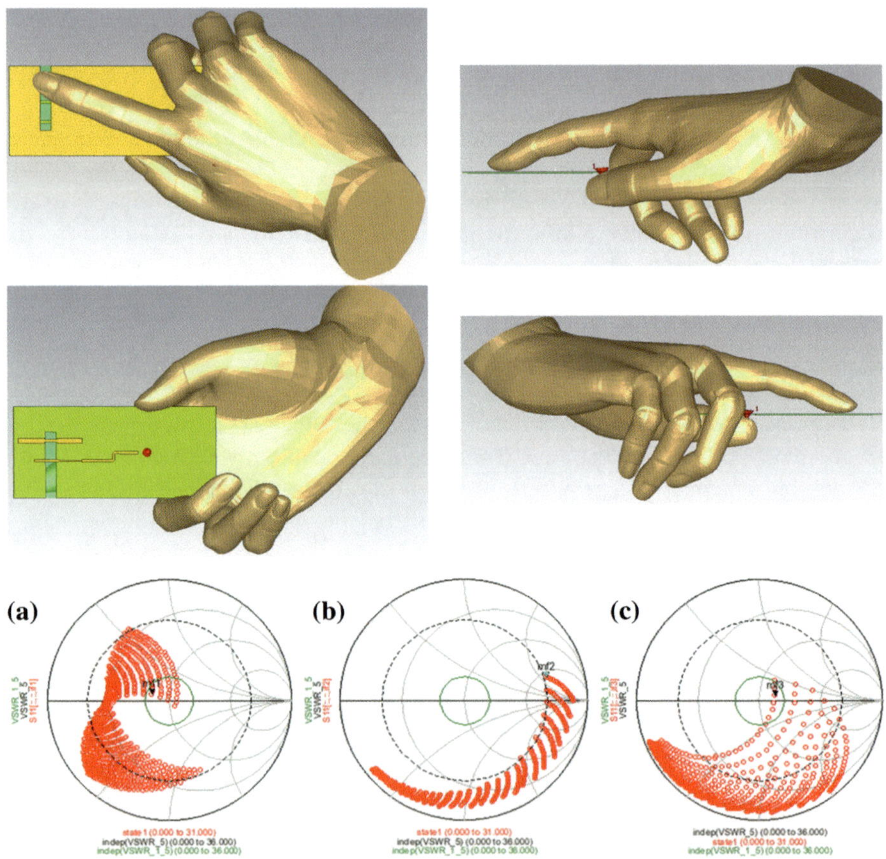

Fig. 2.31 CAD views of the Hand#2 position (case 3) and simulation of the 1,024 impedance states of the TMN-Antenna module at **a** 1.7 GHz, **b** 2.2 GHz, and **c** 2.7 GHz

Table 2.4 Performance summary of the TMN-Antenna module in case 3

	1.7 GHz				2.2 GHz				2.7 GHz			
	State #		Performance (dB)		State #		Performance (dB)		State #		Performance (dB)	
	C1	C2	S_{11}	S_{21}	C1	C2	S_{11}	S_{21}	C1	C2	S_{11}	S_{21}
No TMN	–	–	−3.1	−4.4	–	–	−3.8	−3.9	–	–	−3.4	−4.2
TMN	31	28	−16.5	−3.4	0	0	−3.3	−5.5	2	0	−17.3	−2.7
Gain	–	–	–	+1.0	–	–	–	−1.6	–	–	–	+1.5

impedance states are well into the VSWR = 1.5 circle. At 2.2 GHz, there is no impedance state in the VSWR = 5 circle; the reflection coefficient is slightly degraded and the transmission coefficient is reduced by 1.6 dB due to the insertion losses of the TMN. A further investigation based on a larger set of hand positions and test frequencies will lead to a better assessment of the occurrence of this issue, i.e. to determine if there is an actual weakness of the design at 2.2 GHz or if there are rare cases like this one. Nevertheless, considering the randomness and wide range of antenna perturbation occurring in actual applications, it is not expected to reach a net benefit in all the cases and all the frequencies. The design optimization may be done at the level of the series inductance (which can be easily changed) or the switched capacitors (which would require another CMOS fabrication runs).

The three antenna cases presented above show promising performances and lead to the following main conclusions:

- When the antenna is correctly matched, the TMN slightly degrade the overall efficiency of the system due to its own insertion losses [34, 35]. This is why it is of upmost importance to design such circuits in advanced technologies with minimum losses such as CMOS-SOI, RF MEMS, or Silicon on-Sapphire. This degradation can be acceptable if limited to about 0.5 dB on average, which is the case here at least in the range 1.7–2.2 GHz (Table 2.2).
- In critical configurations such as cases 2 and 3 presented above, the TMN can result in an improvement of the radiated power by 1–2.4 dB, which corresponds to an efficiency improvement of 25–73 %.
- In some cases of antenna configuration and frequency, the TMN does not provide an acceptable matching state. This can be improved by (i) further optimization of the inductance/capacitance values, (ii) reducing the targeted bandwidth to achieve a better optimization, (iii) increasing the number of states at the cost of increased complexity.

It is important to note that another factor of improvement not evaluated here is the benefit for the power amplifier efficiency of the better output impedance matching achieved thanks to the TMN in an integrated terminal. This topic is beyond the scope of this work, since it would require a co-design and close integration of the power amplifier and TMN.

2.4 Conclusion

This chapter presents a design approach for the RF front-end in multi-mode future emerging handsets, and engineered according towards energy saving, miniaturization, flexibility and re-configurability. The chapter mainly addresses two topics: the design of a multi-standard transceiver and the design of energy efficient miniature antennae co-designed with a matching network to overcome the perturbation caused by user's hand, face or metal close to the antenna.

At the transceiver level, a major challenge is to implement integrated architectures addressing multiple standards in different bands with a minimum of off-chip components for better miniaturization and lower energy consumption. We provided a generic design for a multi-standard RF transmitter architecture, with specific use-cases for TETRA, TETRAPOL and LTE. The performance results show the feasibility of innovative architectures with a high level of integration in state-of-the-art silicon CMOS technologies.

Pivotal to the RF front end in mobile handset, is the requirement for miniature antennae. We presented investigated candidate energy efficient design along with a matching network, to study the perturbation by and sensitivity of miniature antenna to metal sheets, user's hand or the human head. The miniature antennae embedded into wireless terminals are quite sensitive to their environment and exhibit wide impedance variations, resulting in high reflection losses between the RF front-end and the antenna. The antenna has been designed and demonstrated with an excellent agreement between the simulations and the measurements. It covers the GSM band as well as the 1.7–3.6 GHz band with a reflection coefficient better than −6 dB and efficiency higher than 70 %. The sensitivity of this antenna to its environment was investigated and it was found that the presence of a metallic plane or the user's hand close to the antenna would lead to reflection coefficients as high as −2 dB, which corresponds to −4.3 dB of losses.

The tunable matching network is based on a CLC pi-network with switched capacitors fabricated in CMOS-SOI technology. This circuit has been designed for the 1.7–2.7 GHz band.

A sensing unit was designed based on an integrated coupler and two power detectors measuring the incident and reflected power levels. This circuit was tested and operates adequately over the 1.7–2.7 GHz band with a good accuracy. An advantage of this implementation is the very low power consumption of the detectors (<10 mW). A drawback is that only the magnitude of the reflection coefficient is measured. Other sensing systems reported in the literature allow the measurement of the complex impedance using power and phase measurements. The power consumption of the TMN-Antenna module is quite low; it is estimated at <30 mW (10 mW per detector, 2.5 mW for the TMN). Finally, the performance of the TMN-Antenna module was evaluated by simulations in several cases of antennae with or without perturbation. The results indicate an efficiency improvement of more than 25 % in cases of strong perturbations.

It is worth noting that the insertion losses of the TMN always result in some performance reduction in specific cases where the antenna is not perturbed, which is expected due to the adding of circuits [34, 35]. On the other hand, it provides great benefits in cases of strong perturbations. Therefore, an accurate evaluation of its benefit to the overall power consumption of the terminal would require a statistical investigation of the terminal usage to determine the long-term average gain. To the best of the author's knowledge, such investigation has never been realized but it is interesting to note the recent announcements of leading terminal manufacturers of the integration of TMN systems in their terminals.

References

1. Wong, A., Kathiresan, G., Chan, C., et al.: A 1 V wireless transceiver for an ultra-low-power SoC for biotelemetry applications. IEEE J. Solid-State Circuits **43**(7), 1511 (2008)
2. Quinlan, P., Crowley, P., Chanca, M., et al.: A multimode 0.3–200-kb/s transceiver for the 433/868/915-MHz bands in 0.25-_m CMOS. IEEE J. Solid-State Circuits **39**(12), 2297 (2004)
3. Qi, Z., Kuang, X., Wu, N.: An ultra-low-power RF transceiver for WBANs in medical applications. J. Semiconductors **32**(6), 065008 (2011)
4. Peiris, V., Arm, C., Bories, S., et al.: A 1 V 433/868 MHz 25 kb/s-FSK 2 kb/s-OOK RF transceiver SoC in standard digital 0.18 _m CMOS. IEEE ISSCC Digest of Technical Papers, vol. 48, p. 258 (2005)
5. Guofeng, L., Nanjian, W.: A low power flexible PGA for software defined radio systems. J. Semiconductors 33(5), 055006 (2012)
6. Telecommunications Industry Association: APCO Project 25 System and Standards Definition. Telecommunications Industry Association, Arlington, TIA/EIA-102.A (1995)
7. Telecommunications Industry Association: Project 25 FDMA Common Air Interface New Technology Standards Project Digital Radio Technical Standards. Telecommunications Industry Association, Arlington, TIA/EIA-102.BAAA (1998)
8. Cayla, G.: TETRA: the new digital professional mobile radio. In: Proceedings of 5th Seminar on Digital Mobile Radio Communications, pp. 113–118 (1992)
9. http://www.tetrapol.com/
10. Skrivervik, A.K., Zurcher, J.-F., Staub, O., Mosig, J.R.: PCS antenna design: the challenge of miniaturization. IEEE Antennas Propag. Mag. **43**, 12–27 (2001)
11. Lee, C.S., Tseng, K.-H.: Size reduction of microstrip antennas. Electron. Lett. **37**, 1274–1275 (2001)
12. Shackelford, A.K., Lee, K.F., Luck, K.M.: Design of small-size wide bandwidth Microstrip patch antennas. IEEE Antennas Propag. Mag. **45**, 75–83 (2003)
13. Deshmukh, A.A., Kumar, G.: Half U-slot loaded rectangular microstrip antenna. In: IEEE Antennas Propagation Society International Symposium, vol. 2, pp. 876–879, Columbus OH (2003)
14. Chair, R., Mak, C.-L., Lee, K.-F., Luk, K.-M., Kishk, A.A.: Miniature wide-band half U-slot and half E-shaped patch antennas. IEEE Trans. Antennas Propagat. **53**, 2645–2651 (2008)
15. Guo, L., Wang, S., Chen, X., Parini, C.: Miniaturised antennas for UWB communications. In: Proceedings of the European Conference on Antennas and Propagation, pp. 3774–3778. Berlin, Germany, 23–27 Mar 2009
16. Gianvittorio, J.P., Rahmat-Samii, Y.: Fractal antennas: a novel antenna miniaturization technique, and applications. IEEE Antennas Propag. Mag. **44**(1), 20–36 (2002)

17. Byndas, A., Hossa, R., Bialkowski, M.E., Kabacik, P.: Investigation into operation of single and multi-layer configurations of planar inverted-F antenna. IEEE Antenna Propag. Mag. **49** (4), 22–33 (2007)
18. Razali, A.R., Bialkowski, M.E., Tsai, F.-C.E.: Multi-band planar inverted-F antenna with microstripline coupling to open-end ground slots. In: Proceedings of Asia Pacific Microwave Conference, pp. 2471–2474, Dec 2009
19. Razali, A.R., Bialkowski, M.E.: Coplanar inverted-F antenna with open-end ground slots for multi-band operation. IEEE Antenna Wirel. Propag. Lett. **8**, 1029–1032 (2009)
20. Hossa, R., Byndas, A., Bialkowski, M.E.: Improvement of compact terminal antenna performance by incorporating open-end slots in ground plane. IEEE Microw. Wirel. Compon. Lett. **14**(6), 283–285 (2004)
21. Alam, M.S., Islam, M.T., Misran, N.: A novel compact split ring slotted electromagnetic bandgap structure for microstrip patch antenna performance enhancement. Prog Electromagnet. Res **130**, 389–409 (2012)
22. Yu, A., Yang, F., Elsherbeni, A. Z.: A dual band circularly polarized ring antenna based on composite right and left handed metamaterial. Prog Electromagnet. Res **78,** 73–81 (2008)
23. Sayem, A.T.M., Ali, M.: Characteristics of a microstrip-fed miniature printed Hilbert slot antenna. Prog Electromagnet. Res **56**, 1–18, (2006)
24. Niamien, M.A.C., Dussopt, L., Delaveaud, C.: A Compact dual-band notch antenna for wireless multi-standard terminals. IEEE Antennas Wirel. Propag. Lett. **11**, 877–880 (2012)
25. Karimullah, K., Nyquist, D., Chen, K.: Interaction of thin wire antennas with conducting, polarizable bodies—theory and experiment. In: Proceedings of IEEE International Symposium on Antennas and Propagation Society (APS '78), vol. 16, pp. 219–222. College Park, Md, USA, May 1978
26. Norklit, O., Teal, P.D., Vaughan, R.G.: Measurement and evaluation of multi-antenna handsets in indoor mobile communication. IEEE Trans. Antennas Propag. **49**(3), 429–437 (2001)
27. Jensen, M.A., Rahmat-Samii, Y.: Performance analysis of antennas for hand-held transceivers using FDTD. IEEE Trans. Antennas Propag. **42**(8), 1106–1113 (1994)
28. Toftgard, J., Hornsleth, S.N., Andersen, J.B.: Effects on portable antennas of the presence of a person. IEEE Trans. Antennas Propag. **41**(6), 739–746 (1993)
29. de Mingo, J., Valdovinos, A., Crespo, A., Navarro, D., Garcia, P.: An RF electronically controlled impedance tuning network design and its application to an antenna input impedance automatic matching system. IEEE Trans. Microw. Theory Tech. **52**(2), 489–497 (2004)
30. Vicki Chen, L.-Y., Forse, R., Chase, D., York, R. A.: Analogtunable matching network using integrated thin-film BST capacitors. In: IEEE MTT-S International Microwave Symposium Digest, vol. 1, pp. 261–264. FortWorth, Tex, USA, June 2004
31. Moritz, J.R., Sun, Y.: Frequency agile antenna tuning and matching. In: Proceedings of 8th International Conference on HF Radio Systems and Techniques (IEE Conf. Publ. No. 474), pp. 169–174. Guildford, UK, July 2000
32. Leenaerts, D.M.W.: Low power RF IC design for wireless communication. In: Proceedings of International Symposium on Low Power Electronics and Design, pp. 428–433. Seoul, Korea, Aug 2003
33. Abidi, A.A.: Low-power radio-frequency IC's for portable communications. Proc. IEEE **83**(4), 544–569 (1995)
34. Van Bezooijen, A.: Antenna tuner for hand-sets. In: Proceedings of Advancements in Front End Modules for Mobile and Wireless Applications workshop, 2012 IEEE MTT-S Int. Microwave Symposium, June 2012
35. Boyle, K.R., Spits, E., de Jongh, M.A., Sato, S., Bakker, T., Van Bezooijen, A.: A self-contained adaptive antenna tuner for mobile phones. In: Proceedings of 6th European Conference on Antennas and Propagation (EuCAP 2012), pp. 1804–1808, Mar 2012

Chapter 3
Energy Efficient Power Amplifier Design

Abubakar Sadiq Hussaini, Issa Elfergani, Ayman Radwan, Jonathan Rodriguez, Laurent Dussopt, Alexandre Giry, Michael Pelissier, Sami Aissa, Frederic Fraysse and Dany Lenox

Abstract Legacy mobile handsets already incorporate multimode features, with dual, tri and quad band functionality already the norm, and the very latest smart phone having WiFi connectivity. However, complete interoperability is still in its infancy with several hurdles still to overcome, and energy efficiency is one of them. Unless mobile handsets are smart enough to be energy aware and flexible in switching between several standards, the 5G slogan of anytime, anywhere connectivity could remain a "vision" rather than close to market. To attain true energy efficiency lays down stringent design requirements on the RF system that in today's handset is a key consumer of power. In fact the power amplifier is the main culprit for stealing energy from the battery supply, and the reason why mobile handset become hot after a lengthy call. This effect will multiply when ubiquitous broadband connectivity becomes the standard, leading to significant research on active cooling, considering approaches similar to fan techniques in laptop and desktop devices. However, rather than trying to reduce the effects of overheating, we aim to tackle the problem at the bud, by considering energy efficient designs. In fact, this chapter is the sequel to the former that was dedicated to the study of new designs for hardware components that can provide truly flexible and energy efficient multi-standard transceivers. This chapter completes this study by focusing on energy efficient power amplifier design. We consider three innovative designs, to operate at varying operating frequencies, covering the frequency bands 400 MHz, 1,920 MHz and 3.5 GHz. The proposed RF Power Amplifiers show considerable enhancements compared to the state-of-the-art existing technologies.

A.S. Hussaini (✉) · I. Elfergani
Instituto de Telecomunicações, Aveiro, Portugal
e-mail: ash@av.it.pt

L. Dussopt · A. Giry · M. Pelissier
CEA-LETI, Grenoble, France

S. Aissa · F. Fraysse · D. Lenox
CASSIDIAN, Élancourt, France

A. Radwan · J. Rodriguez
Instituto de Telecomunicações, Campus Universitário de Santiago, 3810-193 Aveiro, Portugal

A. Radwan and J. Rodriguez (eds.), *Energy Efficient Smart Phones for 5G Networks*, Signals and Communication Technology,
DOI 10.1007/978-3-319-10314-3_3

3.1 Introduction

The research within this book addresses a critical problem facing current and upcoming mobile handsets, namely hungry power applications. The foreseen increase in data over the next decade, and the ever growing market trends towards rich broadband services will require smart, configurable and energy efficient mobile handsets that are able to deliver these services on demand. Today′s phones are already heading towards this direction with multimode features, providing connectivity to other frequency bands on the same technology, and with WiFi connectivity. Moreover, applications are power hungry that a simple GPS application running on your smart phone has the potential to run down your battery power. However, with more sophisticated and rich content services becoming available, demand will increase providing the impetus for more stringent designs that can go that step further toward energy saving. This chapter looks at one most responsible culprits for consuming power on today′s handset devices, and for this we need to dive into the RF world again and investigate the power amplifier design. In the previous chapter, innovative architectures for energy efficient multi-standard transceivers were presented, together with an adaptive antenna front-end. This chapter completes the whole picture of a multi-standard front-end radio by analyzing the Power Amplifier (PA), and addressing energy efficient designs. With a power-added efficiency in the range of 20–40 %, power amplifiers are major contributors to the power consumption of wireless terminals, especially for recent non-constant envelope modulations used in LTE imposing severe linearity constraints. In this regard, the development of new power amplifier architectures with recent solid-state technologies is a very active research topic [1]. Two architectures are investigated here, the Doherty [2] approach and the Envelope-Tracking [3] approach at different frequency bands.

In the first part, we address two power amplifiers based on the Doherty topology, operating at different frequencies for different applications. The first Doherty design addresses PMR mobile applications to match the transceiver architectures given in the previous chapter, in the 400 MHz band with a maximum output power of 10 W. The second design is engineered for mobile WiMAX user terminal applications in the 3.4–3.6 GHz band with a maximum output power of 1 W. The second part presents energy efficient power amplifier design based on the envelope-tracking technique, designed for LTE hand-held applications in the 1,920−1,980 MHz band with a maximum output power of 1 W. The results obtained from the different PA designs in this chapter show promising enhancements in terms of the energy efficiency and linearity trade-off, and confirm the benefits of these technologies for future mobile terminals.

3.2 Doherty Power Amplifier for LTE 400 MHz Band

This section focuses on a Power Amplifier topology often used in base stations for modulation with complex envelope: Doherty Topology [2]. The efficiency enhancement technique in the linear region of operation of power amplifiers is implemented, for better efficiency and a low-level output power. Several efficiency enhancement techniques are available, (such as Chireix outphasing [4] or Linear amplification using Non-linear Components (LINC), Doherty Configuration, Kahn Envelope Elimination, Restoration (EER) and Envelope Tracking (ET) [5]). Most of these techniques involve complex architecture and external control circuits and signal processing, apart from Doherty configuration which does not require any additional components, which together with its self-managing characteristics, makes its implementation more attractive. The targeted application is a 5 W mobile LTE Modem in 400 MHz band. Currently, the solution implemented in the PMR uses Class AB amplifier based on discrete transistors.

3.2.1 Target Specifications

The main objective of this work is to evaluate the interest of a Doherty PA for vehicular PMR applications, as compared to a classic class AB topology, with LTE uplink modulation at 5 MHz bandwidth. The figures of merit of the targeted PA are listed below:

- Frequency range: 380–400 MHz
- Average output power (Pout): ≈40 dBm
- Power Added Efficiency (PAE): >40 %
- Adjacent Channel Leakage Ratio (ACLR): <−30 dBc
- Error Vector Magnitude (EVM): <17.5 % in QPSK
- 3rd order Intermodulation distortion (IMD3) <−30 dBc

The main figure of merit of power amplifiers is the Power Added Efficiency (PAE), defined as

$$PAE = 100\frac{P_{out} - P_{in}}{P_{Supply}} \text{ expressed in\%}$$

where P_{out} and P_{in} are the output and input signal power (Watt) and P_{supply} equals to Vcc*Icc.

The linearity of a PA can be evaluated by two main measurements:

- Adjacent Channel Leakage Ratio (ACLR), expressed commonly in dB, which is defined as the ratio of the transmitted power to the power in the adjacent radio channel;

- The error vector magnitude (EVM), often expressed in percentage, which measures the distortion introduced on the signal emitted by an RF transmitter.

IMD3 refers to the 3rd order intermodulation distortion of a system. This value is expressed in dBc. This measurement is done by injecting a 2 continuous-wave (CW) tones (one at Frequency F1 and the other at frequency F2) with a frequency spacing (F2–F1) between the two carriers at the input of the PA. In addition to the two amplified carriers at F1 and F2, the non-linearity of the PA creates some intermodulation products at frequencies 2F1-F2 and 2F2-F1. The level difference between carrier and intermodulation product at 2F1-F2 or 2F2-F1 is the IMD3.

The different phases of the presented work are listed below:

- Simulation of Doherty structure on AGILENT ADS;
- Printed Circuit Board layout and realization;
- Power Amplifier Optimization and Measurement on PCB.

The measurement results will be presented for a Doherty Amplifier structure compared to a topology commonly used for a Private Mobile Radio market.

3.2.2 ADS Simulations

3.2.2.1 Circuit Description

The transistor selected for designing the Doherty structure is a discrete 25 W transistor from ST Microelectronics PD85025. Two of these transistors are used to create the Doherty topology: one as main amplifier and a second as auxiliary amplifier. Each amplifier is matched to 50 Ω at input and output.

The input signal is split with a 3 dB hybrid coupler and drives the main and auxiliary transistors. The main transistor is biased in class AB and the quiescent

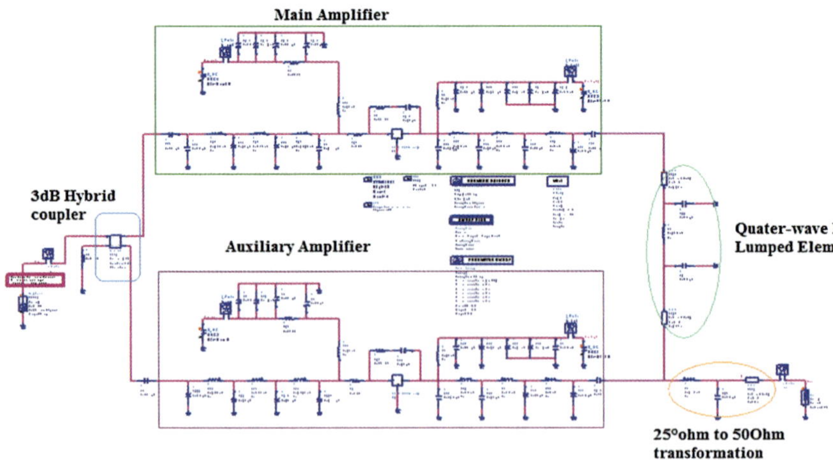

Fig. 3.1 Doherty schematic

current of the auxiliary transistor is optimized following the PAE and linearity performances. At 400 MHz frequencies, lumped components replacing the quarter wave line are used to realize the impedance transformation. This change allows reducing the size of the Doherty structure.

A model of this transistor is available and allows simulating the circuit on CAD tools. The schematic of simulation is shown in Fig. 3.1.

3.2.2.2 Simulations Results

The simulations have been performed in harmonic balance mode. Input and output powers, current and power consumption are calculated. The Doherty topology is compared with a common class AB structure for an output power of 40 dBm at three frequencies. The curves are presented in Figure.

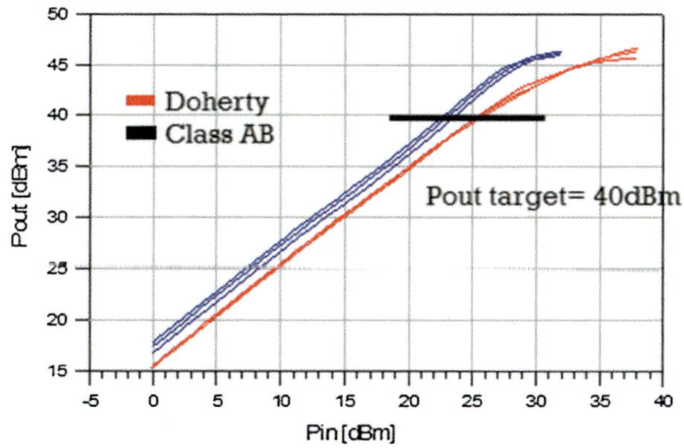

Fig. 3.2 Output power versus input power for Doherty and class AB topologies at 380–390–400 MHz

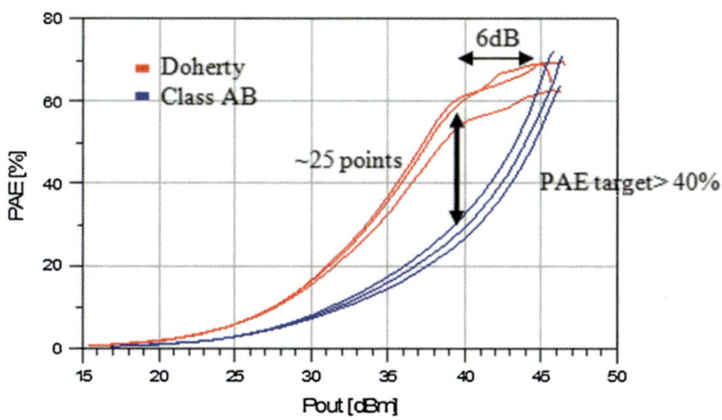

Fig. 3.3 PAE versus output power for Doherty and class AB topologies at 380–390–400 MHz

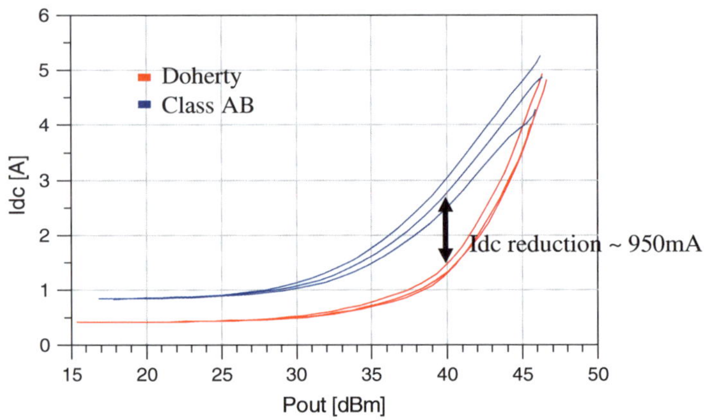

Fig. 3.4 Current consumption versus output power for Doherty and class AB topologies at 380–390–400 MHz

Table 3.1 IMD3 measurements at 40 dBm with 2-tones

IMD3			
Frequency (MHz)	Spec (dBc)	Doherty (dBc)	AB–AB (dBc)
380	−30	−38	−46
390	−30	−37	−45
400	−30	−37	−48

In Fig. 3.2, the gain of the designed Doherty structure is 2 dB lower than a class AB topology. The saturation output power is close to 47 dBm on the two topologies. This indicates that the power recombination is good after the transistors with the quarter-wave line.

The PAE curves shown in Fig. 3.3 are representative of a Doherty structure. The saturation of the efficiency on 6 dB of output power back-off is observed in simulation. The PAE improvement is clearly significant. The enhancement is between 25 and 30 points of PAE for Pout = 40 dBm following the frequency range.

The PAE depends directly on Current Consumption (Idc); hence the reduction of current consumption achieved with the Doherty topology is around 950 mA (Fig. 3.4) and decreases the total dissipated Power by 12 W. This avoids a high thermal elevation which could be a root cause of failure for the amplifier in the long term.

The last figure of merit evaluated in simulation is the third order Intermodulation Products (IMD3) at 40 dBm Peak Envelope Power (PEP). This measurement is done by injecting a 2-tone signal with 5 MHz frequency spacing at the input of the Power Amplifier.

With a Doherty topology, the linearity is degraded by 8 dB, but IMD3 simulation is aligned with the specification. Based on the simulation results (Table 3.1), a board has been built, which is detailed in the next subsection.

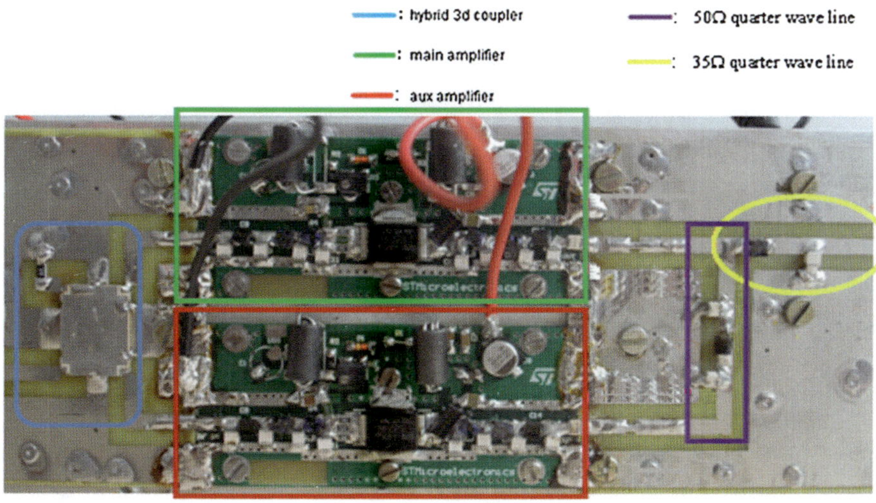

Fig. 3.5 Printed circuit board

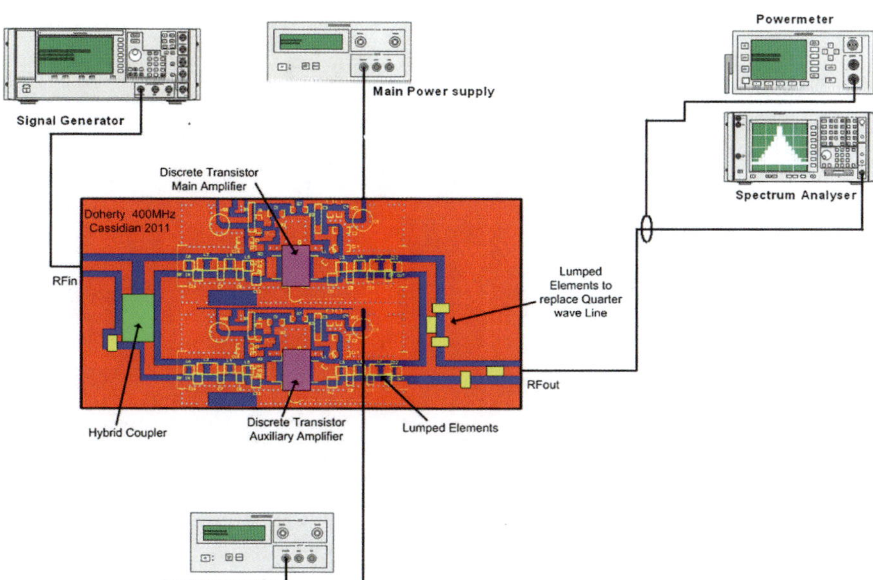

Fig. 3.6 RF test bench

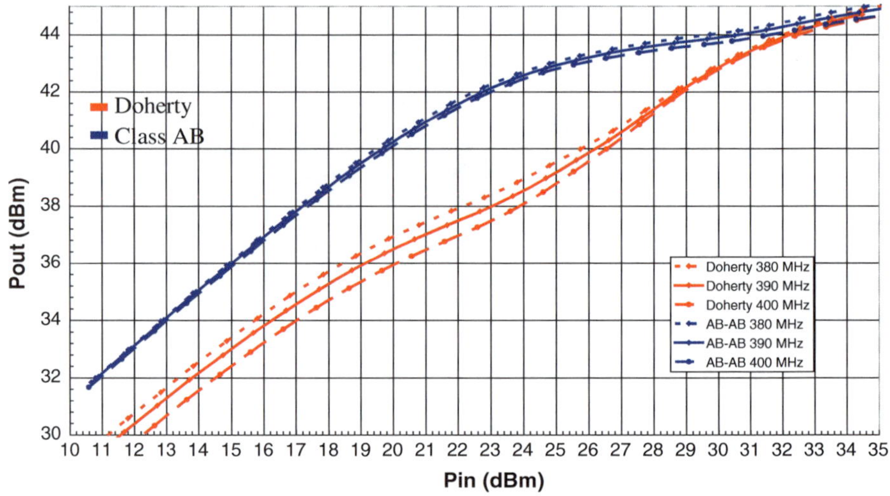

Fig. 3.7 Pout versus pin for a Doherty and class AB topologies at 380–390–400MHz

3.2.3 10 W Doherty Evaluation Board

3.2.3.1 Printed Circuit Board

The realized Printed Circuit Board (PCB) is presented in Fig. 3.5. Its size is 15 cm × 7 cm = 105 cm^2. The realized mockup has not been optimized in term of size. Taking into account that it is just the first iteration on the board, it seems

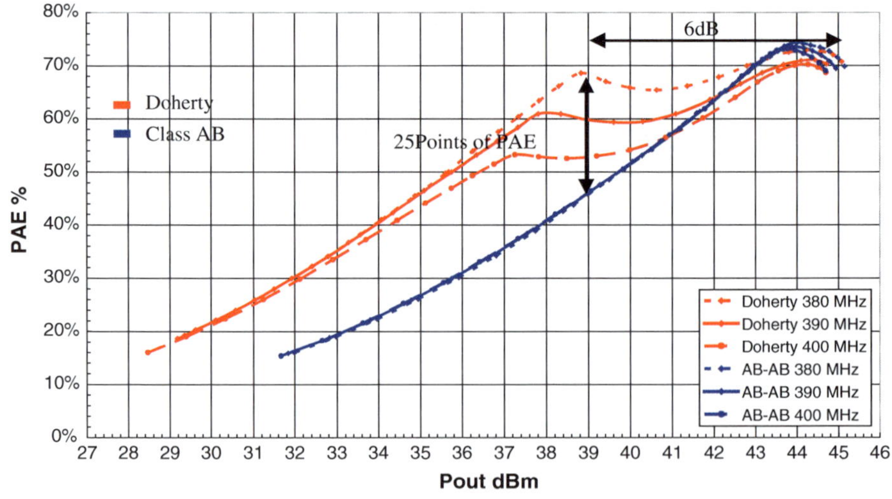

Fig. 3.8 PAE versus Pout for a Doherty and class AB topologies at 380–390–400 MHz

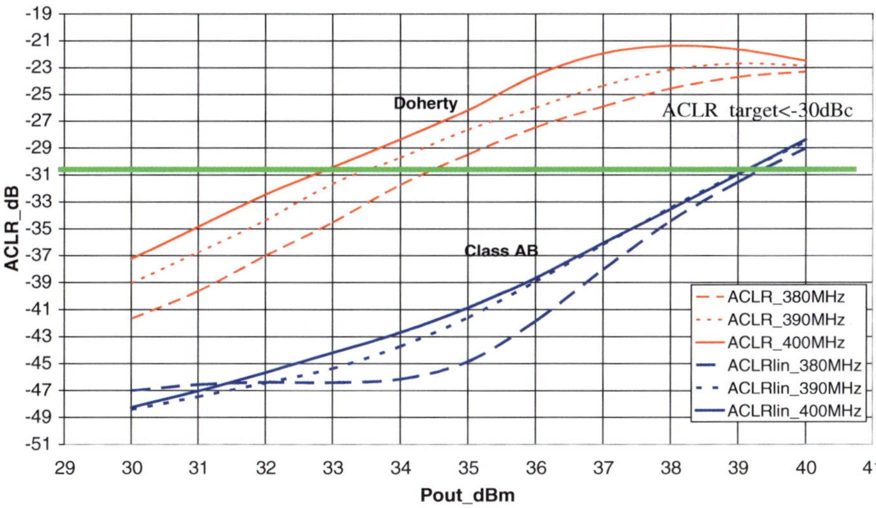

Fig. 3.9 A CLR versus Pout for a Doherty and class AB topologies at 380–390–400 MHz

possible to reduce the total Power Amplifier area within 70 cm^2. Comparatively, the size of an equivalent class AB topology design is integrated in 35 cm^2. But the gain in PAE could compensate the size drawback.

The RF test bench to evaluate the performance of the Doherty PA is sketched in Fig. 3.6.

3.2.3.2 Measurement Results

Different measurements have been performed after the first round of optimization of the Doherty matching. Figure 3.7 shows the Pout versus Pin curves for the 3 frequencies. The main observation is that the output power saturation (45 dBm) is lower than the simulation results (46–47 dBm). This also impacts the 1 dB compression point. The gain difference of 2 dB between the 2 topologies is also observed on measurements. The ripple on the band is lower than 1 dB.

The next curves in Fig. 3.8 are PAE results over frequency range for the Doherty and class AB structures. The results show a PAE improvement of 6 dB output

Table 3.2 EVM measurement at 40 dBm with LTE uplink signal in QPSK

IMD3			
Frequency (MHz)	Spec (%)	Doherty (%)	AB–AB (%)
380	17, 50	6.74	4.36
390	17, 50	6.95	4.17
400	17, 50	6.96	4.42

power back-off as on the simulations. But due to the difference in saturation power, the peak of PAE is at 39 dBm and not at 40 dBm. The best enhancement is observed at 380 MHz with 25 points improvement for 39 dBm. However, the improvement is not homogeneous on frequency range (between 7 and 25 points). At 40 dBm, the gain in PAE is not significant.

As observed in simulation, the Doherty structure degrades (Fig. 3.9) the linearity of the power amplifier (7 dB), but the class AB topology is not fully optimized in linearity. The ACLR is not compliant at 40 dBm (upper than −30 dBc). As an equivalent, matching of class AB PA has been used to design the main amplifier, this explains the non optimum results obtained with the Doherty structure. Investigations are on-going to have a better trade-off between PAE and linearity. Table 3.2 presents the EVM measurement obtained with an LTE uplink signal.

3.3 1-W Load Modulation Power Amplifier for 3.4–3.6 GHz Mobile WiMAX

The last technical subsection of the chapter describes the characterization and design of energy efficient user terminal load modulation power amplifier. The core of the design is based on the combination of Class B and Class C that includes quarter-wavelength transformer at the output to perform the load modulation. The user terminal load modulation power amplifier is designed to operate over the frequency range of 3.4–3.6 GHz band. The performances of the load modulation

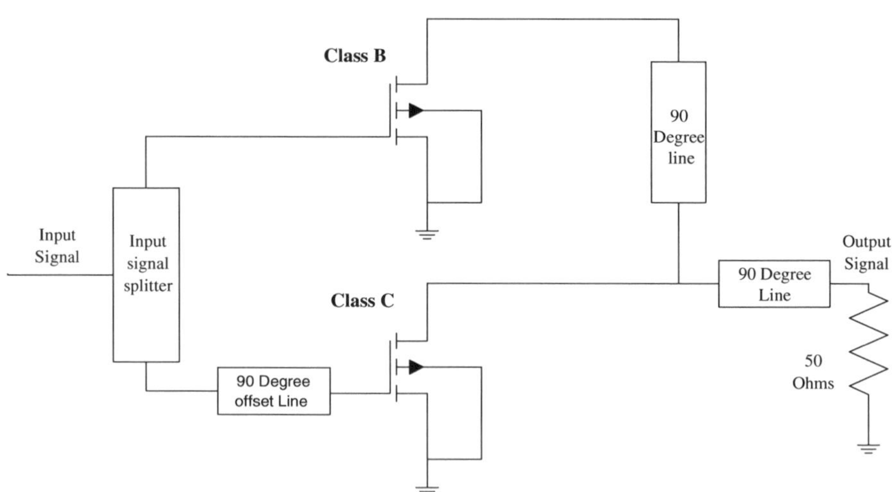

Fig. 3.10 The schematic diagram of load modulation power amplifier

amplifier are compared with a conventional Class B amplifier. The schematic diagram is presented in Fig. 3.10.

3.3.1 Target Specifications

The main objective of this work is to design energy efficient user terminal load modulation power amplifier and compare its performance to that of a conventional Class B amplifier. The figures of merit of the targeted PA are listed below:

- Frequency range: 3.4–3.5 GHz
- Average output power (Pout): >30 dBm
- Power Added Efficiency (PAE): >40 %

As in the above PA design, the main figure of merit is the Power Added Efficiency (PAE), defined as $PAE = 100\frac{P_{out}-P_{in}}{P_{Supply}}$ expressed in% where P_{out} and P_{in} are the output and input signal power (Watt) and P_{Supply} equals to Vcc*Icc.

3.3.2 Circuit Design

A 3.5 GHz, 30 dBm Mobile WiMAX handset load modulation RF power amplifier, has been designed using the TOM3 large signal model and FPD1500 transistor. The FPD1500SOT89 is a packaged depletion model AlGaAs/InGaAspseudomorphic High Electron Mobility Transistor (pHEMT). It contains double recessed gate structure, which minimizes parasitics and optimizes performances.

This design comprised several design steps, as each optimization is applied in order to obtain global high performances of the entire load modulation RF power amplifier. Initially, the design of carrier and peak amplifiers, input 3 dB 90° hybrid coupler designs, Output 90° offset line and impedance transformer designs were performed.

However, it is important to note that in the design of carrier and peak amplifiers, the DC simulation should be done first in order to find the optimal bias point and bias network based on the class of operation and power requirements. In this work, the bias circuit was designed based on Class B and Class C of the carrier peaks. Class B was used to improve the efficiency and linearity instead of Class AB, which is widely used in the combination of Doherty amplifier. The DC quiescent current for Class B is at threshold while for Class C, it is below the threshold. In theory, the quiescent current of Class B is zero but for the current work the quiescent current was increased to an order of 8 % of the peak drain current that is resulting in 0.046 mA. The reason for this is to minimize the cross-over distortion and increase the efficiency. The peak drain current is 0.587 mA when the VGS is at 0 V and VDS is 5 V. 5 V was chosen for VDS since it is located between cut-off and saturation of the transistor. 0.046 mA is 8 % of the peak drain current, which gives VGS of −0.9 V, while the overall power consumption is 0.228 W. Moreover,

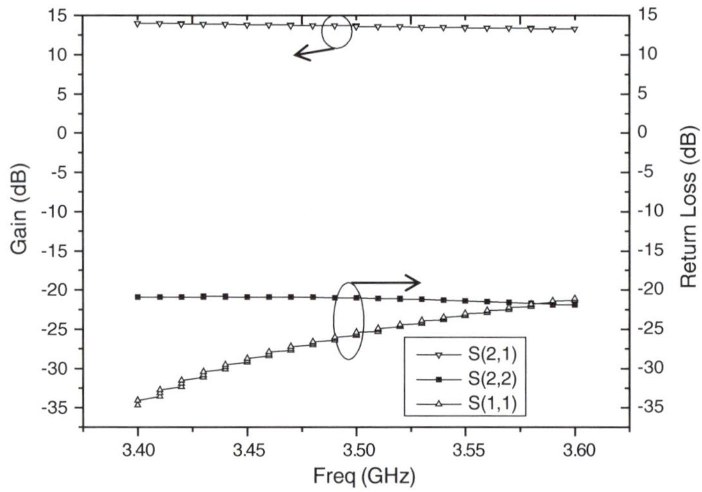

Fig. 3.11 Linear simulation: flat gain and return loss

the same supply power was applied to class C but the drain current was 0.004 mA and the power consumption was 0.019 W.

Having obtained the DC quiescent current, the next step is to determine the load line impedance to design the output and input matching of Class B amplifier, to obtain the performance regarding the output power and efficiency. The output matching network was designed for optimum output power performance with load pull technique based on input matching S-parameters.

The transistor parasitic elements are included in the load-pull analysis in order to optimize the output matching network. The results obtained from the load pull simulation show that the transistor needs to see an impedance of 20.492 + j3.775 at the output. Therefore the target of the matching network was to transform the

Fig. 3.12 Power splitter

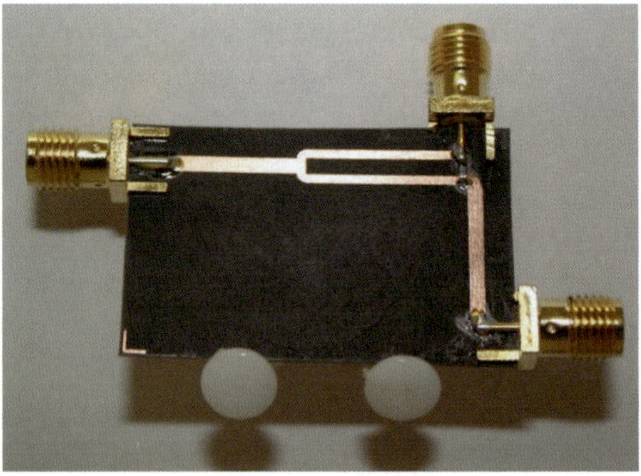

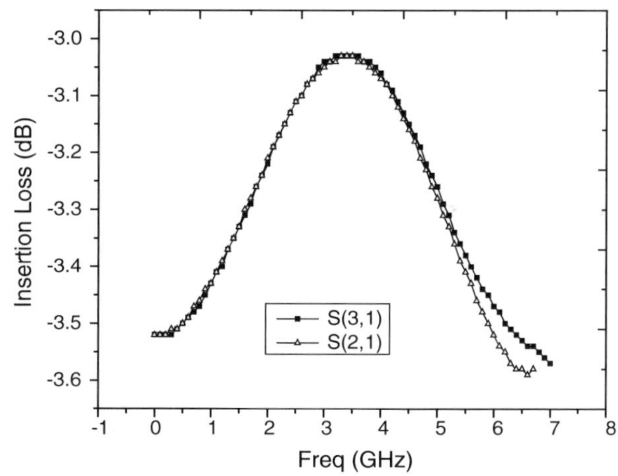

Fig. 3.13 Insertion loss

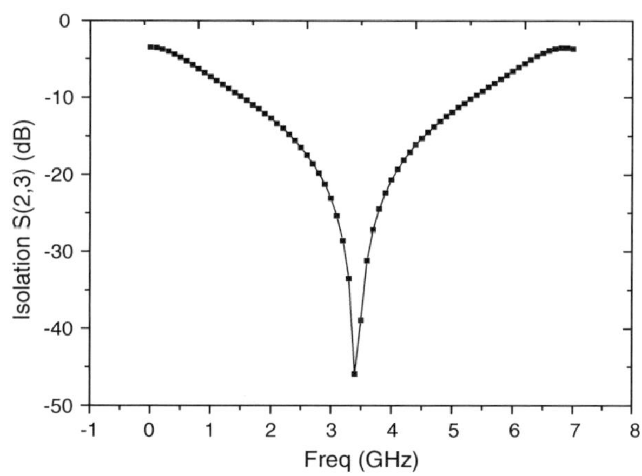

Fig. 3.14 Isolation

impedance from 49.393 − j1.776 to 20.492 + j3.775. This impedance is the optimal load value, which compromises the efficiency of 40.39 %, and 27.05 dBm output power at 1-dB compression point of single Class B alone. The load reflection coefficient was used to design the output matching network, while for the input matching network S-parameter was employed and conjugated the input reflection coefficient for maximum power transfer. Figure 3.11 shows the linear results obtained from matched Class B power amplifier, the gain is flat over the range of 3.4–3.6 GHz, with excellent matching at the input and output return losses.

The non-linear simulation of Class B was performed and the performance of the design in terms of output power and efficiency was observed. The 26.98 dBm output power was achieved at 1-dB compression point and 39 % efficiency. From this nonlinear simulation, the results show a clear compromise of the load pull measurement values. The same was applied to Class C but using different bias point.

Fig. 3.15 Implemented prototype of energy efficient load modulation PA

Table 3.3 Bias point setting for load modulation

Drain voltage (V)	Carrier VGS (V)	Peaking VGS (V)
5	−0.9	−1.1

A 3 dB quadrature splitter is part of load modulation and if properly set can contribute a lot to the total efficiency of the system. Our investigation shows that the operation of this technique is strongly influenced by the coupling factor of the input splitter. In fact, in this research 3 dB quadrature splitter have been designed (Fig. 3.12) and tested in terms of the operated frequency and bandwidth, and it provides good results as shown in Figs. 3.13, 3.14. It should be noted that this splitter is at the input of amplifier which divides the input signal equally between the carrier and peaking amplifiers. The splitter, the Carrier Class B, the peaking Class C, and impedance transformer at the output are combined to form a load modulation RF power amplifier.

3.3.3 Implementation and Results

Figure 3.15 shows the prototype diagram of the proposed load modulation RF power amplifier with offset transmission line at both output and input circuit, which maximizes the overall system's efficiency with the configuration of Class B amplifier. The FPD1500SOT89 transistors with 27.5 dBm output power is used for both Class B and Class C amplifiers and produces a load modulation amplifier with 30 dBm and Efficiency of 53 %. Table 3.3 summarizes the recommended bias setting. The bias condition for the Class B carrier amplifier are Vgs = −0.9 V

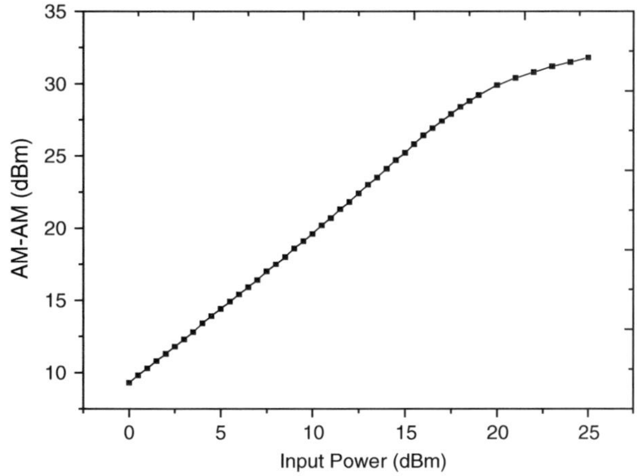

Fig. 3.16 AM-AM characteristics of load modulation amplifier

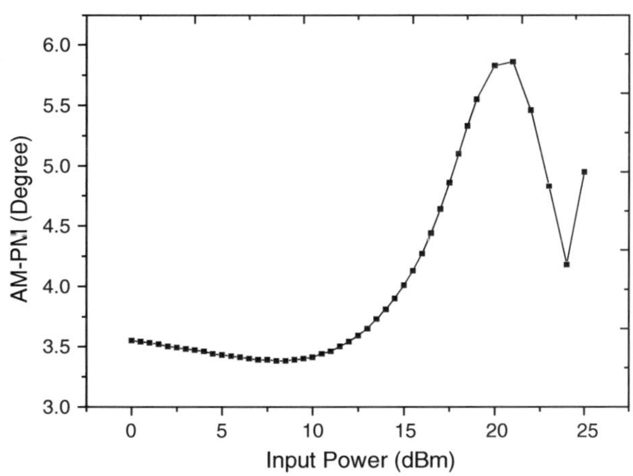

Fig. 3.17 AM–PM characteristics of load modulation amplifier

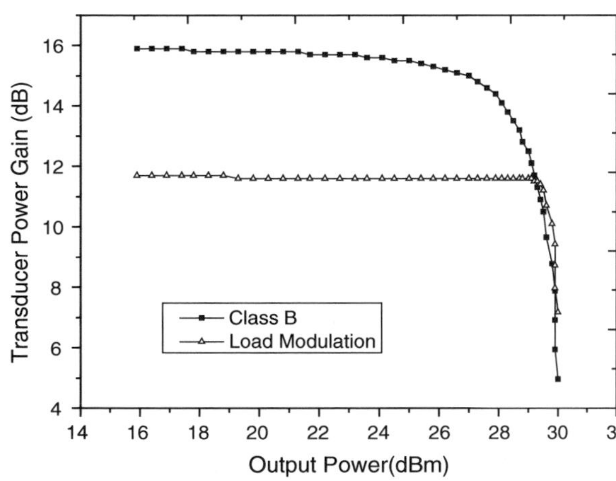

Fig. 3.18 Gain characteristics

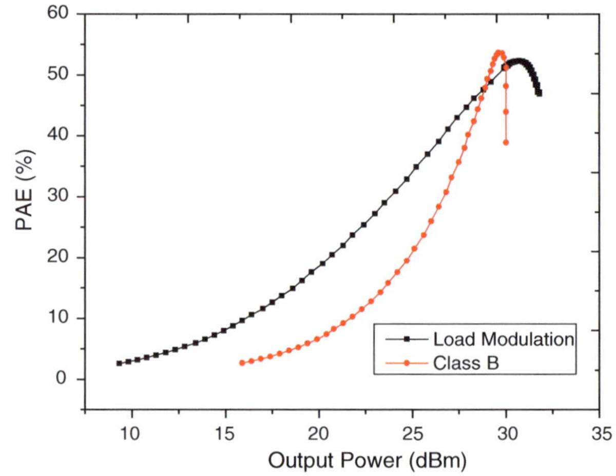

Fig. 3.19 Power-added efficiency

Table 3.4 Comparison performance of class B and load modulation at Pout 1 dB compression point

Amplifier	Gain (dB)	Pout (dBm) at 1 dB	PAE (%) at P1 dB
Class B amplifier	15.4	27.5	37
Load modulation	11.8	30	53

(Ids = 46 mA), and for the Class C peaking amplifier, Vgs = −1.1 V (Ids = 4 mA). Both of the amplifiers use the same drain voltage (5 V).

The load modulation is initially characterized for AM–AM and AM–PM responses as well as output power and efficiency. The performance comparisons between the load modulation amplifier and Class B amplifier are performed and the output power increases to 30 dBm at 1-dB compression point, while the efficiency increases to 53 %. Figure 3.16 represents the variation of the input power versus the output power of the load modulation. It clearly shows that 30 dBm output power is

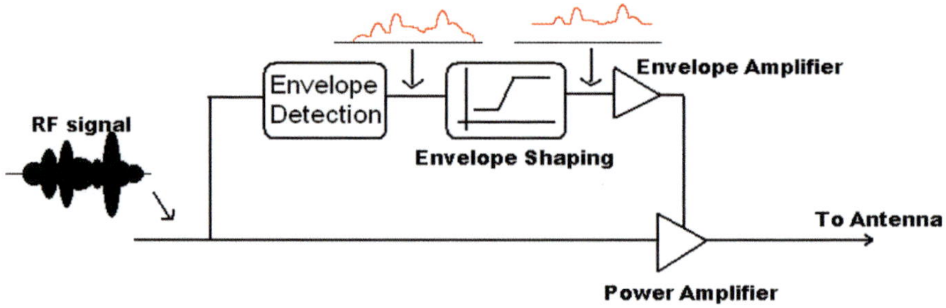

Fig. 3.20 Envelope tracking principle

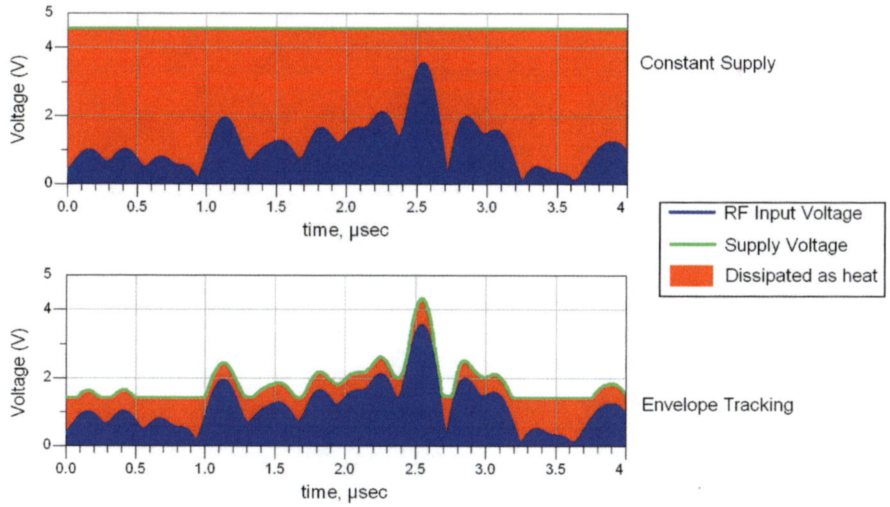

Fig. 3.21 Biasing voltage modulation introduced by envelope tracking compared to constant biasing voltage

at the linear region of the amplifier and is achieved due to the characteristic of gain compression and expression of the load modulation (Fig. 3.17). The peaking amplifier Class C late gain expansion can compensate the carrier Class B amplifier gain compression. Figure 3.18 represents the gain characteristic versus output power. The graph shows that the power gain of load modulation amplifier is degraded compared to Class B, due to the arrangement of lower biasing.

Figure 3.19 shows the power added efficiency versus output power. The load modulation amplifier have higher efficiency over the range of wide output power levels compared to Class B amplifier (Table 3.4).

3.4 200-mW Envelope-Tracking Power Amplifier for LTE in the 1,920–1,980 MHz Band

The efficiency enhancement technique studied within this section is the Envelope-Tracking (ET) technique. The concept of the ET system involves the detection of the envelope of the RF input signal and the control of the linear RF power amplifier supply voltage accordingly [3, 6], as represented by Fig. 3.20.

In this configuration, the supply voltage varies so that the power amplifier operates linearly, close to the 1-dB compression point and at high efficiency regime in order to preserve power. The Envelope Tracking is a great enhancement technique for such signals as uplink LTE signals, which have a high PAPR value and lead to a large dissipated power in traditional PA (Fig. 3.21).

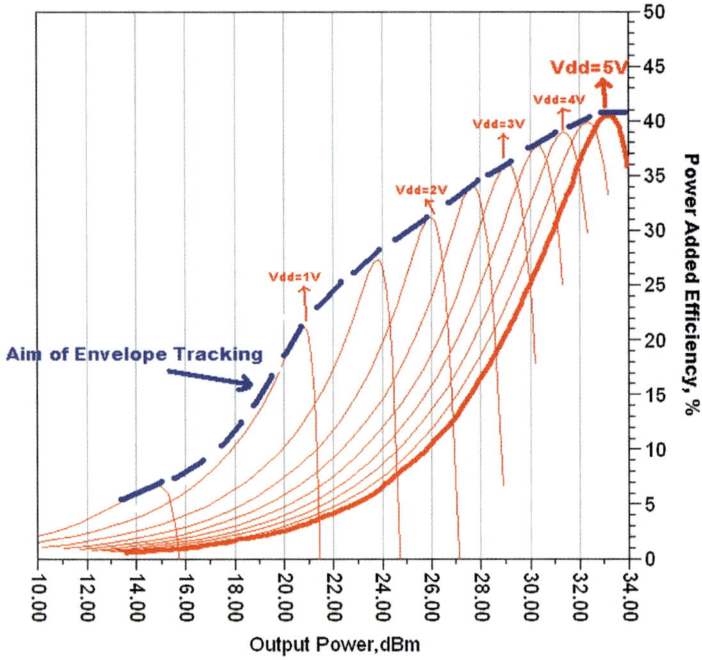

Fig. 3.22 Simulation results with a two tone signals (center frequency: 1960 MHz, spacing: 10 MHz)

Figure 3.22 shows the PAE of the power amplifier under consideration as a function of the output power. The maximum of PAE is different for each supply voltage. The aim of the Envelope-Tracking is to vary the supply voltage according to the output power to maintain the operating point close to the maximum PAE.

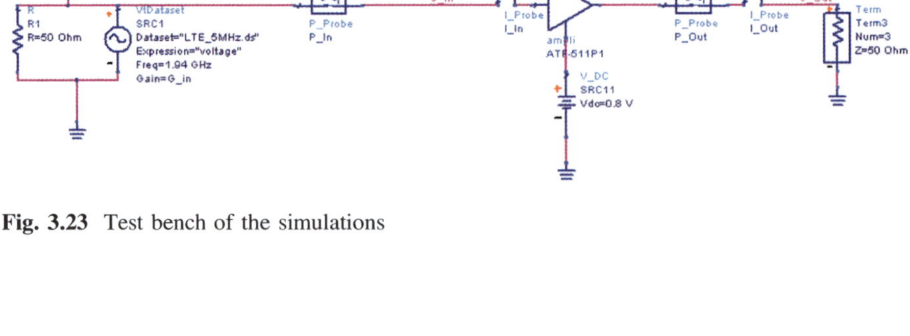

Fig. 3.23 Test bench of the simulations

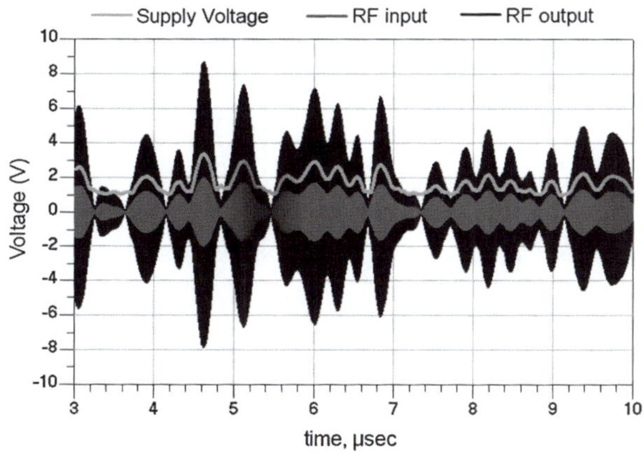

Fig. 3.24 Waves of the signals simulations: supply voltage, RF input and output

In these simulations, the PAE can be enhanced by up to 20 % between 20 and 28 dBm of output power by modulating the DC voltage.

3.4.1 Target Specifications

The figures of merit targeted for our system with the Envelope tracking and PA are listed below:

- Frequency range: 1,920–1,980 MHz
- Average output power (Pout): ≈23 dBm
- Power Added Efficiency (PAE): >40 %
- Adjacent Channel Leakage Ratio (ACLR): <−30 dBc
- Error Vector Magnitude (EVM): <17.5 % in QPSK

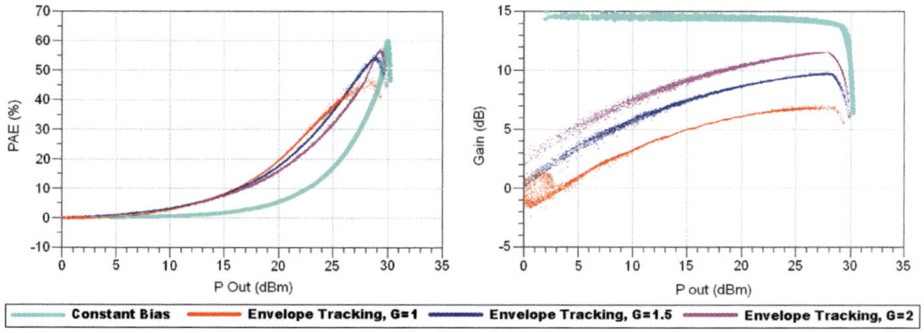

Fig. 3.25 PAE and gain versus Pout with and without envelope tracking system for different values of envelope gain

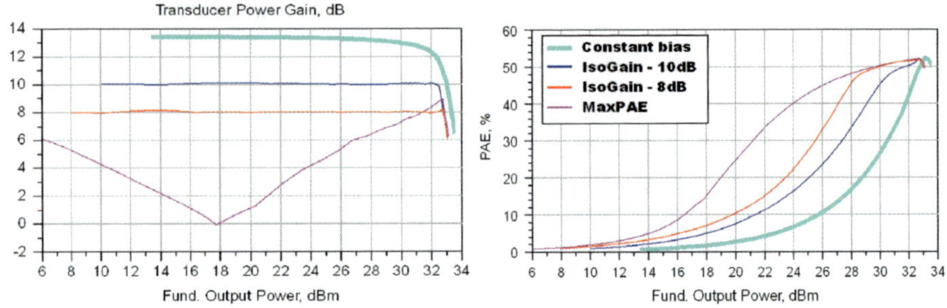

Fig. 3.26 Harmonic balanced simulation (F = 1,960 MHz)

Table 3.5 Simulated PAE of the PA with and without envelop-tracking with different optimization goals Simulations performed with a continuous-wave (not modulated) signal

PAE (%)	Constant bias (%)	IsoGain 10 dB (%)	IsoGain 8 dB (%)	MaxPAE (%)
At −7 dB of max compression point	10	24	40	48
At max compression point	50	50	50	50

3.4.2 ADS Simulations

The Envelope-Tracking system has been designed and simulated with Agilent-ADS, as shown in Fig. 3.23. The source signal is an up-link LTE signal with a 5 MHz bandwidth and a 64-QAM modulation. Power-meters, current-meters and voltage-meters are used to monitor the power consumption and the main figures of merit (PAE, gain, ACLR). An envelope detector and modulator supply the matched power amplifier. The supply voltage follows the input signal envelope with a minimum of 1 V in this example [7], as can be seen in Fig. 3.24.

From these simulations, the instant PAE and gain can be extracted. The PAE and gain are plotted versus the output power for each calculated point with 4 different voltages in Fig. 3.25: one with a constant 5 V supply, and 3 others, supplying the Power Amplifier with the envelope amplified by different gains G = 1, 1.5 or 2.

These curves show how important the control of the envelope is in order to enhance the linearity or the PAE (Fig. 3.26). In this example, three optimizations were done with the following goals:

- a 10 dB gain, high linearity (dark blue curve),
- a 8 dB gain, high linearity (red curve),
- a maximum efficiency (purple curve).

The light blue curves are the simulation results of the stand-alone PA without ET system. In all cases, the PAE increases significantly. For the maximum efficiency, the gain is not constant. This shape of gain degrades the linearity of the PA and

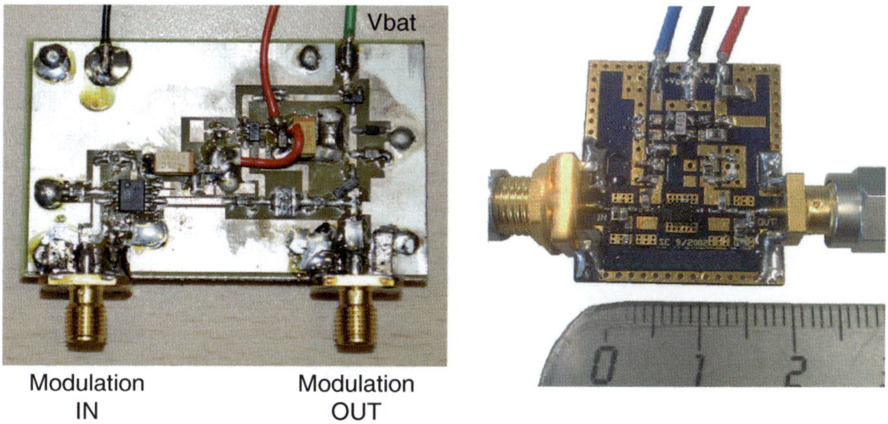

Fig. 3.27 Photographs of the voltage modulator (4 × 6 cm) (*left*) and the PA evaluation board (*right*)

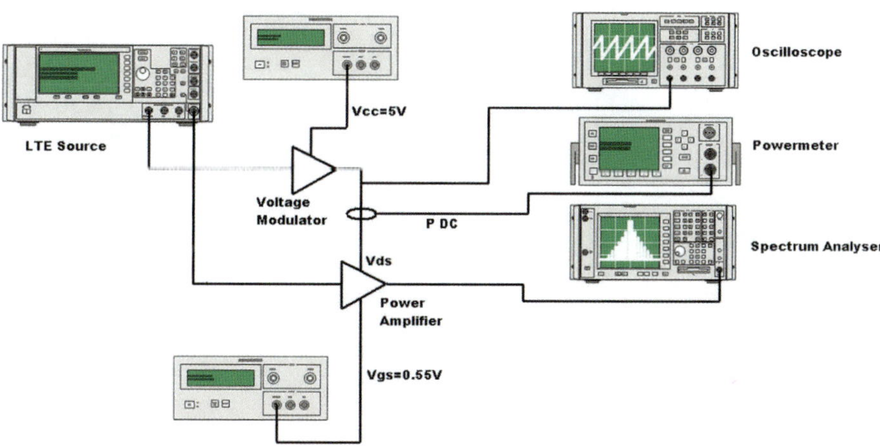

Fig. 3.28 The test bench

increases EVM and ACPR. A trade-off exists between PAE improvement, Linearity target and Gain, which can be optimized based on the specific application (Table 3.5).

3.4.3 Measurements

The prototype was fabricated as two separate boards for the voltage modulator and the power amplifier (Fig. 3.27). The test bench is presented in Fig. 3.28.

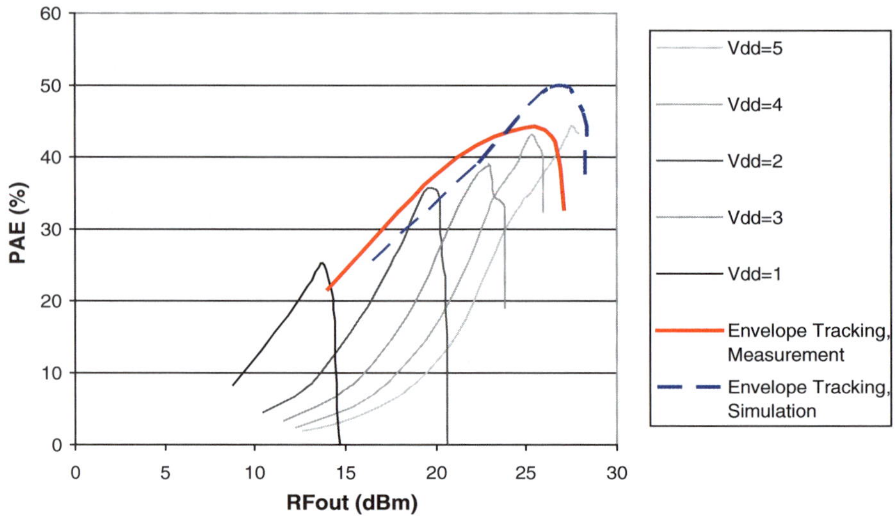

Fig. 3.29 PAE of the PA as a function of the output power comparison between fixed supplies and the envelope tracking system

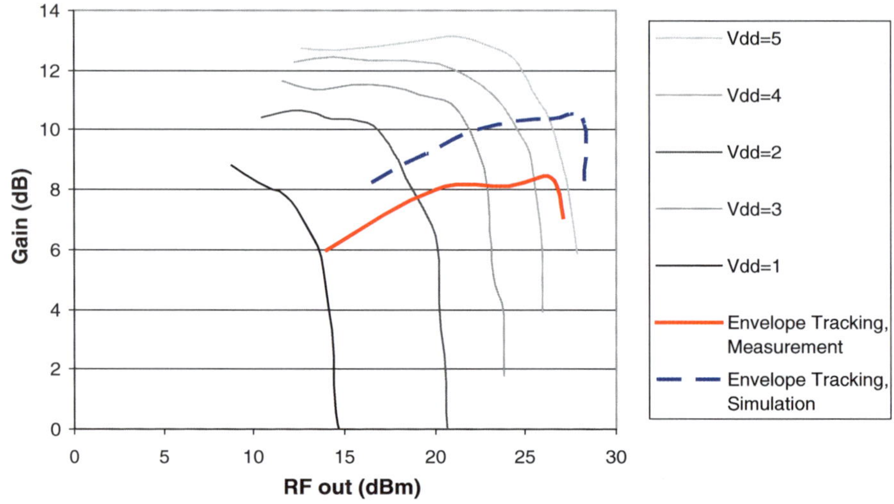

Fig. 3.30 Gain of the PA; comparison between fixed supplies and the envelope tracking system

The source signal is an uplink LTE signal with 5 MHz of bandwidth, a 4 QAM or 16 QAM modulation and a carrier at f_0 = 1,960 MHz. The generator delivers the RF signal to the PA input and the envelope of this signal to the modulator. The Power-meter provides the information on the output power. In our application, the targeted average output power is 23 dBm. The spectrum analyzer allows to measure

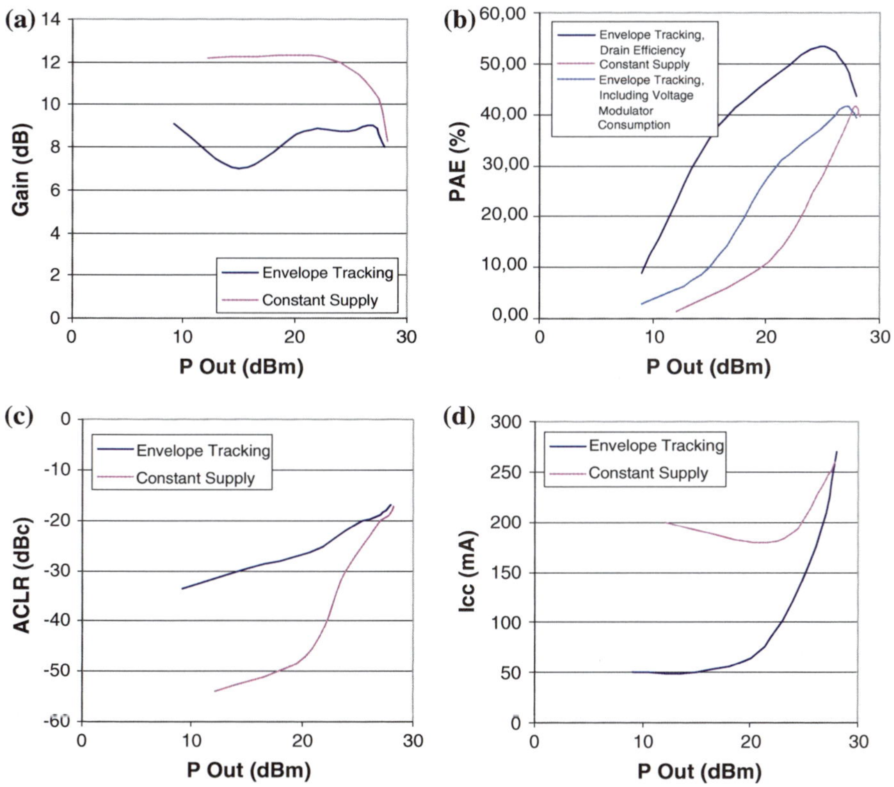

Fig. 3.31 Gain, PAE, ACLR and Icc versus output power comparison between a 5 V constant supply and envelope tracking

Table 3.6 Main figures of merit measured at PA output power of 23 dBm for LTE uplink signal with and without envelope tracking system

Values at 23 dBm of average output power					
	Vcc (V)	Icc (A)	Gain (dB)	PAE (%)	ACLR (dBc)
Envelope tracking	5	0.1	8.7	34	−25
Constant supply	5	0.18	12	20	−33

the spectrum at the PA output and to evaluate the linearity of the PA and its conformance to the LTE standard.

The first tests have been performed for an optimization of PAE. Figure 3.29 presents the PAE of the Power Amplifier supplied by several constant voltages from 1 to 5 V, and also with Envelope-Tracking, measured and simulated. The PAE is calculated from the RMS values of the power amplifier supplies.

At 23 dBm of average output power, the PAE with ET is upper than 40 %, which is twice higher than with a single 5 V supply. The corresponding gain values are presented in Fig. 3.30. The experimental gain is lower than expected because of a lower impedance matching; this issue is the subject of further investigations.

In order to increase the performance of the designed system, some SMD components have been changed, whereas the topology of the voltage modulator is kept unchanged. With this configuration of Envelope Tracking, the power consumption has been divided by two. The obtained results are presented in Fig. 3.31.

Table 3.6 summarizes the figures of merit at 23 dBm. With the envelope tracking system, the efficiency is enhanced by 14 points. The gain is decreased by around 3dB and the ACLR measurement is −25 dBc. This is higher than the target (−30 dBc) for Pout = 23 dBm, and the ACLR result is not perfectly aligned with 3GPP standard of −30 dBc.

3.5 Conclusion

This chapter presented energy efficient designs targeting power amplifiers for next generation handsets. The design of the power amplifier represents great challenges both in terms of power consumption and integration for wireless terminals since their efficiency can be rather limited. In this respect, three PA implementations were investigated at 400 MHz, 1,920 and 3.5 GHz. Two designs are based on a Doherty topology, while the third design is based on an Envelope Tracking technique. Interesting results have been shown:

- 400-MHz Doherty PA demonstrated a power-added efficiency of 50–70 % at 39 dBm (7.9 W) of average output power, which represents an improvement of 7–25 % as compared to a classical class-A PA;
- 3.5 GHz Doherty (or Load-Modulation) PA demonstrated a power-added efficiency of 53 % at 1 W of output power as compared to 37 % efficiency for a class-C PA.
- Envelope-Tracking PA designed for the 1,920–1,980 MHz LTE band demonstrated the promise of the envelope-tracking technique to improve the efficiency of PA. With the designed circuit, the PAE increased from 20 to 34 % at 23 dBm of average output power.

It is worthwhile to mention, that in the Envelope-Tracking approach, the 34 % PAE performance value can be improved up to at least 40 % by optimizing the voltage modulator power consumption. Furthermore, the quoted PAE (34 %) is a combination of the final stage PA efficiency (52 %) and the voltage modulator efficiency (65 %). Furthermore, the power lost in the voltage modulator contributes to the increase in the ACLR, which is about 25 dBc only at 23 dBm of output power.

This chapter represents a continuation of the previous chapter, tackling the issue of power consumption within the front end radio. In the previous chapter, the design of multi-standard transceiver and miniature antenna was addressed. The latter also took into account the impedance mismatch caused by perturbation due to the human hand/head or metal close to antenna. A sensing unit and a matching network have been designed to overcome such perturbations, and enhance energy efficiency further, even under extreme operating conditions. To complete the picture of a full front end radio design, the current chapter provided three designs of Power Amplifier targeting Doherty technology and the envelope tracking approaches. All these elements in working in synergy can lead to significant energy enhancements, and start to shift the perception of the 5G from a vision to pragmatic reality.

References

1. Zhang, G., Khesbak, S., Agarwal, A., Chin, S.: Evolution of RFIC handset PAs. IEEE Microwave Mag. **11**, 60–69 (2010)
2. Doherty, W.H.: A new high efficiency power amplifier for modulated waves. Proc. Inst. Radio Eng. **24**(9), 1163–1182 (1936)
3. Institute of Electrical and Optical Communications Engineering, http://www.unistuttgart.de/int/forschung/PA/Grundlagen/Architekturen/VSM.en.html
4. Chireix, H.: High power outphasing modulation. Proc. IRE **23**(11), 1370–1392 (1935)
5. Kahn, L.R.: Comparison of linear single-sideband transmitters with envelope elimination and restoration single sideband transmitters. Proc. IRE **44**(12), 1706–1712 (1956)
6. Moon, J., Kim, J., Kim, I., Kim, B.: A wideband envelope tracking Doherty amplifier for WiMAX systems. IEEE Microwave Wirel. Compon. Lett. **18**(1), 49–51 (2008)
7. Kenington, P.B.: High-linearity RF Amplifier Design. Artech House, Norwood (2000)

Chapter 4
Energy-Efficient Roaming for Heterogeneous Wireless Networks

Dionysis Xenakis, Nikos Passas, Lazaros Merakos and Christos Verikoukis

Abstract Heterogeneous networks enable multi-standard terminals to provide increased Quality of Experience with the ability to always stay connected. Nevertheless, the price paid for multiplicity of active radio interfaces and increased data rates is the rise in energy consumption of multi-standard terminals. This factor coupled with the slow progress in battery technology creates the so called energy trap, which at times can limit the mobile user to the nearest electricity socket creating a real paradox to the headlines of 4G and future generations in terms of unrestricted mobility. Therefore real solutions need to be in place that can support unrestricted roaming, but in an energy efficient manner whilst maintaining Quality of Service (QoS). Today's mobile networking environment is already considering a HetNet (Heterogeneous networking) environment, with LTE release 12 considering the ABS (almost blank space) approach for coordinated interference management between macro and femto cells (3GPP, Technical report on UMTS heterogeneous networks, 2013, [1]). Therefore, specific user case that can handle mobility management between femto and macro cells, and in a more broader sense vertical handovers will be aligned with current technology trends in 3GPP, and ease the stringent requirements on interference management as well promote the coexistence and dense deployment of heterogeneous small cells, that can include WLAN hotspots among others. In this chapter, we discuss the key aspects of vertical handover (VHO); focusing on the IEEE 802.11 and 3GPP LTE-Advanced network and propose a novel VHO decision algorithm. Following a similar approach, we discuss the key challenges for (horizontal) handover in macrocell–femtocell network, and propose a novel interference-aware decision algorithm as well. The key features of both algorithms is that they are fully compatible with the IEEE and the 3GPP systems, while they focus on minimizing the energy consumption of mobile terminals; a critical requirement for 4G heterogeneous systems and the foreseen 5G system.

D. Xenakis (✉) · N. Passas · L. Merakos
Department of Informatics and Telecommunications, University of Athens,
Athens, Greece
e-mail: nio@di.uoa.gr

C. Verikoukis
Telecommunications Technological Centre of Catalonia, Barcelona, Spain

A. Radwan and J. Rodriguez (eds.), *Energy Efficient Smart Phones for 5G Networks*, Signals and Communication Technology,
DOI 10.1007/978-3-319-10314-3_4

4.1 Introduction

Current mobile devices enable multi-standard operation for jointly utilizing the existing IEEE and 3GPP Radio Access Network (RAN) infrastructure, either to offload the user traffic, or to mitigate the interference in the cellular network, or to reduce the energy consumption of the mobile devices. IEEE has recently updated the baseline version of the IEEE 802.11 Standard for wireless local area access [2], providing technical corrections/clarifications to the baseline operation of the previous standard and incorporating various enhancements to existing medium access and physical layer functions. These modifications include higher data rates, use of power control, and provision for advanced radio measurements. Seamless mobility through Vertical Handovers (VHO) between the 3GPP cellular systems and IEEE-based WLAN systems has always been challenging [3]. Aiming to realize the so-called heterogeneous networking paradigm, 3GPP provided several functional enhancements to the Evolved Packet Core (EPC) of the LTE-Advanced (LTE-A) system [4], such as the deployment of the Access Network Discovery and Selection Function (ANDSF) [5, 6]. The ANDSF can assist the multi-mode mobile terminals (MMTs), i.e. devices equipped with both 3GPP and non-3GPP radio interfaces, to discover and select an appropriate Point of Attachment (PoA) in the heterogeneous network, i.e. either a 3GPP cell or a WLAN Access Point (AP). Specifically, the ANDSF can respond to the MMT requests for access network discovery using the pull mode and can initiate data transfer to the MMT using the push mode. Besides, the ANDSF can provide the MMT with access network discovery information, an inter-system mobility policy and an inter-system routing policy.

On the other hand, the existing cellular infrastructure undergoes major changes towards providing seamless connectivity and supporting diverse QoS requirements to the end user. Over the past few years, the deployment of low-power and low-cost cellular access points, such as femtocells, has drawn a lot of attention from both the industrial and academia stakeholders [7]. However, due to their unplanned deployment and their operation within the licensed band, the smooth integration of femtocells into the predominant macro-cellular network layout is a key challenge, mainly owing to the need for advanced mobility management. Femtocells can improve the energy saving potential for the network nodes and enhance the QoS perceived by the end users. In fact, existing reports foresee that the number of deployed femtocells will surpass that of macrocells by up to six times within the next few years [8] and thus, the development of femtocell-specific solutions for (horizontal) handover between the cellular stations will be essential for recent 3GPP technologies, such as the LTE-Advanced system [9].

To resourcefully facilitate the mobility and the radio resource management procedures, the Release 10 series of the standards of the LTE-Advanced system includes a wide set of signal quality measurements for both the LTE-Advanced access network and UEs. Table 4.1 summarizes part of the measurements capabilities in the LTE access network and UEs [10], while it additionally includes notation for a tagged LTE UE u and LTE cell c. Note that $R_{c,DL}$ and $R_{c,UL}$

Table 4.1 LTE UE and network measurements [10]

Measurement	Definition	Performed by	Notation
Reference signal received power (RSRP)	The linear average over the power contributions (in [W]) of the resource elements that carry cell-specific reference signals within the considered measurement frequency bandwidth. For RSRP determination the cell-specific reference signals R0 shall be used while if the UE may use R1 in addition to R0 if it is reliably detected. The reference point for the RSRP shall be the antenna connector of the UE	UE	$RSRP(c)$
E-UTRA carrier Received signal strength indicator (RSSI)	The linear average of the total received power (in [W]) observed only in OFDM symbols containing reference symbols for antenna port 0, over $R_{c,DL}$ number of RBs by the UE from all sources, including co-channel serving and non-serving cells, adjacent channel interference, thermal noise etc. RSSI is not reported as a stand-alone measurement rather it is utilized for deriving RSRQ	UE	$RSSI(c)$
Reference signal received quality (RSRQ)	The ratio $R_{c,DL}$ × RSRP/(E-UTRA carrier RSSI) where $R_{c,DL}$ is the number of RB's of the E-UTRA carrier RSSI measurement bandwidth. The measurements in the numerator and denominator shall be made over the same set of RBs. The reference point for the RSRQ shall be the antenna connector of the UE	UE	$RSRQ(c)$
Downlink reference signal transmitted power (DL RS Tx)	The linear average over the power contributions (in [W]) of the resource elements that carry cell-specific reference signals which are transmitted by a tagged cell within its operating system bandwidth. For DL RS TX power determination the cell-specific reference signals R0 and if available R1 can be used. The reference point for the DL RS TX power measurement shall be the TX antenna connector	E-UTRAN	$P_{RS}(c)$

(continued)

Table 4.1 (continued)

Measurement	Definition	Performed by	Notation
Received interference power	The uplink received interference power, including thermal noise, within one physical RB's bandwidth of N_{sc}^{RB} resource elements. The reported value shall contain a set of Received Interference Powers for all the uplink physical RBs. The reference point for the measurement shall be the RX antenna connector.	E-UTRAN	I(c)

correspond to the number of utilized Resource Blocks (RB) within the operating bandwidth of cell $c \in C$. The LTE measurements presented in Table 4.1 will be utilized in the following section for the VHO decision phase as well.

At the moment, we focus on the radio measurement capabilities included in IEEE 802.11-2012 [2], which allow enhanced mobility management solutions in the integrated 3GPP and IEEE WLAN environment. The radio resource measurement (RRM) service includes measurements that extend the capability, reliability, and maintainability of WLANs by providing standard measurements across vendors. The measurement data are commuted to upper layers of the communications stack [2]. An IEEE 802.11-2012-conformant station (STA) can make local measurements, request a measurement from another STA, or be requested by another STA to make one or more measurements and return the respective results. The radio measurement data are made available to STA management and upper protocol layers, where data may be used for a wide range of applications, such as radio resource and mobility management. Both the measurement requests and reports are conveyed through IEEE 802.11 management frames, which are as follows: beacon, frame, channel load, noise histogram, STA Statistics, Location Configuration Information (LCI), neighbor report, link measurement, and transmit stream/category measurement. A measurement pause mechanism (request-only type mechanism) and a measurement pilot (report-only mechanism) are also provisioned. Table 4.2 summarizes some of the basic STA measurement capabilities in IEEE 802.11-2012, while it also includes the adopted notation for two tagged IEEE 802.11-2012-conformant STA s, s'. Note that both the MMTs and PoAs are referred to as STAs according to [2]; thus the measurement capabilities in Table 4.2 apply to both the MMT and the WLAN PoA sides.

In the remainder of this chapter, we focus on utilizing the above measurement capabilities of both the IEEE and LTE-A compatible MMTs towards enhancing the VHO and HO decision procedure and reducing the energy consumption at the mobile terminals [11–13].

Table 4.2 IEEE 802.11-2012 measurement capabilities [2]

Measurement	Definition	Notation
Received channel power indicator (RCPI)	An indication of the total channel power (signal, noise, and interference) of a received IEEE 802.11 frame, transmitted from STA s to STA s′, measured on the channel and at the antenna connector used to receive the frame at STA s′. The RCPI value can be translated in dBm according to the respective formula in [2]	$RCPI(s, s')$
Average noise power indicator (ANPI)	A MAC indication of the average noise plus interference power measured by STA s, when the channel is idle as defined by three simultaneous conditions: (1) the virtual CS mechanism indicates idle channel, (2) the STA s is not transmitting a frame, and (3) the STA s is not receiving a frame. The ANPI value can be translated in dBm according to the respective formula in [2]	$ANPI(s)$
Received signal to noise indicator (RSNI)	An indication of the signal to noise plus interference ratio of a received IEEE 802.11 frame, transmitted from STA s to STA s′. RSNI is defined by the ratio of the received signal power (RCPI-ANPI) to the noise plus interference power (ANPI) as measured on the channel and at the antenna connector used to receive the frame	$RSNI(s, s')$
Max transmit power (MTP)	The max transmit power field is a 2's complement signed integer and is 1 octet in length, providing an upper limit, in units of dBm, on the transmit power as measured at the output of the antenna connector to be used by STA s on the current channel. The maximum tolerance for the value reported in max transmit power field shall be 5 dB. The value of the max transmit power field shall be less than or equal to the max regulatory power value for the current channel.	$P_{\max}(s)$
Transmit power used (TPU)	The transmit power used field is a 2′s complement signed integer and is 1 octet in length. It shall be less than or equal to the max transmit power and indicates the actual power used as measured at the output of the antenna connector, in units of dBm, by STA s when transmitting the frame containing the Transmit Power Used field. The Transmit Power Used value is determined any time prior to sending the frame in which it is contained and has a tolerance of ±5 dB	$P_{used}(s)$

4.2 Energy-Efficient Vertical Handover for the Integrated IEEE—3GPP Heterogeneous Network

Based on the 3GPP Access Network Discovery and Selection Function (ANDSF) and the advanced measurement capabilities provided by the IEEE 802.11-2012 [2] and the 3GPP Long Term Evolution—Advanced (LTE-A) Standards [10], in this section we discuss an ANDSF-assisted energy-efficient vertical handover decision algorithm for the heterogeneous IEEE 802.11-2012/LTE-A network. The proposed algorithm enables a multi-mode mobile terminal to select and associate with the network PoA that minimizes its average power consumption and guarantees a minimum supported quality of service for its ongoing connections. System-level simulation is used to evaluate the performance of the proposed algorithm and compare that to other competing solutions.

4.2.1 System Model

A heterogeneous wireless network (HWN) is considered, consisting of a set of 3GPP LTE PoAs *c* a set of IEEE 802.11-2012-conformant WLAN PoAs *W*, and a set of Multi-mode Mobile Terminals (MMTs) *U* equipped with both LTE and IEEE 802.11-2012-conformant radio access interfaces (Fig. 4.1). The ANDSF module is assumed to facilitate the VHO decision at the MMTs, by providing an appropriate candidate PoA set (network discovery information) and an inter-system mobility policy with respect to the candidate PoA set (described in the following), through the standardized S14 interface [5]. Each PoA within the HWN is assumed to perform standardized network measurements and convey them to the ANDSF as described below.

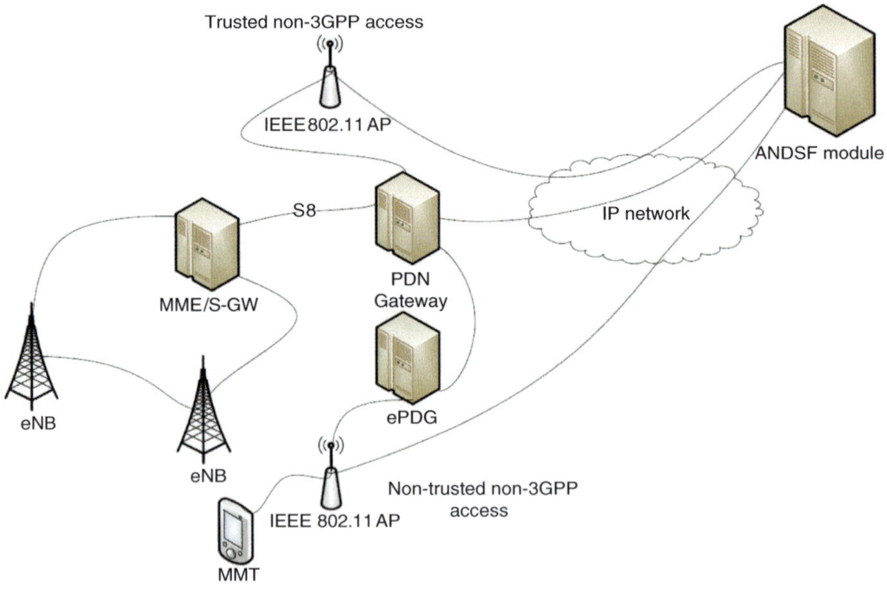

Fig. 4.1 Heterogeneous wireless network

Depending on the particular characteristics of the ongoing user service, each MMT $u \in \boldsymbol{U}$ is assumed to have a mean SINR target $\overline{\gamma}_{LTE}$ and bandwidth requirement B_{LTE} on service reception from the LTE network, and a mean SINR target $\overline{\gamma}_{WLAN}$ and bandwidth requirement B_{WLAN} on service reception from the WLAN network. Note that different mean SINR target and bandwidth requirements are allowed for the two systems, owing to their heterogeneous communication characteristics, e.g. incorporation of MIMO and allowed AMC schemes. It is also assumed that an active radio access interface of a tagged MMT (i.e. an interface which is switched-on and connected to a PoA) can be in one of the communication states in $\boldsymbol{S} := \{T, R, I\}$, where T corresponds to the transmit, R to the receive, and I to the idle state, respectively.

4.2.1.1 Power Consumption Model

Let us now focus on the power consumption model of an active MMT $u \in \boldsymbol{U}$. For a tagged time interval *TTT* the expected power consumption of the LTE interface is given as follows:

$$P_{LTE} = \sum_{s \in \boldsymbol{S}} p_{LTE}(s) \cdot P_{LTE}(s) \tag{4.1}$$

where $P_{LTE}(s)$ corresponds to the average power consumption in state $s \in \boldsymbol{S}$ for the LTE interface, and $P_{LTE}(s)$ represents the probability of the LTE interface being in state $s \in \boldsymbol{S}$, both of which are measured within the decision time interval *TTT*. Following a similar approach and notation, the expected power consumption corresponding to the WLAN interface is given as follows:

$$P_{WLAN} = \sum_{s \in \boldsymbol{S}} p_{WLAN}(s) \cdot P_{WLAN}(s) \tag{4.2}$$

In (4.1) and (4.2), the parameters $p_{LTE}(s)$ and $p_{WLAN}(s)$ depend on the particular characteristics of the ongoing user service, while they can also vary with respect to the channel state and the RAT type of the target PoA. For instance, a bad channel increases the probability of retransmissions, which in turn alters the sojourn time per interface states. These probabilities can be derived either through theoretical analysis, such as the work in [14] which includes some references for VoIP modeling with Markov chains, or through simulation and analysis, or by utilizing PHY or MAC interface monitoring techniques. In the work in this chapter, these probabilities are assumed to be known and fixed (but not equal) for both RATs.

Let us now focus on the power consumption parameters per interface state and RAT type. Current literature includes numerous reports for estimating the power consumption values per interface state with respect to the RAT type of the MMT interface [14, 15]. However, the efforts of [15] show that the power consumption at the MMT side is highly correlated to a plethora of statically and dynamically changing parameters such as supported standard, distance between the MMT and the target PoA, interface manufacturer, underlying computer engine, adopted interface state set, current channel state, and current traffic pattern characteristics. Even though the power consumption at the receive and idle state can be assumed fixed [16, 17], the power consumption at the transmit state is closely related to the mean SINR target, the interference and noise at the PoA site, as well as the channel gain between the MMT and the target PoA [16, 17]. Herein, the power consumption for the transmit state is allowed to vary with respect to the MMT and the target PoA status, for both the LTE and the WLAN interfaces.

Following a similar approach as the one in [17], the power consumption of the transmit interface state is decomposed to the base power consumption of the radio access interface and the transmitted power of the interface. As a result, for a tagged time interval T and MMT $u \in \boldsymbol{U}$, the expected power consumption of the LTE and WLAN transmit state is given by (4.3) and (4.4), respectively, where $P_{B,LTE}$ and $P_{B,WLAN}$ correspond to the base power consumption of the LTE and WLAN interfaces, while $P_{t,LTE}$ and $P_{t,WLAN}$ correspond to the respective transmitted power per interface.

$$P_{LTE}(T) = P_{B,LTE} + P_{t,LTE} \tag{4.3}$$

$$P_{WLAN}(T) = P_{B,WLAN} + P_{t,WLAN} \tag{4.4}$$

The following analysis is pursued to express the transmitted power per RAT type with respect to signal quality measurements, which can be estimated based on MMT or network measurements. For a tagged LTE PoA $c \in \boldsymbol{C}$, MMT $u \in \boldsymbol{U}$ and time interval TTT, the expected mean SINR in the LTE uplink direction, i.e. the mean SINR as seen by the LTE PoA c within the time interval TTT is given as follows:

$$\gamma_{LTE}(c) = \frac{P_{t,LTE}(c) \cdot h_{LTE}(u,c)}{\left(I_{LTE}(c) + (\sigma_{LTE}(c))^2\right)} \tag{4.5}$$

where $P_{t,LTE}(c)$ corresponds to the average transmitted power at the MMT for the LTE interface, $h_{LTE}(u,c)$ represents the average channel gain from MMT u to the LTE PoA c, $I_{LTE}(c)$ is the average interference power at the LTE PoA c, and $(\sigma_{LTE}(c))$ represents the average noise power at the LTE PoA c, all averaged within the time interval TTT. Using (4.5) and taking into account the mean SINR target $\overline{\gamma}_{LTE}$ for sustaining the MMT services, it can be readily shown that the expected power transmission of MMT u within the time interval TTT is given as follows:

$$P_{t,LTE}(c) = \frac{\overline{\gamma}_{LTE} \cdot \left(I_{LTE}(c) + (\sigma_{LTE}(c))^2\right)}{h_{LTE}(u,c)} \tag{4.6}$$

Following a similar approach for a tagged WLAN PoA $w \in \boldsymbol{W}$, it can be shown that the expected power transmission of MMT u within the time interval T is given as follows:

$$P_{t,WLAN}(w) = \frac{\bar{\gamma}_{WLAN} \cdot \left(I_{WLAN}(w) + (\sigma_{WLAN}(w))^2\right)}{h_{WLAN}(u, w)} \tag{4.7}$$

where $I_{WLAN}(w)$ corresponds to the average interference power at the WLAN PoA w, $(\sigma_{WLAN}(w))^2$ corresponds to the average noise power at the WLAN PoA w, and $h_{WLAN}(u, w)$ represents the channel gain from the MMT u to the WLAN PoA w all averaged within the time interval *TTT*.

By carefully examining (4.6) and (4.7), it can be seen that the transmitted power per interface, can be estimated with respect to the mean SINR target which is known to the MMT, the received interference and noise power at the PoA site, and the channel gain between the MMT and the PoA set. Next section presents a novel MMT power consumption minimization inter-system mobility policy, which is based on standardized LTE and WLAN measurements commuted to the ANDSF. This policy is conveyed to the MMT through the standard 3GPP S14 interface, which subsequently utilizes it to perform a VHO (if needed).

4.2.2 Proposed VHO Algorithm

This section describes a methodology for estimating the power consumption of a tagged MMT connected to a candidate PoA, based on standard measurements at the PoA sites (Tables 4.1 and 4.2). Accordingly, a novel MMT power consumption minimization policy is proposed, which is formed in the ANDSF module and conveyed to the MMTs through the standard S14 interface.

4.2.2.1 Overall Power Consumption Estimation for the LTE System

Taking into account the measurements in Table 4.1, and assuming a symmetric channel gain between user u and the LTE cell c, $h_{LTE}(u, c)$ can be estimated as follows:

$$h_{LTE}(u, c) \cong h_{LTE}(c, u) \cong \frac{RSRP(c)}{P_{RS}(c)} \tag{4.8}$$

Using (4.4) and taking into account the measurement definition in Table 4.2, it can be readily shown that the transmitted power for a tagged user u and LTE PoA c, is given as follows:

$$P_{t,LTE}(c) = \frac{\bar{\gamma}_{LTE} \cdot P_{RS}(c) \cdot I(c)}{RSRP(c)} \tag{4.9}$$

where the measurements are assumed to be derived for the tagged time interval TTT and over the operating bandwidth of the respective LTE nodes. Accordingly, the expected overall power consumption for sustaining a connection between a tagged MMT u a and LTE PoA c, can be estimated as follows:

$$\begin{aligned} P_{LTE}(c) = & \sum_{s \in S - \{T\}} p_{LTE}(s) \cdot P_{LTE}(s) \\ & + p_{LTE}(T) \cdot \left(P_{B,LTE} + P_{t,LTE}(c)\right) \end{aligned} \tag{4.10}$$

where the values of $p_{LTE}(s), s \in \boldsymbol{S}$ are fixed, known and adapted with respect to the ongoing user service [14], the values of $P_{LTE}(s), s \in \boldsymbol{S} - \{T\}$ are fixed, known and adapted with respect to the LTE interface characteristics [15], and finally the SINR target $\bar{\gamma}_{LTE}$ is fixed, known and adapted with respect to the rate requirements of the ongoing MMT service as in [18].

4.2.2.2 Overall Power Consumption Estimation for the WLAN System

Following a similar approach, for a tagged WLAN PoA $w \in W$, MMT $u \in U$, and time interval T, $h_{WLAN}(u, w)$ can be estimated as follows:

$$h_{WLAN}(u, w) \cong h_{WLAN}(w, u) \cong \frac{RCPI(w)}{P_{used}(w)} \tag{4.11}$$

Based on the measurements definition in Table 4.2 and using (4.6), it can be readily shown that the transmitted power for a tagged user $\bar{u}$ and WLAN PoA w, is given as follows:

$$P_{t,WLAN}(c) = \frac{\bar{\gamma}_{WLAN} \cdot P_{used}(w) \cdot ANPI(w)}{RCPI(w)} \tag{4.12}$$

where the measurements are assumed to be derived for the tagged time interval TTT and over the operating bandwidth of the respective LTE nodes. Accordingly, the expected overall power for sustaining a connection between a tagged MMT u and WLAN PoA w, can be estimated as follows:

$$\begin{aligned} P_{WLAN}(w) = & \sum_{s \in S - \{T\}} p_{WLAN}(s) \cdot P_{WLAN}(s) \\ & + p_{WLAN}(T) \cdot \left(P_{B,WLAN} P_{t,WLAN}(w)\right) \end{aligned} \tag{4.13}$$

where the values of $p(s) s \in S$ are fixed, known and adapted with respect to the ongoing user service [14], the values of $P_{WLAN}(s), s \in S - \{T\}$ are fixed, known and adapted with respect to the WLAN interface characteristics [15], and the SINR target $\overline{\gamma}^{u}_{WLAN}$ is fixed, known and adapted with respect to the rate requirements of the ongoing MMT service as in [18].

4.2.2.3 VHO Decision Policy for MMT Power Consumption Minimization

As described above, the ANDSF can provide the MMT with an inter-system mobility policy, access network discovery information, and an inter-system routing policy. The inter-system mobility policy is provided in terms of rules [5], which are applicable under particular validity conditions. This section describes a novel inter-system mobility policy for minimizing the power consumption at the MMT side by performing a VHO. This policy is conveyed to the MMT in terms of rules applicable under particular validity conditions, while it is based on estimating the expected overall power consumption for a tagged MMT and candidate PoA with respect to standardized LTE and WLAN measurements. Note that the proposed policy is used only for VHO decision making, i.e. moving from the LTE system to the WLAN system and vice versa. To further reduce the power consumption at the MMT, in the final part of this chapter we discuss how the selection of the least power consuming PoA of the same RAT type can be performed, aiming to provide a complete decision algorithm for both vertical and horizontal HO decision making.

VHO decision making for the integrated LTE-WLAN system can be broken down into two scenarios: the MMT is connected to the LTE system and investigates the possibility of switching to the WLAN system, and vice versa.

Let us focus on the first scenario, where the MMT $u \in \boldsymbol{U}$ is connected to and receives service from a tagged LTE PoA $c \in \boldsymbol{C}$. Let $\boldsymbol{L}_{WLAN} \subseteq \boldsymbol{W}$ be the candidate WLAN PoA set for the tagged MMT $u \in \boldsymbol{U}$, as identified by the ANDSF module. Then, a VHO execution towards a tagged WLAN PoA $w \in \boldsymbol{L}_{WLAN}$ is preferable in terms of overall power consumption at the MMT, if and only if:

$$P_{WLAN}(w) < P_{LTE}(c) \tag{4.14}$$

$$\begin{aligned} &\sum_{s \in \boldsymbol{S}-\{T\}} p_{WLAN}(s) \cdot P_{WLAN}(s) + p_{WLAN}(T) \cdot \left(P_{B,WLAN} + P_{t,WLAN}(w)\right) \\ &\quad < \sum_{s \in \boldsymbol{S}-\{T\}} p_{LTE}(s) \cdot P_{LTE}(s) + p_{LTE}(T) \cdot \left(P_{B,LTE} + P_{t,LTE}(c)\right) \end{aligned} \tag{4.15}$$

$$\sum_{s\in \boldsymbol{S}-\{T\}} p_{WLAN}(s)\cdot P_{WLAN}(s) + p_{WLAN}(T)\cdot\left(P_{B,WLAN} + \frac{\bar{\gamma}_{WLAN}\cdot P_{used}(w)\cdot ANPI(w)}{RCPI(w)}\right)$$
$$< \sum_{s\in \boldsymbol{S}-\{T\}} p_{LTE}(s)\cdot P_{LTE}(s) + p_{LTE}(T)\cdot\left(P_{B,LTE} + \frac{\bar{\gamma}_{LTE}\cdot P_{RS}(c)\cdot I(c)}{RSRP(c)}\right) \tag{4.16}$$

$$p_{WLAN}(T)\cdot\frac{\bar{\gamma}_{WLAN}\cdot P_{used}(w)\cdot ANPI(w)}{RCPI(w)}$$
$$<\left\{\sum_{s\in \boldsymbol{S}-\{T\}} p_{LTE}(s)\cdot P_{LTE}(s) + p_{LTE}(T)\cdot\left(P_{B,LTE} + \frac{\bar{\gamma}_{LTE}\cdot P_{RS}(c)\cdot I(c)}{RSRP(c)}\right)\right\}$$
$$-\left\{\sum_{s\in \boldsymbol{S}-\{T\}} p_{WLAN}(s)\cdot P_{WLAN}(s) + p_{WLAN}(T)\cdot P_{B,WLAN}\right\} \tag{4.17}$$

$$RCPI(w) > \frac{p_{WLAN}(T)\cdot\bar{\gamma}_{WLAN}\cdot P_{used}(w)\cdot ANPI(w)}{\left\{\sum_{s\in S-\{T\}} p_{LTE}(s)\cdot P_{LTE}(s) + p_{LTE}(T)\cdot\left(P_{B,LTE} + \frac{\bar{\gamma}_{LTE}\cdot P_{RS}(c)\cdot I(c)}{RSRP(c)}\right)\right\} - \left\{\sum_{s\in S-\{T\}} p_{WLAN}(s)\cdot P_{WLAN}(s) + p_{WLAN}(T)\cdot P_{B,WLAN}\right\}} \tag{4.18}$$

where Eq. (4.15) is derived by substituting (4.3) and (4.4) in (4.14), Eq. (4.16) is found by substituting (4.6) and (4.7), while Eqs. (4.17) and (4.18) are obtained by rearranging (4.16).

Let us now focus on the case where a tagged MMT u is connected to and receives service from a tagged WLAN PoA $w \in W$. It is assumed that $L_{LTE} \subseteq C$ is the candidate LTE PoA set for the tagged MMT $u \in U$, as identified by the ANDSF module. Then, following a similar approach, a VHO execution towards a tagged LTE PoA $c \in L_{LTE}$ is preferable in terms of overall MMT power consumption, if and only if:

$$RSRP^{c\to u,T} > \frac{p_{LTE}(T)\cdot\bar{\gamma}_{LTE}\cdot P_{RS}(c)\cdot I(c)}{\left\{\sum_{s\in S-\{T\}} p_{WLAN}(s)\cdot P_{WLAN}(s) + p_{WLAN}(T)\cdot\left(P_{B,WLAN} + \frac{\bar{\gamma}_{WLAN}\cdot P_{used}(w)\cdot ANPI(w)}{RCPI(w)}\right) - \left\{\sum_{s\in S-\{T\}} p_{LTE}(s)\cdot P_{LTE}(s) + p_{LTE}(T)\cdot P_{B,LTE}\right\}\right\}} \tag{4.19}$$

It is important to notice that the right side of the Eqs. (4.18) and (4.19) includes standard WLAN and LTE network measurements known to the ANDSF, fixed and known values regarding the service characteristics of MMT u, i.e., $p_{LTE}(s)$, $p_{WLAN}(s)$, $\bar{\gamma}_{LTE}$, and $\bar{\gamma}_{WLAN}$, as well as fixed and known values regarding the power consumption of the MMT interfaces in the receive and idle states. On the other hand, the left side of Eqs. (4.18) and (4.19) correspond to a link quality measurement locally derived at the MMT, which is not known at the ANDSF site given that it is strongly related to the current MMT status and position. However, Eqs. (4.18) and (4.19) can serve as validity conditions for inter-system mobility rules corresponding to handing over from the current serving to the tagged PoA for the tagged MMT. It is important to note here that this set of rules are auxiliary to the final VHO decision taken by the MMT, and do not correspond to a VHO command to the MMT.

To summarize, deploying the proposed VHO decision policy for MMT power consumption minimization, referred to as the MMT Power Consumption Minimization (MPCM) policy, the ANDSF provides the following set of inter-system mobility rules and validity conditions for a tagged MMT u:

> **if** serving PoA is the serving LTE PoA $c \in C$
> **Rule Set**: {if $RPCI(w) > RCPI_{MPCM}(w, c)$ and $w \in L_{WLAN}$ => Prioritized access to w}
> **else if** serving PoA is the serving WLAN PoA $w \in W$
> **Rule Set**: {if $RSRP(c) > RSRP_{MPCM}(c, w)$ and $c \in L_{LTE}$ =>Prioritized access to c}

where $RCPI_{MPCM}(w, c)$ and $RSRP_{MPCM}(c, w)$ are adapted with respect to the WLAN and LTE network measurements, and are given as follows:

$$RCPI_{MPCM}(w,c) = \frac{p_{WLAN}(T) \cdot \bar{\gamma}_{WLAN} \cdot P_{used}(w) \cdot ANPI(w)}{\left\{\sum_{s \in S-\{T\}} p_{LTE}(s) \cdot P_{LTE}(s) + p_{LTE}(T) \cdot \left(P_{B,LTE} + \frac{\bar{\gamma}_{LTE} \cdot P_{RS}(c) \cdot I(c)}{RSRP(c)}\right)\right\} - \left\{\sum_{s \in S-\{T\}} p_{WLAN}(s) \cdot P_{WLAN}(s) + p_{WLAN}(T) \cdot P_{B,WLAN}\right\}} \tag{4.20}$$

$$RSRP_{MPCM}(c,w) = \frac{p_{LTE}(T) \cdot \bar{\gamma}_{LTE} \cdot P_{RS}(c) \cdot I(c)}{\left\{\sum_{s \in S-\{T\}} p_{WLAN}(s) \cdot P_{WLAN}(s) + p_{WLAN}(T) \cdot \left(P_{B,WLAN} + \frac{\bar{\gamma}_{WLAN} \cdot P_{used}(w) \cdot ANPI(w)}{RCPI(w)}\right) - \left\{\sum_{s \in S-\{T\}} p_{LTE}(s) \cdot P_{LTE}(s) + p_{LTE}(T) \cdot P_{B,LTE}\right\}\right\}} \tag{4.21}$$

4.2.2.4 HO Decision Policy for Intra-System MMT Power Consumption Minimization

By using the MPCM VHO decision policy for inter-system mobility, the MMT can choose between switching to either the LTE or the WLAN systems. In this subsection, we describe how the MMT can utilize standard measurements and the inter-system mobility rules, provided by the ANDSF module, to switch to the least power consuming PoA of the same RAT type. This procedure can be broken down into two decision scenarios as well: (a) how to choose the least power consuming WLAN PoA and (b) how to choose the least power consuming LTE PoA.

Let us focus on the case where an active MMT $u \in U$ is connected to and receives service from a tagged LTE PoA $c \in C$. Let $L_{NS,WLAN} \subseteq L_{WLAN}$ denote the WLAN PoA set consisting of candidate PoA in L_{WLAN}, for which the MMT has measured the pilot signals and has evaluated that they satisfy the validity conditions specified by the MPCM inter-system mobility policy. For $w, w' \in L_{NS,WLAN}$, a VHO towards the WLAN PoA W is considered more preferable compared to a VHO towards the WLAN PoA w', if and only if

$$P_{WLAN}(w) < P_{WLAN}(w') \tag{4.22}$$

$$\sum\nolimits_{s\in S-\{T\}} p_{WLAN}(s)\cdot P_{WLAN}(s) + p_{WLAN}(T)\cdot\left(P_{B,WLAN}+P_{t,WLAN}(w)\right) < \sum\nolimits_{s\in S-\{T\}} p_{WLAN}(s)\cdot P_{WLAN}(s) + p_{WLAN}(T)\cdot\left(P_{B,WLAN}+P_{t,WLAN}(w')\right) \tag{4.23}$$

$$P_{t,WLAN}(w) < P_{t,WLAN}(w') \tag{4.24}$$

$$\frac{\bar{\gamma}_{WLAN}\cdot P_{used}(w)\cdot ANPI(w)}{RCPI(w)} < \frac{\bar{\gamma}_{WLAN}\cdot P_{used}(w')\cdot ANPI(w')}{RCPI(w')} \tag{4.25}$$

$$RCPI(w) > RCPI(w')\cdot\frac{P_{used}(w)\cdot ANPI(w)}{P_{used}(w')\cdot ANPI(w')} \tag{4.26}$$

$$RCPI(w) > RCPI(w')\cdot\frac{RCPI_{MPCM}(w,c)}{RCPI_{MPCM}(w',c)} \tag{4.27}$$

where (4.23) is derived by substituting (4.13) into (4.22), Eq. (4.24) can be obtained by rearranging (4.23), while by substituting (4.12) in (4.24) Eq. (4.25) is obtained. Equation (4.26) is formed by rearranging (4.25), and Eq. (4.27) is derived by taking into account that the denominator in (4.20) is common for both parameters $RCPI_{MPCM}(w,c)$ and $RCPI_{MPCM}(w',c)$.

Following a similar approach and notation, for the case where the tagged MMT receives service from a tagged WLAN PoA $w \in W$, $L_{HO,LTE} \subseteq L_{LTE}$ then denotes the WLAN PoA set compromising the candidate PoA in L_{LTE}, for which the MMT has measured the RS and has evaluated that they satisfy the validity conditions specified by the MPCM inter-system mobility policy. Then, a VHO towards the LTE PoA c is considered more preferable compared to a VHO towards the LTE PoA s c', if and only if:

$$RSRP(c) > RSRP(c')\cdot\frac{P_{RS}(c)\cdot I(c)}{P_{RS}(c')\cdot I(c')} \tag{4.28}$$

$$RSRP(c) > RSRP(c')\cdot\frac{RSRP_{MPCM}(c,w)}{RSRP_{MPCM}(c',w)} \tag{4.29}$$

From the above derivations, it can be concluded that the validity conditions which are provided by the ANDSF, apart from facilitating the MMT to choose between the LTE and WLAN systems, can also be utilized for selecting the least power consuming PoA of the target RAT type, i.e. Eqs. (4.27) and (4.29). On the other hand, selecting the least power consuming PoA of the same RAT can be performed at the MMT by using standard LTE and WLAN measurements as shown in Eqs. (4.26) and (4.28). Both these features are subsequently exploited in Sect. 4.2.2.5 to enable the proposed VHO decision algorithm to support all feasible Vertical and Horizontal HO decision scenarios.

4.2.2.5 Proposed VHO Decision Algorithm

Based on the VHO decision policy for MMT power consumption minimization, described in Sect. 4.2.2.3, and the HO decision policy for intra-system MMT power consumption minimization, described in Sect. 4.2.2.4, in this section we propose an energy-efficient HO decision algorithm for the integrated LTE—WLAN network. The proposed algorithm utilizes ANDSF context, built on standard measurements provided by both LTE and WLAN PoAs, which is sent to the MMT by using standard ANDSF pull procedures. By using this context, the proposed VHO decision algorithm enables the MMT to always choose the least power consuming RAT interface and PoA. The proposed algorithm is illustrated by a flowchart in Fig. 4.2.

Upon VHO decision triggering, the proposed algorithm investigates whether the current serving PoA of the MMT is an LTE cell or a WLAN AP. In the former case, the proposed algorithm follows the steps illustrated in the upper part of the flowchart, whereas in the latter case, it follows the steps in the lower part of the flowchart.

When the serving PoA is an LTE cell, it initiates an ANDSF request to acquire the $RCPI_{MPCM}(w, c)$ parameters for all the WLAN AP in proximity. The proposed algorithm then checks whether a WLAN AP, that satisfies Eq. (4.18 (least power consuming compared to the serving LTE cell) and Eq. (4.27) (least power consuming WLAN AP), exists. If such a WLAN AP (denoted by w) exists, it subsequently checks whether another LTE cell exists with lower power consumption requirements. If an LTE cell does exist, a HHO request is initiated, while if such LTE cell does not exist, a VHO procedure is performed towards the least power consuming WLAN AP w. In the scenario where the current LTE serving cell results in lower power consumption compared to any of the WLAN APs in proximity (upper right case), the proposed algorithm initiates an ANDSF request to acquire the parameters in Eq. (4.28), aiming to identify LTE cells that reduce the MMT power consumption. If such a cell exists, a HO request is sent to the current serving LTE cell, whereas in the opposite case, no action is taken and the VHO algorithm is terminated.

Focusing on the scenario where the current serving PoA is a WLAN AP (lower part of the flowchart) and following a similar approach, the VHO algorithm initiates an ANDSF request to acquire the $RSRP_{LTE}(c, w)$ parameters for all LTE cells in proximity. The proposed algorithm checks if an LTE cell exists, that satisfies (4.19) (least power consuming compared to the serving WLAN AP) and (4.29) (least power consuming LTE cell). If such an LTE cell exists, denoted by c, it subsequently checks whether another WLAN AP with lower power consumption

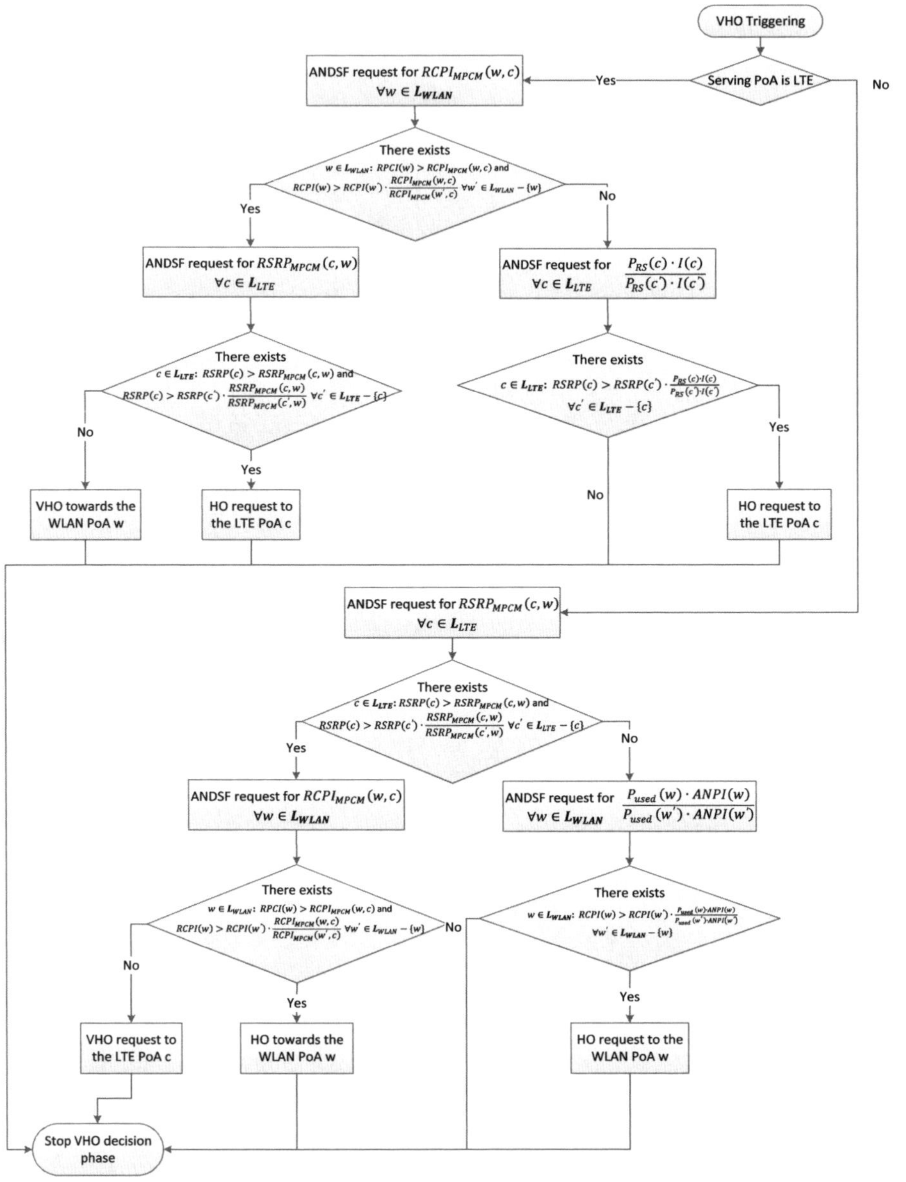

Fig. 4.2 Proposed VHO decision algorithm

requirements exists. If another WLAN AP is present, the MMT initiates a HHO towards the least power consuming WLAN AP w, while if such AP does not exist, a VHO procedure is performed towards the least power consuming LTE cell c. In the scenario where the current serving WLAN AP results in lower power consumption compared to any LTE cells in proximity, the proposed algorithm initiates an ANDSF request to acquire the parameters in Eq. (4.26), aiming to identify WLAN APs that reduce the MMT power consumption. If such a WLAN AP exists, a HO request is sent to the current serving WLAN AP, whereas in the opposite case, no action is taken and the VHO algorithm is terminated.

4.2.3 Performance Evaluation

The performance evaluation of the proposed VHO algorithm is built on an extended version of the system-level methodology described by the Small Cell Forum [19]. A hexagonal LTE-A network is considered with a main cluster composed of 7 eNBs, where each eNB consists of 3 sectors. The wrap-around technique is used to extend the network by copying the main cluster symmetrically on its 6 sides. A set of blocks of apartments, referred to as *wlanblocks*, are uniformly dropped within the main cluster area with respect to a WLAN block deployment density parameter, denoted by d_{WB}, which indicates the percentage of the main cluster area covered with wlanblocks. Wlanblocks are modeled in line with the dual stripe model for dense urban environments in [19], while the path loss models are adapted accordingly. The WLAN APs are deployed with respect to a deployment ratio $r_{wc} = 0.2$, indicating the probability of having an AP deployed inside the apartment. APs and users are uniformly dropped inside the apartments, whereas each AP initially serves one user. Each macrocell sector initially serves ten users, which are uniformly distributed within it. Aiming to guarantee a minimum QoS, all studied algorithms are evaluated by using a mean SINR target of 2.88 dB for all user connections and RATs. The remaining simulation parameters are summarized in Table 4.3.

The proposed algorithm is compared to a VHO algorithm that always prioritizes WLAN over LTE-A access, referred to as the baseline WLAN algorithm, and a VHO algorithm that always prioritizes LTE-A over WLAN access, referred to as the baseline LTE-A algorithm. The performance of all algorithms is averaged over both the LTE-A and WLAN systems/interfaces while it is plotted for increasing d_{WB}.

Figure 4.3 demonstrates the average number of users per RAT for all algorithms. Note that the total number of users in the system increases linearly for increasing d_{WB}, provided that one additional user is introduced for every additional AP. As expected, the baseline LTE-A algorithm forces all users to connect to LTE-A system until the maximum cell capacity for all LTE-A PoAs: for $d_{WB} = 0.5$ the number of users is close to 350. Above this density, all the remainder MMTs inevitably connect to the WLAN system. In contrast, the number of WLAN users for the baseline WLAN algorithm increases rapidly as the density of wlanblocks increases in the system, until all users are finally served by the WLAN system. Interestingly, the proposed VHO balances the number of users between the LTE-A and WLAN systems and sustains a roughly fixed number of LTE-A users by keeping the MMT transmit power and the number of LTE-A interferers low.

Table 4.3 System-level simulation model and parameters

System operating parameters		
Parameter	Macrocell	WLAN
Carrier frequency	Uniformly picked from set {1990, 2000, 2010} MHz	Uniformly picked from the bands 1, 6, 11, 14
Channel bandwidth	10 MHz	22 MHz
Maximum Tx power	46 dBm	20 dBm
Transmit power on the pilot signals	Normally distributed with a mean value of 23 dBm and standard deviation 3 dB	Normally distributed with a mean value of 20 dBm and Standard deviation 5 dB
Maximum number of users	50	20
Simulation model		
Link-to-system mapping	Effective SINR mapping (ESM) [16]	
Path loss	Models for urban deployment in [16]	
Mobility model	Similar to the one in [10]	
Traffic model	Full buffer similar to [16] with $q_l(T) = q_w(T) = 0.4$, $q_l(R) = q_w(R) = 0.3$ and $q_l(I) = q_w(I) = 0.3$	
Power consumption WLAN	$P_w(T) = 924$ mW, $P_w(R) = 594$ mW, $P_w(I) = 80$ mW [12]	
Power consumption LTE-A	$P_l = 1{,}550$ mW, $P_l(R) = 1{,}420$ mW, $P_l(I) = 160$ mW [11]	

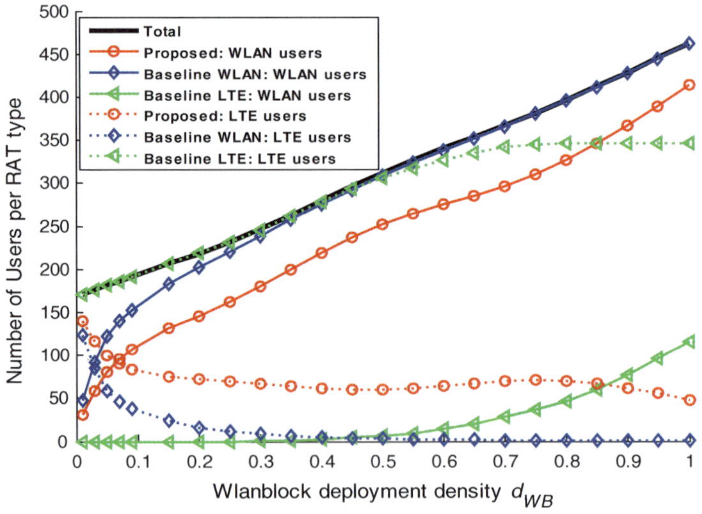

Fig. 4.3 Number of users per RAT type

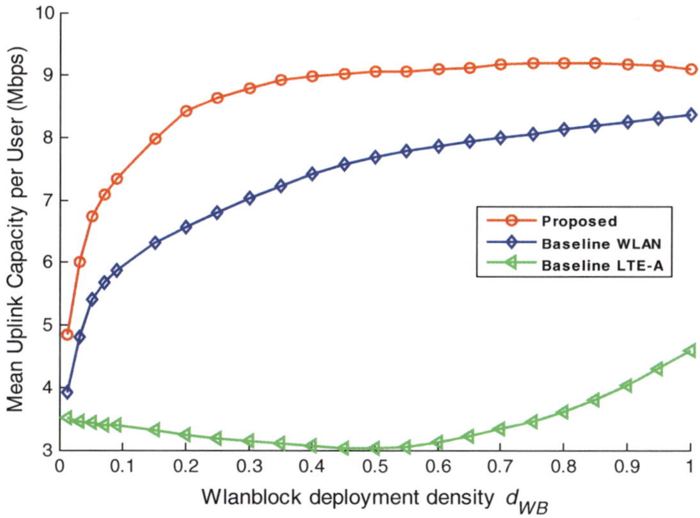

Fig. 4.4 Mean uplink capacity per user (Mbps)

Figure 4.4 depicts the mean uplink (UL) capacity per user for all algorithms. As expected, for $d_W \leq 0.5$ the UL capacity performance for the baseline LTE-A algorithm is deteriorated, due to the increasing number of interferers in the LTE-A UL. Above this density, however, the performance of the baseline LTE-A algorithm improves due to the (inevitable) utilization of the WLAN infrastructure. Due to the shorter transmit-receive range of the WLAN APs, the baseline WLAN algorithm attains improved performance compared to the baseline LTE-A algorithm, whereas, the smart utilization of the LTE-A and WLAN systems (Fig. 4.3) enables the proposed VHO algorithm to attain higher UL capacity per user compared to both baseline algorithms. In particular, the capacity gain per user reaches up to 6Mbps compared to the baseline LTE-A and up to 2 Mbps compared to the baseline WLAN algorithm.

Figure 4.5 demonstrates the average overall MMT energy consumption per bit for all algorithms. Compared to the baseline WLAN, the baseline LTE-A algorithm results in higher MMT energy consumption per bit due to the increased requirements for power consumption per interface state (Table 4.3). On the other hand, the performance of the baseline LTE-A algorithm is improved for $d_{WB} \geq 0.5$, since above this density the LTE-A system reaches its maximum capacity and the surplus of users utilize their WLAN interface. Compared to the baseline LTE-A and WLAN algorithms, the proposed VHO algorithm is shown to reduce the average overall MMT energy consumption per bit by up to 93 and 82 %, respectively.

These performance gains follow from the fact that the proposed VHO algorithm accounts for the power consumption per RAT interface state, while it is based on standard LTE-A and WLAN measurements to estimate the expected transmit power of each interface. On average, the energy per bit gain for the proposed VHO is 80 % compared to the baseline LTE-A and 71 %, compared to the baseline WLAN algorithm.

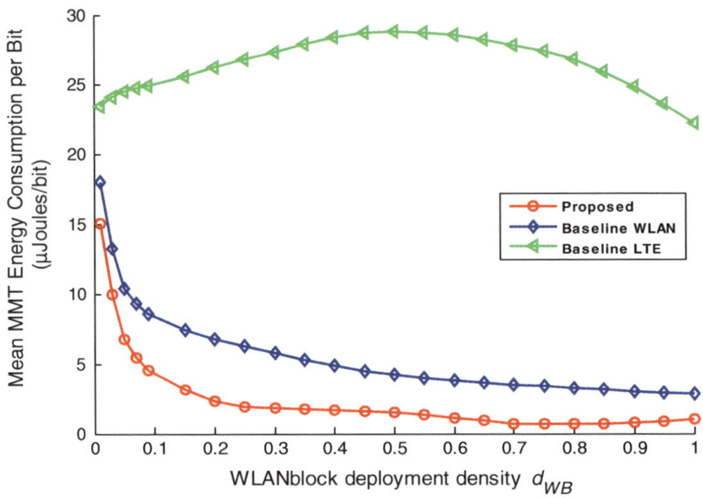

Fig. 4.5 Overall MMT energy consumption per bit (μJ/bit)

Figure 4.6 concludes our simulation study by depicting the signaling rate, i.e. the number of signals exchanged in the system per second, in both the LTE-A and the WLAN systems. Note that the ANDSF signaling rate is plotted only for the proposed algorithm, which necessitates the exchange of the inter-system mobility policy parameters to/from the ANDSF. As expected, the baseline LTE-A algorithm results in the highest signaling rate in LTE-A, as it avoids the utilization of the WLAN infrastructure. On the other hand, the LTE-A network signaling rate for the baseline WLAN algorithm reduces rapidly for increased wlanblock deployment density, owing to the prioritization of the WLAN access. Interestingly, even though the proposed VHO asks for increased ANDSF signaling, it also results in significantly reduced LTE-A signaling compared to the baseline LTE-A algorithm due to the load balancing between the LTE-A and WLAN systems. Nevertheless, the exchange and utilization of standard WLAN measurements at the ANDSF also result in increased WLAN signaling for the proposed VHO algorithm compared to the baseline WLAN algorithm. Nevertheless, the consequential ANDSF and WLAN signaling load for the proposed VHO algorithm, increases linearly for increasing wlanblock deployment density, allowing the system engineers to easily anticipate for the signaling requirements of the proposed algorithm.

4.3 Energy-Efficient Handover for the 3GPP LTE-a Femtocell Network

In this section, we present an interference-aware handover decision algorithm for the LTE-Advanced femtocell network, which utilizes standard signal measurements to select the candidate cell that a) attains the minimum required channel gain for

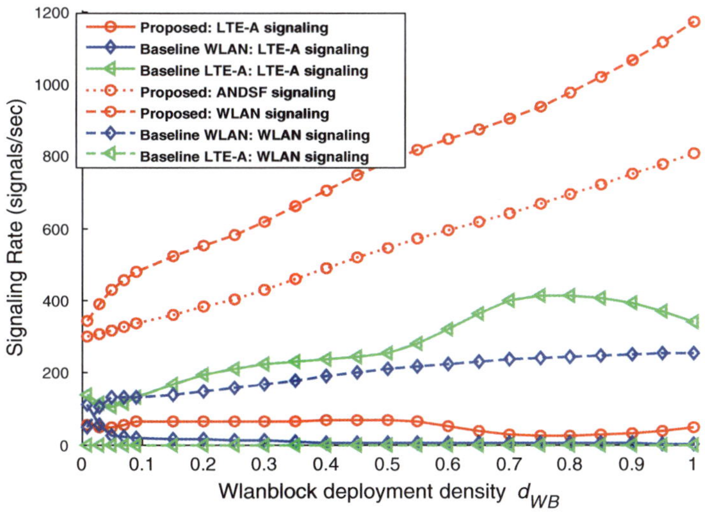

Fig. 4.6 Signaling rate (signals per second)

sustaining service continuity and b) minimizes the mean UE transmit power for a pre-determined mean SINR target. The proposed algorithm attains backwards compatibility with the LTE-Advanced system, as it is deployed by using the private mechanism for non-standard use. Based on the evaluation methodology of the Small Cell forum, we validate the performance of the proposed algorithm and compare it against that of other state-of-the-art algorithms.

4.3.1 System Model

We consider an LTE-A network consisting of a macrocell and a femtocell tier. Macrocells are referred to as evolved Node B (eNB), while femtocells as Home eNB (HeNB). Both tiers operate in the band set denoted by $N := \{1, \ldots, N\}$. Let C_n denote the set of cells operating in band n, including both eNBs and HeNBs, and U_n the set of users served from a cell in C_n. We focus our analysis on the HO decision phase for a tagged LTE-A user $u \in U_n$, which is served from cell $s \in C_n$ and is in proximity of the (candidate and accessible) cell set $L \subseteq \bigcup_{n \epsilon N} C_n$. To sustain service continuity, the tagged user is assumed to have a minimum requirement in terms of Reference Signal Received Power (RSRP) [13], denoted by $RSRP_{th}$, while to support its ongoing services, it is assumed to have a prescribed mean SINR target, denoted by γ_t.

Table 4.4 summarizes the notation used for the system model parameters and measurements. Note that all these parameters are assumed to be derived in the context of the HO decision phase, i.e. averaged within the operating bandwidth of the target cell over the time interval Time To Trigger (TTT) [8, 20].

Table 4.4 System model parameters and LTE-A measurements

	Notation
System model parameters	
Mean transmit power of node x (LTE-A cell or user)	$P(x)$
Mean noise power in node x (LTE-A cell or user)	$\sigma(x)^2$
Mean channel gain from node x to node y (LTE-A cells or users)	$h(x, y)$
Maximum allowed transmit power of node x (LTE-A cell or user)	$P_{\max}(x)$
Handover hysteresis margin (HHM) for cell c	$HHM(c)$
LTE-A signal quality measurements [12]	
Reference signal received power (RSRP) in cell c	RSRP (c)
Received interference power (REP) in cell c	$I(c)$
Downlink reference signal transmit power (DL RS Tx) of cell c	$P_{RS}(c)$

For the tagged time interval TTT, the mean uplink (UL) SINR of user u is given by

$$\gamma(s) = \frac{P(u) \cdot h(u, s)}{\sum_{u' \in \boldsymbol{U}_n - \{u\}} P(u') \cdot h(u', s) + \sigma(s)^2} \tag{4.30}$$

where the numerator corresponds to the signal strength from user u in cell s, while the denominator represents the interference caused by in-band operating users plus the noise power in cell s. Given the prescribed mean SINR target γ_t, the mean UE transmit power of user u in the candidate cell $c \in L$, denoted by $P(u, c)$, can be derived as follows:

$$P(u, c) = \frac{\gamma_t \cdot \left(\sum_{u' \in \boldsymbol{U}_n - \{u\}} P(u') \cdot h(u', c) + \sigma(c)^2\right)}{h(u, c)} \tag{4.31}$$

Note that the positive impact of handing over to cell c in terms of lower interference is included in Eq. (4.31), as we omit the interference caused by the user connection with cell s, i.e. $P(u) \cdot h(u, s)$. Assuming that the transmit power is a primary contributor to the UE power consumption, Eq. (4.31) can be used to estimate the mean power consumption of user u as well.

4.3.2 Handover Decision Algorithm

The proposed HO decision algorithm utilizes standard LTE-A measurements to mitigate the negative impact of user mobility and lower the mean UE transmit power for the prescribed mean SINR target. The former is achieved by excluding cells with channel gain lower than that required to sustain service continuity, while the latter by estimating the mean UE transmit power for the remaining cells and handing over to the one with the minimum required power.

In this section, two criteria are described. The first criterion, in Sect. 4.3.2.1, is a novel criterion for excluding cells with channel gain lower than the one required for sustaining service continuity, while the second described in Sect. 4.3.2.2, is a criterion for handing over to the cell with the minimum required mean UE transmit power. Both these criteria are integrated in the proposed algorithm in Sect. 4.3.2.3, which also discusses the network signaling for deploying the algorithm.

4.3.2.1 Criterion for Sustained Service Continuity

Sustaining service continuity is of critical importance in the presence of femtocells, if we consider the short-range nature of communications, the denser network layout and the fast varying radio environment. Taking into account the definition of the RSRP and the DL RS Tx measurements in Table 4.2, it can be readily shown that they are related as follows:

$$RSRP(c) = P_{RS}(c) \cdot h(c,u) \tag{4.32}$$

In macrocell deployments, higher RSRP typically results in improved channel gain, i.e. comparable RS transmit powers are radiated among the cells. In the presence of femtocells, however, the RSRP is biased in favor of the cells with the higher RS transmit powers (Eq. 4.32). As a result, handing over to the strongest cell does not necessarily improve the channel gain or the SINR performance. Based on this observation, we propose a criterion that can be used to avoid cells with poor channel conditions, i.e. compromised service continuity.

Taking into account the minimum required RSRP for sustaining service continuity, i.e. $RSRP_{th,}$ and the maximum allowed transmit power for user u and cell c, i.e. $P_{max}(u)$ and $P_{max}(c)$, respectively, we can estimate the channel gain threshold, denoted by $h_{min}(c)$, above which the service continuity between cell c and user u is sustained:

$$h_{\min}(c) = \frac{RSRP_{th}}{\min(P_{\max}(u), P_{\max}(c))} \tag{4.33}$$

Note that the $P_{max}(u)$ and $P_{max}(c)$ constraints can be owed to either an interference limitation for the cell or the user, or to the power class of the cell or the UE. It follows that sustaining service continuity is equivalent to satisfying the condition: $h(c,u) > h_{min}(c)$. Using (4.32) and (4.33), it can be shown that the set of candidate cells that sustain service continuity, denoted by M can be identified by the following criterion:

$$\boldsymbol{M} := \{c|RSRP(c) > RSRP_{th} + P_{RS}(c) - \min(P_{max}(u), P_{max}(c)), \mathrm{c} \in \mathbf{L}\} \quad (4.34)$$

where the parameters in (4.34) are assumed to be taken in decibels (dB).

4.3.2.2 Criterion for Reduced UE Transmit Power

This section describes an LTE-A compliant methodology for estimating the mean UE transmit power of the user on a per candidate cell basis, given a prescribed mean SINR target. The incorporation of the SINR target enhances the supported QoS, while the utilization of standard LTE-A measurements provides an accurate estimation for the mean UE transmit power requirements. By taking into account the definition of the RIP measurement in Table 4.2, it follows that:

$$I(s) = \sum_{u' \in \boldsymbol{U_n} - \{u\}} P(u') \cdot h(u', s) + \sigma(s)^2 \quad (4.35)$$

Assuming a symmetric channel gain, i.e., $h(s,u) = h(u,s)$, and by using Eqs. (4.2), (4.3), and (4.6), it follows that the mean UE transmit power for the current serving cell s can be estimated as

$$P(u) = \frac{\gamma_t \cdot P_{RS}(s) \cdot I(s)}{RSRP(s)} \quad (4.36)$$

Under the same viewpoint, the mean UE transmit power for a candidate cell $\mathrm{c} \in M$ can be estimated as follows:

$$P(u,c) = \begin{cases} \dfrac{\gamma_t \cdot P_{RS}(s) \cdot \left(I(s) - \frac{\gamma_t \cdot P_{RS}(s) \cdot I(s)}{RSRP(s)} \frac{RSRP(c)}{P_{RS}(c)} \right)}{RSRP(c)}, & if \quad c, s \in C_n \\ \dfrac{\gamma_t \cdot P_{RS}(c) \cdot I(c)}{RSRP(c)}, & otherwise \end{cases} \quad (4.37)$$

where the condition $c, s \in C_n$ is introduced to remove the interference caused by the ongoing user link with cell s, i.e. $P(u) \cdot h(u,c)$, if cells c and s operate in the same band.We can argue that a HO to the candidate cell $\mathrm{c} \in M$ is expected to lower the mean UE transmit power if the condition $P(u) > P(u,c)$ is met. Using (4.8) and taking the values in dB, it can be shown that this condition can be rearranged in a relative RSRP comparison with adaptive HHM as follows:

$$RSRP(c) > RSRP(s) + HHM(c) \quad (4.38)$$

where the $HHM(c)$ is adapted according to

$$HHM(c) = \begin{cases} 10\log \frac{P_{RS}(c)\cdot\left(I(c) - \frac{\gamma_t \cdot P_{RS}(s)\cdot I(s)}{RSRP(s)}\cdot\frac{RSRP(c)}{P_{RS}(c)}\right)}{P_{RS}(s)I(s)} & c, s \in C_n \\ 10\log \frac{P_{RS}(c)\cdot I(c)}{P_{RS}(s)\cdot I(s)} & otherwise \end{cases} \tag{4.39}$$

Combined with the criterion for sustained service continuity in (4.34), Eqs. (4.38) and (4.39) can be used to select the candidate cell with the minimum required mean UE transmit power as follows:

$$\arg\max_{c \in \boldsymbol{M}} RSRP(c) := \{c | RSRP(c) > RSRP(s) + HHM(c)\} \tag{4.40}$$

where the RSRP measurements are assumed to be taken in dB.

Note that Eq. (4.40) provides a backward compatible method with the standard HO decision procedure in cellular networks, i.e. the strongest cell criterion. The key differences are that the HO decision criterion in Eq. (4.40) adapts the HHM based on standard LTE-A context, i.e. signal measurements, and it is performed on the candidate cells that sustain service continuity.

4.3.2.3 HO Decision Algorithm

The proposed HO decision algorithm integrates the criteria for sustained service continuity and reduced UE transmit power, to enhance the HO decision efficiency in the integrated LTE-A macro-femtocell network. It is worth noting that the required HO decision context consists of the following information for the candidate cells: (a) operating frequency, (b) current LTE-A measurement status (Table 4.2), and (c) maximum allowed transmit power constraint.

The proposed HO decision algorithm is illustrated in Fig. 4.7. Upon HO decision triggering (step 1), the proposed algorithm acquires the HO context from the candidate cells in L(step 2). The network signaling for performing this step is discussed as well. Based on the acquired HO decision context, the algorithm performs the criterion for sustained service continuity in step 3, and evaluates whether at least one cell that satisfies the criterion exists. If not, the cell search procedure is triggered and the algorithm is completed (step 5). In the opposite case, the HHM per candidate cell in M is calculated (step 6), and the HO decision criterion for reduced UE transmit power is deployed (step 7). If the current serving cell is the one that satisfies the criterion in step 7, the algorithm is terminated and no further action is taken. However, if another candidate cell satisfies the condition, a HO is executed and the algorithm is completed (step 8).

Now we are going to now focus on how the HO context acquisition in Fig. 4.7 can be performed (step 2). The LTE-A standard describes a wide set of signals for the S1 and X2 interfaces which, however, are not provisioned to transfer the entire HO decision context required for deploying the proposed algorithm. Nevertheless,

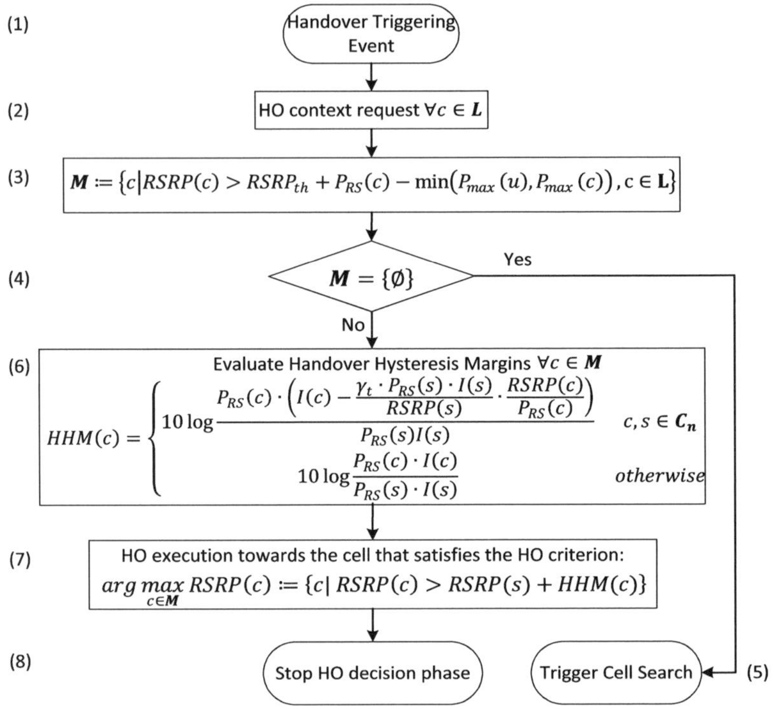

Fig. 4.7 Proposed HO decision algorithm for the LTE-A femtocell network

the HO context acquisition signals can be performed by using the *private message mechanism* for non-standard use [20]. Provided that both the HO context under use and the procedure for exchanging it among the cells, i.e. the private mechanism, are already part of the LTE-A standard, it follows that the proposed algorithm is backward compatible with the LTE-A system and can be performed by using a simple software update at the eNBs and HeNBs.

4.3.3 Performance Evaluation

To evaluate the performance of the proposed algorithm, an extended version of the system level methodology described by the Small Cell forum in [19] is used. A hexagonal LTE-A network is considered with a main cluster composed of 7 eNBs, where each eNB consists of 3 sectors. The wrap-around technique is used to extend the network, by copying the main cluster symmetrically on each of the 6 sides. A set of blocks of apartments, referred to as femtoblocks, are uniformly dropped within the main cluster area with respect to the femtoblock deployment density parameter, denoted by d_{FB}, which indicates the percentage of the main cluster area covered with femtoblocks. Femtoblocks are modeled in line with the dual stripe model for dense urban environments in [19] and the path loss models are adapted accordingly. The deployment of HeNBs within each femtoblock is based on the HeNB deployment density parameter, denoted by r_{fc}, which indicates the percentage of femtoblock apartments with a HeNB installed. Note that a higher d_{FB} corresponds to denser femtoblock layout within the main cluster, whereas a higher r_{fc} refers to denser HeNB deployment within the femtoblocks. HeNBs and users are uniformly dropped inside the apartments, where each HeNB initially serves one user and can serve up to four users. Each eNB sector initially serves ten users, which are uniformly dropped within it. The users are members of up to one Closed Subscriber Group (CSG), where three CSG IDs are used. The rest of the simulation parameters are given in Table 4.5.

The performance of the proposed algorithm is compared against the strongest cell algorithm, referred to as the SC algorithm, and the algorithm in [21], referred to as the Zhang algorithm, which is the HO algorithm that provides the best energy efficiency available in current literature. All the results are derived for increasing

Table 4.5 System level simulation model and parameters

<table>
<tr><td colspan="4">System operating parameters</td></tr>
<tr><td>Parameter</td><td colspan="2">eNB (macrocells)</td><td>HeNB (femtocells)</td></tr>
<tr><td>Carrier frequency</td><td colspan="2">2,000 MHz</td><td>Uniformly picked from the set {1980, 2000, 2020} MHz</td></tr>
<tr><td>Channel bandwidtb</td><td colspan="2">20 MHz</td><td>20 MHz</td></tr>
<tr><td>RS transmit power (DL RS Tx)</td><td colspan="2">Normally distributed with a mean value of 23 dBm and standard deviation 3 dB</td><td>Uniformly distributed within the [0, 20] dBm interval</td></tr>
<tr><td>Link-to-system mapping</td><td colspan="3">Effective SINR mapping (ESM) [15]</td></tr>
<tr><td colspan="4">UE parameters</td></tr>
<tr><td>Mean UL SINR target</td><td colspan="3">$\gamma_t = 3$ dB</td></tr>
<tr><td>Traffic model</td><td colspan="3">Full buffer [15]</td></tr>
<tr><td rowspan="5">Mobility model</td><td rowspan="3">User speed v_t</td><td colspan="2">$v_t = N(\bar{v}, s_u)$ m/s</td></tr>
<tr><td>Mean user speed</td><td>$\bar{v} = 3$ km/h</td></tr>
<tr><td>User speed standard deviation</td><td>$s_u = 1$ km/h</td></tr>
<tr><td>User direction φ_t</td><td colspan="2">$\varphi_t = N\left(\varphi_{t-1}, 2\pi - \varphi_{t-1}\tan\left(\sqrt{\frac{v_t}{2}}\right)\Delta t\right)$</td></tr>
<tr><td colspan="3">where Δt is the time period between two Updates of the model and N(a, b) the Gaussian distribution of mean a and Standard deviation b</td></tr>
<tr><td colspan="4">Other Simulation parameters</td></tr>
<tr><td colspan="3">Overall simulation time</td><td>200 s</td></tr>
<tr><td colspan="3">Simulation time unit</td><td>$\Delta t = 1$ s</td></tr>
</table>

Table 4.6 Number of femtocell users/total number of users

d_{FB}	Number of femtocell users/total number of users					
	$r_{fc} = 0.1$			$r_{fc} = 0.5$		
	Sc	Zhang	Prop	Sc	Zhang	Prop
0.01	2.5/211	3/211	8/211	7.5/218	7.5/218	10/218
0.05	7.5/215	7/215	29/215	14/242	14/242	29/242
0.1	17.5/224	16.5/224	52.5/224	19.5/271	17/271	53.5/271
0.25	27/239	29/239	75.5/239	39.5/359	55.5/359	74/359
0.5	45.5/269	56,5/269	105/269	77.5/500	99/500	129.5/500
0.75	52/303	69.5/303	128.5/303	115.5/641	142.5/641	193.5/641
1	60.5/337	82.5/337	159/337	153.5/782	186/782	257.5/782

d_{FB} and r_{fc}, to investigate the impact of uniform and non-uniform femtocell deployment on the algorithms' performance.

Table 4.6 depicts the number of femtocell users to the number of total users in the main cluster, which is a good measure of the femtocell utilization. At first, we observe that compared to the SC algorithm, the deployment of femtocell-specific HO decision substantially improves the femtocell utilization, i.e. the proposed and Zhang algorithm. Nevertheless, the proposed algorithm substantially increases the femtocell utilization compared to the SC and the Zhang algorithms, where a fourfold increase is observed under low d_{FB} and r_{fc} densities.

Figure 4.8 illustrates the performance of the algorithms in terms of mean received interference at the UEs. For $r_{fc} = 0.1$ the SC and Zhang algorithms show a similar performance, whereas the proposed algorithm lowers the mean received

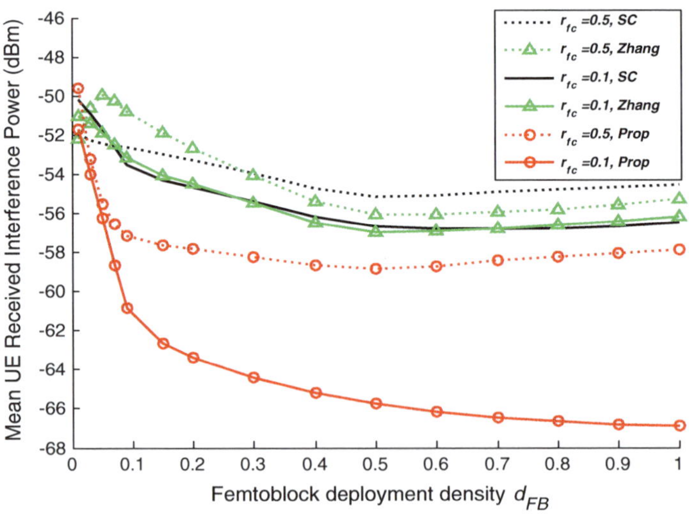

Fig. 4.8 Mean received interference power at the UE versus d_{FB}

interference by up to 10 and 8 dB, respectively. Significant lower interference is shown for the proposed algorithm under $r_{fc} = 0.5$ as well, with the higher gain attained under low to medium femtoblock deployment densities, i.e. $0.05 < d_{FB} < 0.3$. Noticeably, the performance of the proposed algorithm under dense HeNB deployments per femtoblock ($r_{fc} = 0.5$) is better, even compared to that of the competing algorithms in sparser HeNB densities per femtoblock ($r_{fc} = 0.1$). These results follow from the interference awareness of the proposed algorithm, which accounts for the actual channel gain between the UEs and the candidate cells. The importance of this result is clearer if we consider that existing interference mitigation techniques primarily reduce the interference at the cell rather than the UE sites.

Figure 4.9 illustrates the performance of the algorithms in terms of mean UE energy consumption per bit, owing to transmit power. For dense HeNB deployments per femtoblock ($r_{fc} = 0.5$), as the d_{FB} increases, a constantly increasing UE energy expenditure per bit is required for the SC and Zhang algorithms to sustain the mean SINR target γ_t. On the other hand, comparably lower UE energy consumption is observed for the proposed algorithm as the d_{FB} increases, which reaches up to 19 % compared to that of the competing algorithms. Comparably lower UE energy expenditure per bit is also observed for the proposed algorithm in sparser HeNB deployments per femtoblock ($r_{fc} = 0.1$), reaching up to 28 % compared to both the SC and the Zhang algorithms. Note that this improvement results from both lower UE power emissions, as well as the increased femtocell utilization; hence increased resource availability per served user.

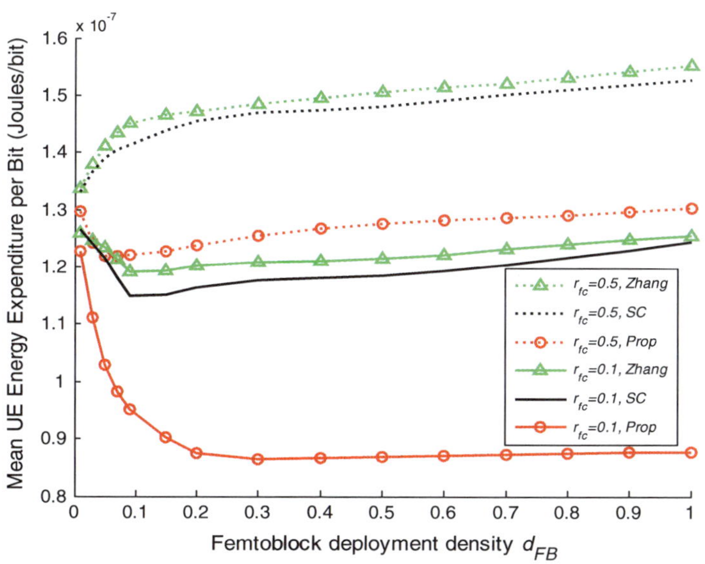

Fig. 4.9 Mean energy consumption per bit at the UE versus d_{FB}

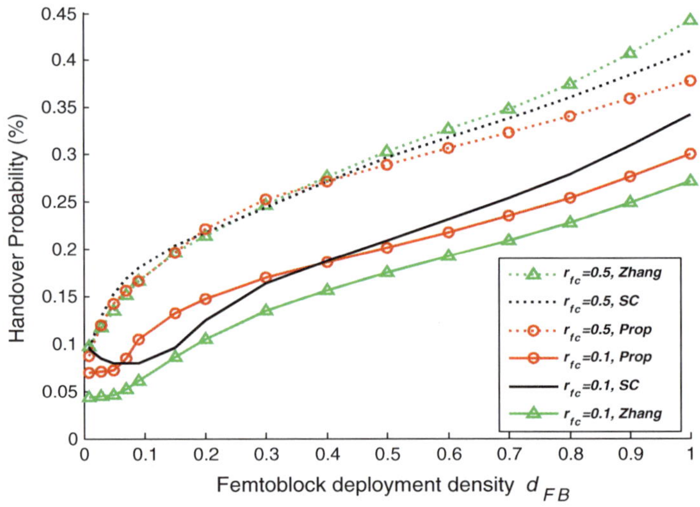

Fig. 4.10 Handover probability versus d_{FB}

Figure 4.10 depicts the HO probability performance for all algorithms. As expected, a higher HO probability is observed for all algorithms as the d_{FB} increases. The same implies for denser HeNB deployments per femtoblock ($r_{fc} = 0.5$), where a comparably lower mean inter-site distance characterizes the HeNB deployment layout. For $r_{fc} = 0.1$ the Zhang algorithm sustains the lowest HO probability, whereas the proposed algorithm attains an improved performance compared to the SC algorithm under very low and medium to high d_{FB}, i.e. for $d_{FB} < 0.1$ and $d_{FB} \geq 0.4$, respectively. On the other hand, for $r_{fc} = 0.5$, all algorithms show similar performance under low to medium deployment densities ($d_{FB} < 0.4$). However, in medium to high femtoblock deployment densities the proposed algorithm attains the lowest HO probability ($d_{FB} \geq 0.4$). It follows that even though the proposed algorithm does not account for the actual UE speed, it attains comparable performance with the speed-based Zhang algorithm due to the incorporation of the criterion for sustained service continuity.

Figure 4.11 depicts the S1 signaling overhead for all algorithms, which equals the number of signals exchanged in the core network over the S1 interface, during the phases of the HO decision and execution. Results show that the S1 signaling load strongly depends on the HeNB deployment density. Specifically, for $r_{fc} = 0.1$, the performance of all algorithms grows almost linearly with respect to the d_{FB} parameter. However, the proposed algorithm necessitates higher S1 signaling overhead due to the employment of the HO context acquisition. For $r_{fc} = 0.5$, a rapidly growing S1 signaling overhead is observed for the SC and the Zhang algorithms, whereas under medium to high d_{FB} the proposed algorithm requires the lowest S1 signaling overhead, owing to the increased use of the X2 interface.

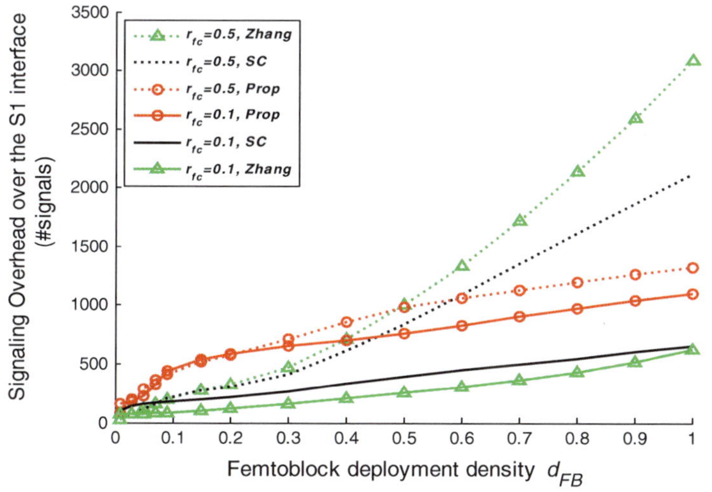

Fig. 4.11 Number of signals over the S1 interface versus d_{FB}

4.4 Conclusion

Vertical and Horizontal Handover decision is a powerful tool for reducing the energy consumption of mobile terminals in both heterogeneous and multi-tier radio access networks. In this chapter, we have presented the enhanced measurement capabilities of the IEEE 802.11-2012 and LTE-Advanced systems and proposed two novel vertical and horizontal handover decision algorithms for the heterogeneous 4G and the multi-tier LTE-Advanced system, respectively. The vertical HO algorithm is proposed for heterogeneous networks comprising LTE-A and IEEE 802.11, while the horizontal HO algorithm is dedicated to LTE-A networks consisting of marcocell APs (eNBs) and Femtocell APs (HeNBs).

System-level simulations have shown that even though the deployment of the proposed VHO algorithm asks for increased ANDSF and WLAN network signaling, it provides efficient load balancing between the heterogeneous RATs, reduced MMT energy consumption per bit and enhanced uplink capacity per user. On the other hand, the system-level simulations results for the macrocell–femtocell handover decision scenario have shown that compared to existing algorithms, the proposed HO algorithm significantly reduces the interference and energy expenditure at the UEs, at the cost of a moderately increased core network signaling.

References

1. 3GPP: Technical report on UMTS heterogeneous networks, TR 25.800 V12.1.0, Dec 2013
2. IEEEStd 802.11-2012 (Revision of IEEE Std 802.11-2007): IEEE standard for information technology-telecommunications and information exchange between systems-local and metropolitan area networks-specific requirements—part 11: Wireless LAN medium access control (mac) and physical layer (phy) specifications, Mar 2012
3. Yan, X., Şekercioğlu, Y.A., Narayanan, S.: A survey of vertical handover decision algorithms in fourth generation heterogeneous wireless networks. Comput. Netw. **54**(11), 1848–1863 (2010)
4. 3GPP: E-UTRA and E-UTRAN overall description, TS 36.300 V10.7.0, Mar 2012
5. 3GPP: Access to the 3GPP evolved packet core (EPC) via non-3GPP access networks, TS 24.302 V10.4.0, Jun 2011
6. 3GPP: Architecture enhancements for non-3GPP accesses, TS 24.402 V10.4.0, Jun 2011
7. Andrews, J.G., Claussen, H., Dohler, M., Rangan, S., Reed, M.C.: Femtocells: past, present, and future. IEEE J. Sel. Areas Commun. **30**(3), 497–508 (2012)
8. Telecoms & Media: Small cell market status, Small Cell Forum, Feb 2012
9. Xenakis, D., Passas, N., Merakos, L., Verikoukis, C.: Mobility management for femtocells in LTE-advanced: key aspects and survey of handover decision algorithms. IEEE Commun. Surv. Tutorials **16**(1), 64–91 (2014)
10. 3GPP: Physical layer; measurements, TS 36.214 V10.1.0, Mar 2011
11. Xenakis, D., Passas, N., Radwan, A., Rodriguez, J., Verikoukis, C.: Energy efficient mobility management for the macrocell–femtocell LTE network. In: Eissa, M. (ed.) Energy Efficiency—The Innovative Ways for Smart Energy, the Future Towards Modern Utilities. InTech Publishers, Winchester (2012). ISBN:978-953-51-0800-9
12. Xenakis, D., Passas, N., Verikoukis, C.: An energy-centric handover decision algorithm for the integrated LTE macrocell–femtocell network computer communications. Comput. Commun. **35**(14), 1684–1694 (2012). Elsevier
13. Xenakis, D., Passas, N., Verikoukis, C.: A novel handover decision policy for reducing power transmissions in the two-tier LTE network. In: Proceedings of IEEE International Communications Conference (IEEE ICC) 2012, Jun 2012
14. Zhang, J., de la Roche, G.: Femtocells: Technologies and Deployment. Wiley, New York, ISBN: 978-0-470-74298-3, 2010
15. Xenakis, D., Passas, N., Di Gregorio, L., Verikoukis, C.: A context-aware vertical handover framework towards energy-efficiency. In: Proceedings of 73rd IEEE Vehicular Technology Conference (VTC), pp. 1–5, May 2011
16. Feeney LM., Nilsson, M.: Investigating the energy consumption of a wireless network interface in an ad hoc networking environment. In: Proceedings of 12th Annual Joint Conference of the IEEE Computer and Communications Societies (INFOCOM 2001), vol. 3, pp. 1548–1557 (2001)
17. Liu, J.-S., Lin, C.-H.R.: ECTP: an energy-efficiency label-switching MAC protocol for infrastructure wireless networks. IEEE Trans. Veh. Technol. **56**(3), 1399–1417 (2007)
18. Xenakis, D., Passas, N., Merakos, L.: Multi-parameter performance analysis for decentralized cognitive radio networks. Wirel. Netw. **20**(4), 787–803 (2013). doi:10.1007/s11276-013-0635-4
19. Small Cell Forum: Interference management in OFDMA femtocells. Small Cell Forum, Mar 2010
20. Xenakis, D., Passas, N., Merakos, L., Verikoukis, C.: Energy-efficient and interference-aware handover decision for the LTE-advanced femtocell network. In: Proceedings of IEEE International Communications Conference (IEEE ICC), pp. 2464–2468, Jun 2013
21. Zhang, H., Ma, W., Li, W., Zheng, W., Wen, X., Jiang, C.: Signalling cost evaluation of handover management schemes in LTE-advanced femtocell. In: Proceedings of 73rd Vehicular Technology Conference (Spring), pp. 1–5, May 2011

Chapter 5
Cooperative Paradigm for Energy Saving

Firooz B. Saghezchi, Ayman Radwan and Jonathan Rodriguez

Abstract Most portable devices are likely to be equipped with a variety of radio technologies, enabling multiple opportunities for wide area access. As the density of these devices increases in typical urban environments, it becomes increasingly possible and desirable to participate or establish cooperation to achieve a common goal. In this context, we consider cooperation among mobile devices within a short range area as a means to save energy at the handset device, but it has been proven that the energy savings can also proliferate to the network side. In particular, we address how cooperative strategies that exploit long range connectivity in synergy with short-range connectivity can lead to significant energy savings. Game theoretical approaches are used as an engineering tool to find the optimum configuration of cooperative clusters to minimize the energy consumption of the whole network, that also includes the mobile handset. Specific use-cases that consider selfish behaviour among mobile users is a crucial impediment hampering cooperation, however cooperative game theory is used in this work to overcome such a problem, by offering credit to cooperative users as incentive. This technique results in rewarding cooperative users, as well as detecting and isolating selfish ones.

5.1 Introduction

Inspired by nature, cooperation is known as an effective strategy to improve the efficiency of wireless networks by forming cooperative groups [1]. Cooperation is a broad term referring to a wide range of techniques throughout all layers of the network protocol stack, from the physical layer up to the application layer. In this

F.B. Saghezchi (✉)
Instituto de Telecomunicações, Aveiro, Portugal
e-mail: firooz@av.it.pt

A. Radwan · J. Rodriguez
Instituto de Telecomunicações, Campus Universitário de Santiago,
3810-193 Aveiro, Portugal

A. Radwan and J. Rodriguez (eds.), *Energy Efficient Smart Phones for 5G Networks*, Signals and Communication Technology,
DOI 10.1007/978-3-319-10314-3_5

chapter, we primarily focus on cooperative techniques for physical and network layers suitable for improving the energy efficiency of multi-mode mobile terminals.

Perhaps, the most natural and traditional form of cooperative communications stems from multihop packet forwarding, primarily employed in wireless ad hoc networks where wireless nodes, without having support from any pre-established infrastructure, forward packets for each other to ensure end-to-end connectivity. This kind of cooperation has widely been investigated to provide reliable connectivity by means of appropriate ad hoc routing algorithms and protocols in wireless ad hoc networks. Recently, this technique has also been employed in wireless infrastructure-based networks to improve the coverage or QoS of the network. The resulting network from the integration of ad hoc multihop communications with the cellular networks is called multihop cellular network (MCN) or hybrid ad hoc network (HANET) [2–4]. There has been considerable interest from both standardization bodies and academia in multihop cellular networks. Opportunity Driven Multiple Access (ODMA) [5] is a multihop relaying protocol, proposed by 3GPP to be applied to UMTS TDD to (i) increase the high data rate coverage in the network; (ii) increase the capacity of the network; (iii) provide distributed network architecture for spot coverage and traffic hotspots. In [6], the authors integrated MANET and GSM and introduced a so called ad hoc GSM (A-GSM) platform, addressing practical issues contributing to the evolutionary changes of GSM in order to enable relaying of calls. Moreover in [7] the authors overview several contributions to Working Group 4 of the Wireless World Research Forum (WWRF) and present several relay-based deployment concepts such as multihop relaying, cooperative relaying, virtual antenna arrays as well as routing and radio resource management.

Cooperative communications have also been widely investigated to improve the efficiency of the physical layer, leveraging the spatial diversity of the cooperative relay channel to combat fading impairment. The basic idea behind wireless relay channel can be traced back to the groundbreaking work of Cover and Gamal [8] on the information theoretic characterization of the relay channel. The work builds upon the three-node (including a source node, a destination node, and a relay node) channel model first introduced by Van der Meulen [9] and examines its channel capacity for the case where the channel is contaminated by additive white Gaussian noise (AWGN). Recent research work on this subject is focused on taking advantage of the underlying spatial diversity introduced by the relay node(s) to combat detrimental effects of wireless channel fading and hence improve the reliability of the wireless communications [10–20]. The spatial diversity has also been exploited to improve the spectral efficiency of the wireless channel through distributed space-time multiplexing techniques [21, 22]. The key benefit of cooperative relaying lies in the fact that mobile terminals can share their antennas and form a virtual multiple-input-multiple-output (MIMO) channel to take advantage of the provided spatial diversity. In contrast to transmit diversity or receive diversity, this form of spatial diversity is referred to as cooperative diversity. Laneman et al. [10] developed different cooperative fixed relaying schemes such as amplify-and-forward (AF) and decode-and-forward (DF), selection relaying schemes that adapt according to the channel quality, and incremental relaying schemes that adapt based on limited feedback from the destination terminal and examined their performance in terms of outage events and associated outage probabilities. Unlike this work where the authors constrained the relay node to operate in half duplex mode and employ TDMA scheme, Sendonaris et al. [12, 13], study the cooperation of mobile users when both users have data to transmit and address practical implementation issues within a CDMA framework. Finally, Nosratinia et al. [20] provides a good tutorial about cooperative diversity techniques.

There are several previous research efforts on enhancing the energy efficiency of multi-mode mobile terminals. In [23], the authors propose a technique where Bluetooth link is exploited to wake up the wireless local area network (WLAN) interface when there is a pending packet, avoiding the unnecessary periodic wake-ups of the WLAN interface. CoolSpots [24] exploits Bluetooth links not only as a wake-up channel for WLAN interfaces, but also as a data link when the application requires low bandwidth. WLAN interface is powered up when the data rate reaches a threshold, allowing the WLAN interface to spend more time on the sleep mode. A cooperative networking protocol (CONET) is proposed in [25] where, in each cluster, a mobile terminal is elected as the cluster head and every cluster member sends its traffic to the cluster head over a Bluetooth link. The cluster head aggregates all the traffic coming from the cluster members and relays the aggregate traffic to the access point over a WiFi link.

Recently, cognitive and cooperative communications have been applied to enhance the energy efficiency of mobile terminals. A cognitive radio is generally defined as a radio that can adapt its transmitter behaviour according to the changes in the environment in which it operates to improve its efficiency in terms of resource utilization. Traditionally, the only concern regarding resource utilization was the efficient spectrum utilization. However, recently, it has been argued that the concept of cognitive radio can also be applied for efficient battery utilization [26], which is a growing concern for mobile users. Lying in this stream of research, C2POWER [27] and Green-T [28] are two European research projects aiming at exploiting cognitive radio and cooperative communications by exploiting short range interfaces to reduce the power consumption of multi-mode mobile terminals. In [29], the authors provide quantitative analysis to study the energy saving performance for different combinations of short range and long range technologies, namely WiMedia-WiFi, WiFi-WiMAX, and WiFi-WiFi. Cognitive and cooperative communications have also been employed in [30, 31] through exploiting short range interfaces to reduce energy consumption of multi-mode mobile terminals.

Interacting as independent agents while cooperating or competing for efficient resource utilization, conflict of interest naturally arises among mobile users. Game theory [32–34] is the mathematical tool that is applied to analyse this conflict and predict the strategic decisions of the players (i.e., mobile users) [35–42]. Specifically, with the emergence of cooperative paradigm in wireless networks, coalitional game theory, also known as cooperative game theory, has received considerable interest in recent years (see [31, 39–42] and references therein). Solution of a non-cooperative game, which is given by the well-known Nash Equilibrium, basically seeks to predict the strategic equilibrium profile of the game where no player tends to change his strategy any more. On the other hand, solving a cooperative game means incentivizing all players by dividing the common payoff that the players obtain by their cooperation in a fair way where all players are satisfied and no player or group of players has incentive to leave the cooperation. Further detail about cooperative game theory is provided in Sect. 4.1.

Another issue related to the cooperation of mobile users is encouraging the users to keep cooperation on while preventing them from acting selfishly, an issue closely related to cooperative game theory approach. There are two main techniques for this purpose, namely credit based schemes [43–46] and reputation based schemes [47–52]. In the former, there is a central trustworthy entity that maintains a credit account for each mobile user, where the cooperation of each user is fully recorded and tracked. When a user cooperates, he/she gains some credit, and when he/she asks for help from others, he/she spends some credit, which is transferred to the account of the helping relay(s). Therefore, cooperative users increase their credit over time while selfish users fail to add any credit to their account. Consequently, if a selfish node, which has no credit in its account, asks for help from other users, they will refuse to help the selfish node. As a result, selfish nodes are isolated and hence excluded from the cooperative groups. In contrast to credit based schemes, reputation based schemes are distributed techniques where every mobile user maintains a reputation record for any of his/her neighbouring users. Mobile users also disseminate the reputation image of their immediate neighbours to other nodes far away as an attempt to help each other to detect selfish users faster. It is worth pointing out that there is an intricate difference between credit based schemes and cooperative game theory techniques. In the former, normally players receive flat credit regardless of their effort or position, while, in the later, the players receive credit proportional to their influence and effort.

In this chapter we consider mobile terminals' partitioning themselves as coalitions. In each coalition, there can be multiple relays or multiple sources pooling their resources (e.g. their batteries, antennas, etc.) and communicating cooperatively. We do not intend to concentrate on performance evaluation of different cooperation techniques; instead, we aim at considering the cooperation as a strategic design and describe how the potential conflict among cooperating players can be settled, using an approach based on coalition formation game. For the purpose of exposition, we consider multihop cooperation; however, the technique is generic and applicable to other cooperation techniques. We define the characteristic function of the game along with appropriate utility function for assessing the profitability of a coalition. The technique indeed seeks to maximize the social welfare (i.e. aggregate energy savings) of mobile terminals while enabling the required QoS. We validate the effectiveness of our approach using simulations. Finally, we conclude this chapter with some design guidelines with regards to cooperative strategies and energy saving.

5.2 Fundamentals on Cooperative Communications

Cooperative communication can take place in different forms. In this section, we describe two main techniques, namely cooperative relaying and multihop communications. The right side of Fig. 5.1 illustrates a three-node cooperative relaying network where the communication takes place in two stages. In the first stage (illustrated by the solid arrows), the source node *S* broadcasts the message to the destination node *D* and the relay node *R*, whereas in the second stage (illustrated by the dashed arrows), the relay node retransmits the received information to the destination node *D*. The relay node indeed establishes another path (*S*–*R*–*D*) in parallel to *S*–*D*, providing spatial diversity. The left side of Fig. 5.1 illustrates signal-to-noise ratio (SNR) variations of the source and relay channels due to multipath fading effect. The shaded parts show the time intervals where the channels are in deep fade. When cooperative relaying is employed, the communications will fail if both the direct link and the relay link are in deep fading (illustrated by dark shades), an event, which is less likely to happen than the case when only the direct link is in deep fading.

Several relaying strategies have been introduced, such as amplify-and-forward (AF), decode-and-forward (DF), and compress-and-forward (CF). In the first strategy, the relay node *R* amplifies the received signal, without trying to detect it, and forwards the amplified signal to the destination node *D*. Despite its simplicity, AF strategy also amplifies the noise embedded in the received signal, which may deteriorate the signal-to-noise ratio (SNR). To overcome this drawback, in the second strategy (DF), the relay node *R* first detects the received signal then encodes and forwards it to the destination node *D*. Finally, in the last strategy, the relay node detects the received signal, and retransmits a quantized or a compressed version of the received message to the destination, exploiting the statistical dependencies between the message received at the relay and that received at the destination. The performance of these strategies essentially depends on the quality of the source-relay channel. When this channel has good quality, DF strategy shows better performance than AF strategy; otherwise, AF strategy outperforms DF strategy. This is due to the fact that, when the source-relay channel quality is poor, the relay fails to

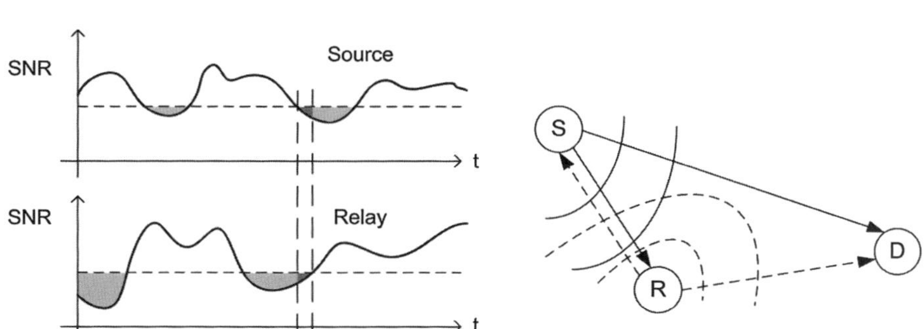

Fig. 5.1 A three-node cooperative relaying model [53]

successfully detect the received signal. In this case, in spite of amplifying the noise, AF strategy demonstrates better performance than DF strategy [17, 18].

At the destination node, signals from both the source and relay paths are combined for a reliable detection. Different combining strategies are used for this purpose. For example, in the selective combining (SC) method, the received signal with higher SNR is chosen for detection, while the weaker signal is ignored. In Maximal ratio combining (MRC) method, both signals are considered for detection; a weighted average of the signal is considered where the weight of each signal is proportional to its SNR. However, in MRC strategy, different phase shifts due to different path lengths of the direct and relay links should be compensated before combining to prevent any destructive addition, which requires that the phase responses of both the direct and relay channels are available at the destination.

In a cooperative relaying setup, if there is no direct link between the source S and the destination D or if the direct signal received at the destination node D is ignored, the cooperative relaying is reduced to multihop communications. The most obvious benefit of cooperative communications is to break a long hop from the source to the destination to shorter hops. The path loss of a radio link is normally proportional to $1/r^{\alpha}$ where r is the distance between the transmitter and the receiver and α is the path loss exponent—which is normally between 2 and 4. Therefore, breaking a long link to several short links can reduce the required transmit power considerably, providing energy saving to the transmitter. The cooperative relaying or multihop communications can also be extended to more than two hops by cascading multiples of the three-node network shown in Fig. 5.1.

Multihop communications are traditionally used in ad hoc networking to ensure connectivity. However, this technique can also be used in cellular networks to enhance the QoS, extend the coverage, or improve the energy efficiency of the network, including the mobile terminals. The integrated network is generally referred to as a multihop cellular network (MCN) or a hybrid ad hoc network (HANET). There are two main approaches for realizing MCN: (a) in the first approach (Fig. 5.2), fixed relay stations (RSs) are deployed by the operator in some strategic locations. The relay stations are indeed small base stations without any

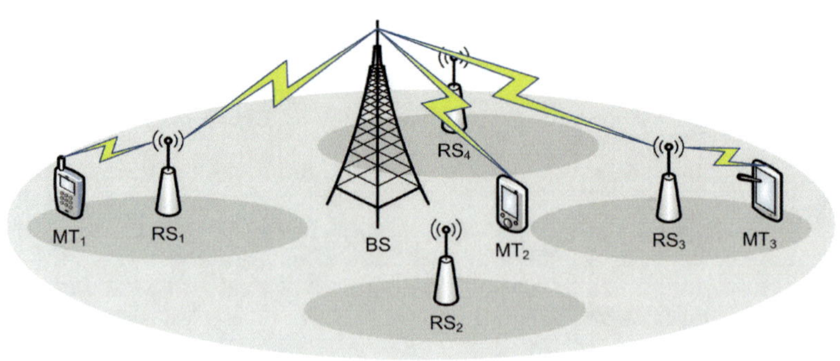

Fig. 5.2 Multihop cellular network with fixed relays

cooling facility, justifying their viability in terms of both operational and capital expenditures, (b) in the second approach, there is no pre-installed relay stations; instead, mobile terminals (MTs) forward traffic on behalf of each other (Fig. 5.3).

Figures 5.2 and 5.3 illustrate two-hop scenarios. As mentioned before, unlike these figures, multihop relaying can be extended to cover more than two hops. However, two-hop relaying has several desirable properties such as avoiding system complexity, reducing routing overhead and collisions, and limiting the embedded latency [54].

In a multihop cellular network, the relaying action can be performed either in the same frequency band as the cellular system or in a different frequency band. In the first approach, which is called in-band relaying, both hops from the source node to the relay node and from the relay node to the destination node are performed over the same frequency channel. More precisely, both links share the same frequency dimension by multiplexing over other dimensions such as time (TDMA) or code (CDMA). Although in-band relaying requires minimal additional complexity for the mobile terminals, the receiver and transmitter at the relay node cannot operate simultaneously. This constrains the relay to operate in half-duplex mode, reducing the link capacity to half. On the other hand, in out-of-band relaying, the first hop (i.e. from the source node to the relay) operates at different frequency band than the second hop (i.e. from the relay to the destination), allowing full-duplex operation for the relay nodes. However, this scheme requires an additional frequency channel as well as good isolation between the transmit and receive signals.

Despite operating in orthogonal channels, in out of band relaying, the transmit signal drowns out the received signal, due to imperfect isolation (typically, the transmit signal is 100–150 dB above the received signal). This requires two radio interfaces at the relay node. Today, any feature phone or smart phone is equipped with a variety of radio interfaces. Therefore, a plausible solution for out-of-band relaying is to exploit prevalent wireless local area networking (WLAN) or personal area networking (PAN) interfaces for relaying purpose. For instance, the first hop from the source to the relay can be performed over a short range (SR) link such as

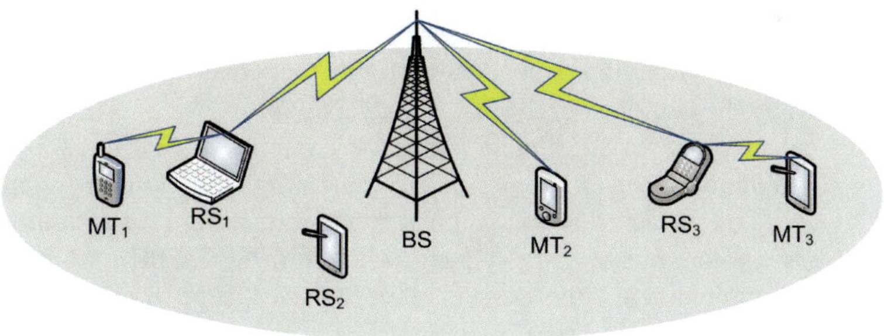

Fig. 5.3 Multihop cellular network with mobile relays

WiFi, Bluetooth, or WiMedia, while the second hop from the relay to the destination can be performed over a cellular link such as UMTS, WiMAX, or LTE.

5.3 Cooperative Energy Saving Strategies

In this section, we describe the application of multihop communications for improving the energy efficiency of mobile terminals in a cellular network. We use cooperative game theory to model strategic interactions of mobile terminals, enabling them to form cooperative groups and share their resources whenever profitable. We address the definition of appropriate utility function to assess the payoff of a cooperative action. Furthermore, we address the threat of selfish players and describe how to tailor the existing credit based schemes for our cooperative game approach to isolate and exclude selfish players from cooperative groups.

5.3.1 Cooperative Game Theory

Cooperative game is a type of game that analyses conflict of interest among rational players who can make coalitions and establish binding agreement among themselves to improve their strength. For example, in a wireless network, MTs may form coalitions and pool their resources (e.g. their batteries, radio interfaces, etc.) and perform their tasks (i.e. communication with their desired destinations) cooperatively whenever they can save energy by doing so. In the following, we introduce the required terminology from coalitional game theory. Those readers who are familiar with coalitional game theory may skip this section.

A coalitional game is defined as an ordered pair $\langle \mathcal{N}, v \rangle$ where $\mathcal{N} = \{1, \ldots, n\}$ is the set of players who seek to form coalitions. Any subset of $\mathcal{N}$ is called a coalition, and the set involving all players (i.e. $\mathcal{N}$) is called the grand coalition. In general, there are 2^n possible coalitions including the empty and the grand coalitions. Furthermore, $v : 2^{\mathcal{N}} \rightarrow \mathbb{R}$ is defined as a function from the coalition space (i.e. a set containing all possible coalitions) to the space of real numbers $\mathbb{R}$). For any coalition S, if the value of this function depends only on the actions of players in S without being affected by the activities of the rest of players who are out of S (i.e. $\mathcal{N} \backslash S$), the function is called characteristic function and the game is referred to as a game in characteristic function form. Otherwise, if the value of the function for a coalition S depends not only on the activities of the players in S, but also on the activities of the rest of players who are out of S (i.e. players who are in $\mathcal{N} \backslash S$), then the function is called a partition function and the game is referred to as a game in partition function form. The value of a coalition S retuned by the characteristic function or the partition function (i.e. $v(S)$) is called the worth of coalition S, and it indicates the maximum payoff that the players in S can achieve by their cooperation; by convention $v(\emptyset) = 0$. Putting it in different words, in a characteristic function game, the worth of any coalition depends only on the members of the coalition and is independent from how the rest of players outside of the coalition organize themselves, while in a partition function game, the worth of a coalition depends not only on the coordination of the members of the coalition but also on how the other players who are outside of the coalition organize themselves. Thus, in a partition function game, forming a coalition might have a kind of side effect on other coalitions, something which is referred to as externalities in coalitional game theory terminology. There are two types of externalities in general: positive externality and negative externality. Positive externality means that forming a coalition increases the worth of other coalitions, while in case of negative externality, forming a coalition reduces the worth of other coalitions. In the rest of our discussion in this chapter, we consider the characteristic function games to avoid blurring important intuitions with too much complexity.

A coalitional game is called super-additive if and only if for any couple of disjoint coalitions S and T (i.e. $S \cap T = \emptyset$),

$$v(S \cup T) \geq v(S) + v(T) \tag{5.1}$$

That is, any disjoint coalitions can always improve their position in the game by merging and forming a bigger coalition. Furthermore, a game is called convex if and only if for any arbitrary coalitions S and T,

$$v(S \cup T) + v(S \cap T) \geq v(S) + v(T) \tag{5.2}$$

Note that any convex game is super-additive also, but the converse may not be true. In fact, in a super-additive environment, all the players tend to join together and form the grand coalition. This type of game where the grand coalition is trivial is also referred to as the canonical coalitional game. In this game, the main problem is only the stability of the coalition. That is, how to divide the common payoff of the grand coalition among the players so as every player is pleased and no one has incentive to leave the coalition. Any distribution of common payoff among the players is called a payoff vector or an allocation vector, denoted by $\boldsymbol{x} = (x_1, \ldots, x_n)$ where x_i is the share of player i from the common payoff (i.e. $v(\mathcal{N})$). While distributing this common payoff among the players, a rational player i always compares the payoff he/she receives from the allocation vector (i.e. x_i) with the payoff that he/she can achieve by acting individually (i.e. $v(\{x_i\})$). The player would prefer to cooperate if $x_i \geq v(\{x_i\})$; otherwise, he/she would decide to leave the coalition as he/she can achieve more payoff by doing so. Hence, the first condition for the stability of a coalition is to offer each player at least the amount that they can obtain individually. A payoff vector that has this property is called individually rational. We say a payoff vector is feasible if and only if

$$\sum_{i \in N} x_i \leq v(N) \tag{5.3}$$

Furthermore, a payoff vector is called efficient or Pareto-optimal if and only if

$$\sum_{i \in N} x_i = v(N). \tag{5.4}$$

For any coalition S, the excess of S under allocation vector $\boldsymbol{x}$ is defined as

$$e(S, \boldsymbol{x}) = v(S) - \sum_{i \in S} x_i \tag{5.5}$$

Therefore, a payoff vector of a canonical game is efficient if the excess of the grand coalition under this allocation vector is zero. A payoff vector is also called group rational if for any coalition $S \subset N$,

$$\sum_{i \in S} x_i \geq v(S) \tag{5.6}$$

Hence, a payoff vector is group rational if the excess of every coalition is either negative or equal to zero. A payoff vector is called an imputation if it is efficient and individual rational. That is

$$I(v) = \{\boldsymbol{x} | \sum_{i \in N} x_i = v(N), \quad x_i \geq v(\{i\})\} \tag{5.7}$$

Finally, the core solution of a coalitional game $\langle \mathcal{N}, v \rangle$ is defined as the set of imputation vectors that are group rational. That is,

$$c(v) = \{\boldsymbol{x} \Bigg| \sum_{i \in N} x_i = v(N), \sum_{i \in S} x_i \geq v(S)\} \tag{5.8}$$

Example 1 Consider a game with three players $\mathcal{N} = \{1, 2, 3\}$ and the following characteristic function.

$$\begin{aligned}
v(\{1\}) &= 0.25 \\
v(\{1,2\}) &= 1.25 \\
v(\{2\}) &= 0 \\
v(\{1,3\}) &= 1.0 \\
v(\{1,2,3\}) &= 2.0 \\
v(\{3\}) &= -0.25 \\
v(\{2,3\}) &= 0.75
\end{aligned}$$

The game is super-additive because:

$$
\begin{aligned}
v(\{1,2\}) &\geq v(\{1\}) + v(\{2\}) \\
v(\{1,3\}) &\geq v(\{1\}) + v(\{3\}) \\
v(\{2,3\}) &\geq v(\{2\}) + v(\{3\}) \\
v(\{1,2,3\}) &\geq v(\{1,2\}) + v(\{3\}) \\
v(\{1,2,3\}) &\geq v(\{1,3\}) + v(\{2\}) \\
v(\{1,2,3\}) &\geq v(\{2,3\}) + v(\{1\})
\end{aligned}
$$

The game is also convex since in addition to the above inequalities:

$$
\begin{aligned}
v(\{1,2,3\}) + v(\{1\}) &\geq v(\{1,2\}) + v(\{1,3\}) \\
v(\{1,2,3\}) + v(\{2\}) &\geq v(\{1,2\}) + v(\{2,3\}) \\
v(\{1,2,3\}) + v(\{3\}) &\geq v(\{1,3\}) + v(\{2,3\})
\end{aligned}
$$

Furthermore, the core solution is any vector $\boldsymbol{x} = (x_1, x_2, x_3)$ that satisfies the following conditions:

$$
\begin{aligned}
x_1 &\geq 0.25 \\
x_2 &\geq 0 \\
x_3 &\geq -0.25 \\
x_1 + x_2 &\geq 1.25 \\
x_1 + x_3 &\geq 1.0 \\
x_2 + x_3 &\geq 0.75 \\
x_1 + x_2 + x_3 &= 2.0
\end{aligned}
$$

For example, $(1, 1, 0)$, $(1, 0.25, 0.75)$, and $(1.25, 0.5, 0.25)$ are all located in the core.

It is clear in this example that the core solution is not unique; rather there can be several core solutions pertaining to a game. The core of a coalitional game can also be empty. However, the core of a convex game is always nonempty. Moreover, the core solution of a coalitional game is a (possibly empty) *convex* polytope in R^n. The core solution provides all the solutions including the edge points of this convex polytope. Other solution concepts exist such as the Nucleolus or the Kernel, which derive interior solutions from this polytope. Further discussion on these solution concepts is out of the scope of this chapter. The interested readers can refer to the textbooks on game theory such as [32].

5.3.2 Cooperative Relaying Game

We define cooperative relaying game as a game in which mobile terminals form coalitions, pool their resources, and relay packets on behalf of each other to reduce their power consumption. There are three main challenges regarding this game. The first problem is how mobile terminals partition themselves in coalitions, that is, whether it is better for mobile terminals to form the grand coalition or they are better to partition themselves in mutually exclusive sets. To answer this question, we need to know whether the game is super-additive or not. If the game is super-additive, it is always beneficial to merge smaller coalitions and make bigger ones; that is, the grand coalition is trivial in such a context. However, if the game is non-super-additive, mobile terminals need to find the best way to partition themselves in order to maximize their social welfare. The second problem regarding the mentioned cooperative relaying game is how to match relays and sources in a coalition to maximize the coalition's payoff—we define the payoff or the worth of a coalition as the maximum energy saving that the coalition can achieve by cooperation. Finally, the third problem regarding this game is how to incentivize relays to cooperate. As the relays are among idle mobile terminals, normally controlled by rational players, if there is no mechanism to prevent selfish behaviour, players will ask others to forward their packets but refuse to forward others' packets to conserve their limited energy for their own needs. Therefore, there should be a mechanism to encourage cooperation while preventing selfish behaviour in the coalition. In the following, we discuss these three problems with further extent.

5.3.2.1 Coalition Structure Generation

In a super-additive environment, any coalition of players can do better by cooperating than by acting individually. Thus, all players are better off forming the grand coalition, sharing their resources and tasks, and cooperating for efficient resource utilization. However, in a non-super-additive environment, forming the grand coalition is not profitable due to the communication or computation burden associated with forming bigger coalitions. In this case, as illustrated by Fig. 5.4, the players first need to be partitioned in mutually exclusive coalitions in a way that maximizes their social welfare. A typical solution to this problem is to search for the optimal coalition structure by performing an exhaustive search among all possible coalition structures. Nonetheless, the number of possible coalition structures—which is given by the Bell number—increases exponentially with the number of players. In fact, as stated by Proposition 1 in [54], for n players, the number of coalition structures is $\mathcal{O}(n^n)$ and $\omega(n^{n/2})$, which means that the number of coalition structures is upper bounded by a constant factor of n^n and lower bounded by a constant factor of $n^{n/2}$. Figure 5.5 illustrates the number of possible coalition structures (i.e. drawn using the Bell number) as the number of players varies between 1 and 50 along with its associated upper and lower bounds.

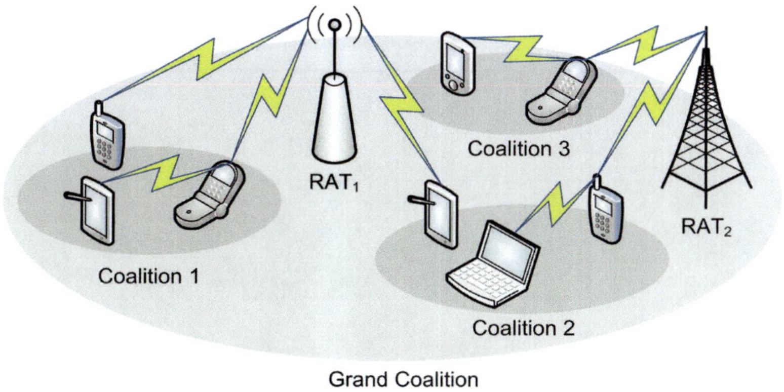

Fig. 5.4 Coalition structure generation

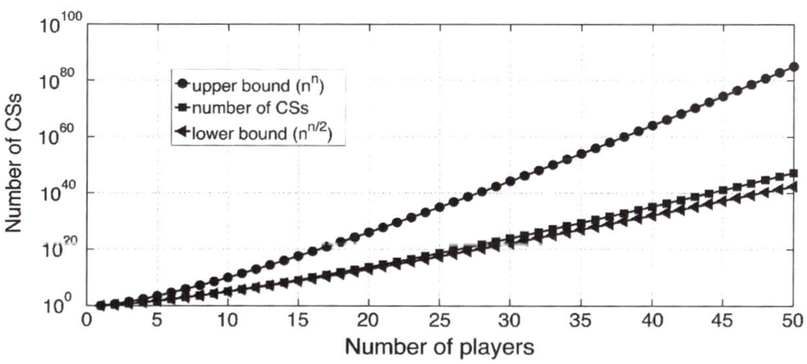

Fig. 5.5 Number of possible coalition structures for different number of players [42]

As apparent from the figure, exhaustive search algorithm is not computationally tractable even for moderate number of players, requiring efficient coalition structure generation algorithms. Further elaboration on designing coalition structure generation algorithms is beyond the scope of this chapter. The interested reader may refer to [54] for more discussion on the topic.

5.3.2.2 Energy Efficient Cooperation

In a coalition, the energy efficient cooperation of mobile terminals can take place in different ways, yet the energy saving performance of the cooperation should outweigh its signalling or computational overhead. Figure 5.6 illustrates a cooperation scenario where every mobile terminal is equipped with two radio interfaces, one for short range and the other for long range communications. For example, Bluetooth and WiMedia can be used as short range interfaces, whereas WiMAX and LTE can

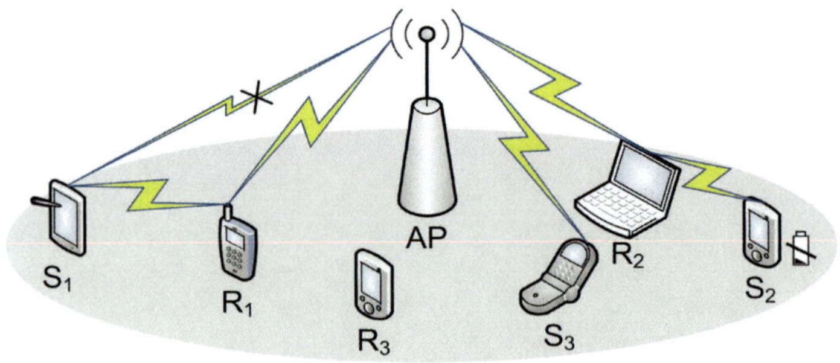

Fig. 5.6 A coalition of MTs under the coverage of a WiFi *AP*

be used as long range interfaces. Note that, although WiFi can generally be used as both short range and long range interfaces, in this chapter we consider it as a long range interface. There are different types of user terminals in the figure such as smartphone, personal digital assistant (PDA), tablet, etc.; we refer to them collectively as mobile terminals (MTs) in general. The scenario is a hybrid combination of long range communications aided by the access point (AP) and short range ad hoc communications. Mobile terminals located in the proximity of each other form a coalition for energy saving. In the scenario depicted by Fig. 5.6, S_1 is experiencing low channel quality, so it starts scanning its neighbourhood and finds R_1 in its short range coverage. It sets up a cooperative short range link with R_1 and relays its traffic to the access point through it. On the other hand, S_2, despite having good channel, suffers from low battery level. Thus, it starts scanning its neighbourhood and discovers R_2 in its coverage, which has good channel and battery level. It sets up a cooperative link with R_2 and relays its traffic through it. Finally, S_3—which has good channel quality and enough battery—communicates with the access point directly, and R_3—which is an idle mobile terminal and a potential relay node—is left unemployed.

A mobile terminal needs to evaluate the energy efficiency (EE) of the cooperation to choose an appropriate strategy, regarding whether to join a coalition or not or whether to communicate over a cooperative link or over the direct one. For this purpose, we can use "Bits/Joule" as the efficiency metric of a link, which can be determined by dividing the data rate of the link by its required power, i.e., $\eta_{EE} = R/P$ [55]. In the following, to illustrate through an example, we model the aforementioned scenario as a coalitional game. The approach is however generic and can be applied to any other cooperative scenario.

For an arbitrary coalition of mobile terminals $S \subseteq N$, we define the characteristic function $v(S)$ as the maximum amount of common utility that the mobile terminals in S can achieve by their cooperation. Let M and $\bar{M}$ denote the sets of relays and sources, respectively. M and $\bar{M}$ are mutually exclusive as we assume that only idle mobile terminals are allowed to act as relays. It is obvious that $v(S) = 0$ if $|S| = 0 \; or \; 1$ because an empty coalition or a coalition of a single player cannot save energy. More generally, any one-sided coalition (i.e. a coalition composed of relays only or a coalition composed of sources only) again leads to no profit. This is due to the fact that in a coalition composed merely of sources (relays), there is no partner relay (source) node to cooperate with. This means that only a mixed coalition of relays and sources can be profitable. Hence, the best a larger coalition can do is to split up into separate cooperative pairs (relays and sources) and pool the profit. The simplest kind of mixed coalition consists of two players—one of each type. For such coalition, we define the utility as a weighted function of energy saving and battery life extension as follows, where we define the battery life extension of a coalition as the amount of energy that the mobile terminal which possesses the least battery level saves with cooperation.

$$u_{ij} = v(\{i,j\}) = \alpha a_{ij} + (1-\alpha)b_{ij} \quad \forall i \in M,\ j \in \bar{M} \tag{5.9}$$

Here, α is the weighting factor, a fraction between 0 and 1. Moreover, a_{ij} and b_{ij} are the normalized energy saving and the normalized battery life extension achieved from cooperation of relay i and source j. Note that, both a_{ij} and b_{ij} are non-negative as the cooperation is performed if it is profitable; otherwise, mobile terminals avoid cooperation and operate individually. In order to compute the characteristic function for a larger mixed coalition, which consists of multiple relays and multiple sources, we must choose an optimal matching of relays and sources that maximizes the coalition's aggregate utility. The evaluation of this maximization problem is generally an assignment problem [56], which can be expressed as the following Linear Programming (LP) problem.

$$\begin{aligned} \max \quad & v(S) = \sum_{i\in M}\sum_{j\in M} x_{ij}u_{ij} \\ s.t.: \quad & \sum_{j\in M} x_{ij} \le 1, \quad \forall i \in M \\ & \sum_{i\in M} x_{ij} \le 1, \quad \forall j \in \bar{M} \\ & x_{ij} \in \{0,1\}, \quad \forall i \in M,\ j \in \bar{M} \end{aligned} \tag{5.10}$$

Here, x_{ij} is a binary decision variable indicating whether relay i should relay source j or not, and u_{ij} is the potential utility from cooperation of relay i and source j. There are two constraint inequalities. The first one indicates that any relay can relay at most one source, while the second one indicates that any source can either communicate directly or utilize at most one relay to reach the access point. The latter reflects the two-hop constraint, where any cooperative link utilizes at most one intermediate relay.

5.3.2.3 Cooperation Enforcement Mechanism

Once the characteristic function is determined, we can solve the game using a solution concept from coalitional game theory, such as the core solution, defined by (5.8), and determine the share of each player from the achieved utility. However, the utility—which is an amount of energy saving—is non-transferable. Only the sources save energy, while the relays are incurred some extra energy consumption.

Consequently, even if the cooperation is socially desirable, relays will be reluctant to cooperate unless they are stimulated by some incentive. Although reciprocal altruism can be adopted to enforce cooperation among mobile terminals, where a relay helps others with the condition of receiving help in the future, it is highly vulnerable to potential free riding attempts from selfish nodes. Therefore, a mechanism should exist to incentivize cooperative nodes while preventing potential threats of the selfish ones; otherwise, selfish nodes will exploit others. Consequently, leaving no motivation to cooperate, the cooperation will collapse soon.

As stated in Sect. 5.2, there are two main techniques to stimulate cooperation and mitigate harmful threats from selfish nodes, namely virtual currency (credit) based techniques and reputation based techniques. We use a credit based approach similar to [46] as it better suits our coalitional game approach. Specifically, we assume that there is a virtual central bank (VCB) that maintains a credit account for each mobile terminal to record its efforts to the common achieved utility of the coalition. One feature that distinguishes this approach from the previous ones is that it offers incentive to each mobile terminal proportional to its impact on achieving the common utility, leaving no incentive for players to break the cooperation. The previous works such as [43–46] provide flat credit to intermediate relays for each packet forwarding regardless of the amount of resources they consume. Here, when relay i forwards a packet for source j, we increase the credit of relay i while decreasing the credit of source j by a value which is equal to the energy cost of packet forwarding for the relay. To incentivize the relay for keeping cooperation on, we provide it an additional credit u_i, the amount of which is determined by the solution of the coalitional game, described in the previous section.

Finally, in any cooperation instance, the VCB checks the credit level of the candidate sources and rejects the ones that have no credit. To enable mobile terminal to start cooperation at the very beginning, the VCB provides an initial credit to each mobile terminal. The cooperative mobile terminals gain credit over time and keep their credit levels increasing, while the selfish ones lose their initial credit rapidly and would soon be isolated from the coalition accordingly.

5.4 Performance Evaluation

For numerical evaluation, we consider the energy saving performance of a single coalition illustrated by Fig. 5.6 where mobile terminals hold two air interfaces: a WiFi (IEEE 802.11g) interface for long range (infrastructure-based) communications as well as a WiMedia interface for short range communications. All parameters used are summarized in Table 5.1. For the WiFi system, the radiation power as well as the transmitter and receiver power consumption values are according to the data sheet of the Cisco Aironet 802.11 a/b/g wireless CardBus adapter [57], and the carrier frequency and the bandwidth values are according to IEEE 802.11g standard [58]. For the WiMedia system, the carrier frequency and the bandwidth values are according to ECMA-368 standard [59], and the transmitter and the receiver power

Table 5.1 System parameters

Parameter	WiFi	WiMedia
Radiation power	13 dBm	−52 dBm/MHz
Transmitter power consumption (mW)	1,749	426
Receiver power consumption (mW)	930	356
Carrier frequency (GHz)	2.445	3.432
Bandwidth (MHz)	20	500
Antenna gain (dBi)	2.2	2.2

consumption values are according to the sample WiMedia transceiver design proposed in [60]. According to Federal Communications Commission (FCC), the Ultra Wide Band (UWB) communications are allowed in the frequency band 3.1–10.6 GHz with maximum allowable radiation power spectral density of −41.3 dBm/MHz to avoid interfering with other licensed radios operating in the same frequency band. According to ECMA-368 standard, which describes the specifications for the PHY and MAC layers of the UWB communication systems, the allocated spectrum for UWB communications (3.1–10.6 GHz) is divided into 14 bands each of which having 500 MHz bandwidth. We assume that the WiMedia system operates in the first frequency band (3.1–3.7 GHz) and its radiation power spectral density is −52 dBm/MHz, which is lower than the maximum allowed radiation power spectral density. Finally, the assumed antenna gains for both systems are 2.2 dBi, which corresponds to a dipole antenna.

We assume non-line-of-sight (NLOS) propagation model for WiFi channels and line-of-sight (LOS) propagation model for WiMedia channels. These assumptions are based on the fact that the WiMedia system operates at higher carrier frequencies comparing to the WiFi system, where the electromagnetic waves tend to propagate in LOS direction. For the case of WiFi, we assume path loss exponent of 5, lognormal shadowing with standard deviation of 8 dB as well as slow flat Rayleigh fading. For the case of WiMedia, we assume path loss exponent of 1.7 without any shadowing or fading defects. Table 5.2 summarizes the channel models. Finally, we consider random way point (RWP) mobility model with maximum speed of 3 m/s and pause time of 5 s to take into account the nomadic mobility in an indoor scenario such as a WiFi hotspot in a coffee shop or a shopping mall.

In the simulations, we assume adaptive modulation and coding (AMC) scheme for both WiFi and WiMedia links. When the receiver and transmitter are close to each other, the path loss is low, so higher order modulations can be used, providing higher reliable data rates. On the other hand, when the receiver and transmitter move away from each other, the path loss increases, deteriorating the channel. In this case, the transmitter has to fall back to lower order modulations, so the reliable

Table 5.2 Assumed channel propagation models

System	Channel model
WiFi (IEEE 802.11g)	$[P_r(d)]_{dBm} = -24.8 - 50\log_{10}(d) + \chi$
WiMedia	$[P_r(d)]_{dBm} = -63.8 - 17\log_{10}(d)$

Table 5.3 Receiver sensitivities of WiFi and WiMedia systems

WiFi		WiMedia	
Rate (Mbps)	Sensitivity (dBm)	Rate (Mbps)	Sensitivity (dBm)
6	−90	53.3	−80.8
9	−84	80	−78.9
12	−82	106.7	−77.8
18	−80	160	−75.9
24	−77	200	−74.5
36	−73	320	−72.8
48	−72	400	−71.5
54	−72	480	−70.4

communication data rate decreases. In order to take into account the AMC technique in the simulations, we first take into account the receiver and the transmitter antenna gains by adding these gains to the transmit power level (in logarithmic scale). Then, we consider the power loss of the wireless channel. As a result, we come up with the received signal strength at the receiver. We compare this signal strength with the receiver sensitivity for different modulation schemes. Based on this comparison, we determine the reliable data rate of the link. Table 5.3 summarizes the receiver sensitivities for both WiFi and WiMedia systems. For example, for the case of WiFi system, when the received signal strength is greater than or equal to −80 dBm, it is possible to have reliable communication with data rate 18 Mbps. The minimum signal strength required for reliable WiFi communication is −90 dBm, which can support reliable communication with data rate 6 Mbps. If the signal strength drops further, an outage event will occur and no reliable communication can take place. Figure 5.7 illustrates the average reliable data rates for both WiMedia and WiFi systems as the distance between the transmitter and the receiver varies between 0 and 20 m. As seen from this figure, the maximum range

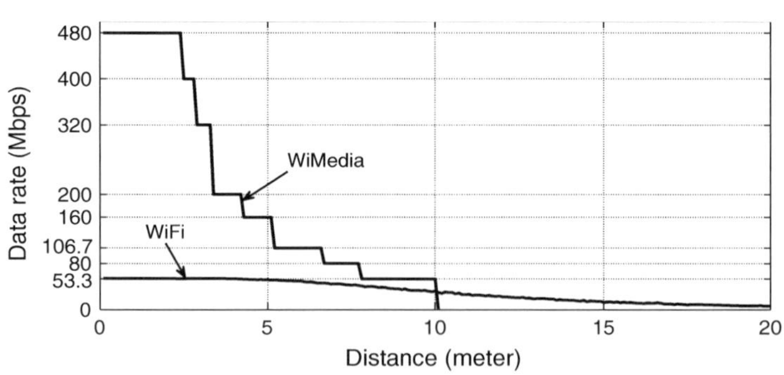

Fig. 5.7 WiFi and WiMedia rates for different distances between transmitter and receiver

of WiMedia and WiFi systems with the considered parameters and channel models are 10 and 20 m, respectively. It should be pointed out that, in the simulations, we consider only the upstream communications from the source mobile terminals to the access point where the first hop is performed over a WiMedia link and the second hop is performed over a WiFi link. Therefore, the receiver for the WiFi link is the access point. For this reason, the receiver sensitivities for WiFi system in Table 5.3 are according to Cisco Aironet 1200 Series Access Point [61].

To evaluate the energy saving performance of the cooperation, we define energy saving gain (ESG) of a coalition as the ratio of the achieved energy saving of the coalition to the required energy for mobile terminals of the coalition to act individually (i.e. communicate directly with the AP). For example, when the common energy saving of the coalition is 1 J and the required energy for all mobile terminals to communicate without cooperation is 2 J, the ESG is 50 %.

In order to evaluate energy saving performance of the coalition, we first conduct a simulation for different coalition sizes and relay densities while the weighting factor of the utility function (α) is constant at 0.5. We vary the coalition size from 10 to 100 mobile terminals. For each coalition size, we repeat the simulation for three different values of the relay density, namely 20, 50, and 80 %. For example, when the coalition size is 50 and the relay density is 20 %, 10 nodes act as relays, while the rest 40 nodes act as sources. Every source node transmits with a constant rate of 10 packets per second with a packet size of 1,024 bytes. Every 10 s, the roles of sources and relays are switched around (the relays become sources and vice versa) to give chance to every mobile terminal to act both as a source and as a relay during the simulation time. This periodic role exchange continues during the whole simulation time, which is 300 s. Figures 5.8 and 5.9 illustrate the average energy saving gain and the corresponding standard deviation, respectively. As it is clear from these figures, when the coalition size increases, the energy saving gain increases, while its standard deviation shrinks. This is due to the fact that the more the number of mobile terminals in a coalition, the more the number of opportunities for more efficient cooperation. In other words, when there are more mobile terminals in a coalition, the chance of finding good relay-source matching increases,

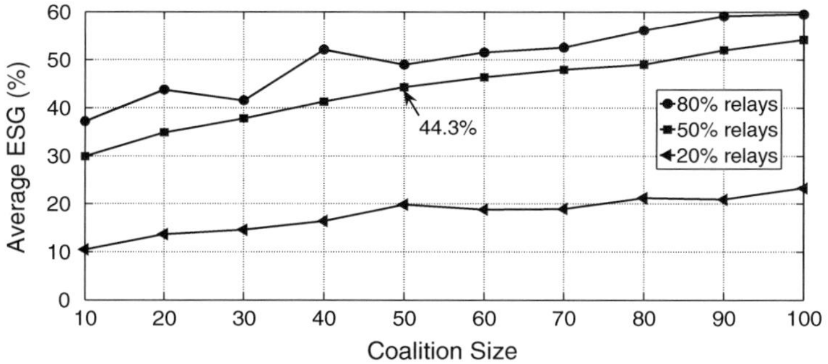

Fig. 5.8 Energy saving gain for different coalition sizes and relay densities [42]

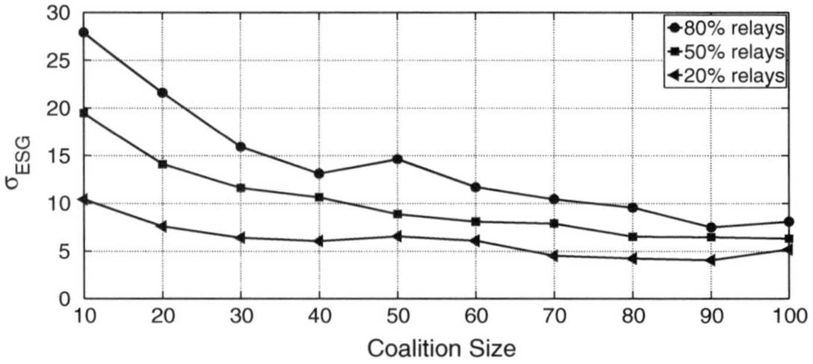

Fig. 5.9 Standard deviation of ESG for different coalition sizes and relay densities

which increases the energy saving. The decrease in standard deviation reflects the fact that the probability of finding a good relay increases as the coalition is populated with more mobile terminals. The standard deviation values in Fig. 5.9 in fact show the precision and reliability of the energy saving gain values of Fig. 5.8. A higher value of standard deviation shows that the energy saving gain cannot be reliable due to high variation from case to case, depending on the topology of mobile terminals and the wireless channel fading. As seen from Fig. 5.8, when the relay density increases from 20 to 50 %, the energy saving gain increases significantly, yet it increases marginally when the density of relays increases further from 50 to 80 %, demonstrating a saturation trend. When the coalition size is 100 and relay density is 80 %, we achieve maximum energy saving gain (60 %) with minimum standard deviation (5 %).The results illustrated by Figs. 5.8 and 5.9 can serve as a guide to choose an appropriate coalition size for a target energy saving gain while avoiding unnecessary bigger coalition sizes to minimize the communication or computation burden.

To study the battery life extension of mobile terminals adopting cooperative communication, we conduct another simulation with a coalition size of 50 mobile terminals and relay density of 50 % (i.e. 25 sources and 25 relays) where all mobile terminals have equal initial battery level of 2 J. The simulation runs until the first battery depletion occurs in the coalition. Figure 5.10 illustrates the battery life extension of the mobile terminal whose battery depletes faster than the battery of other mobile terminals in the coalition. The figure contrasts the battery life extension of this mobile terminal against a baseline scenario, where the short range WiMedia interfaces of all mobile terminals are switched off and every communication is performed over a direct WiFi link. As seen from the figure, cooperation is able to extend the battery life significantly. The figure presents the battery life extension for two different values of the weighting factor: $\alpha = 0.5$ and $\alpha = 1$. As clear from Eq. (5.9), $\alpha = 1$ only takes into account the maximization of the energy saving of the coalition, while $\alpha = 0.5$ gives equal weights to the maximization of the energy saving of the coalition and the maximization of the battery life of the mobile terminal having minimum battery level. As seen from Fig. 5.10, $\alpha = 1$ extends the

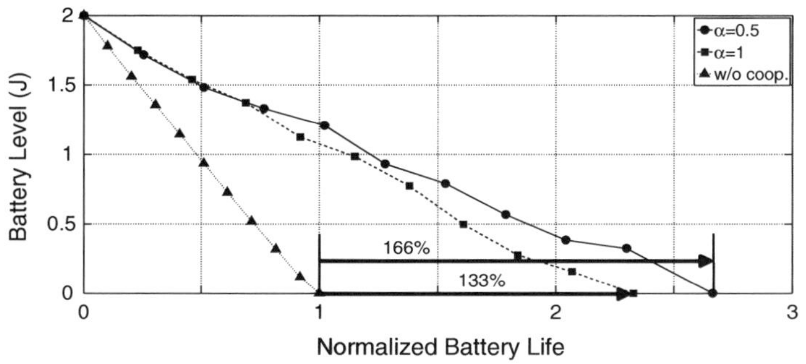

Fig. 5.10 Battery depletion of node whose battery depletes faster than any others [42]

battery life 133 %, comparing to the baseline scenario, while α = 0.5 provides 33 % additional battery life extension. That is, α = 0.5 provides 166 % battery life extension in total comparing to the baseline case. Figure 5.11 illustrates the average energy saving gain for different values of the weighting factor α along with its maximum and minimum values. As seen from this figure, the average energy saving gain starts from 33.9 % and increases gradually until 45.6 %, displaying 11.7 % variation for the whole range variation of α (from 0 to 1), with standard deviation of around 10 %. Several conclusions can be drawn from this figure. First, the energy saving gain demonstrates low sensitivity to the variation of α. Second, even the case of α = 0 (pure battery life extension strategy) results in significant energy saving gain (33.9 %). Finally, as depicted in the figure, the case of α = 0.5 leads to 43.6 % energy saving gain. The overall conclusion that we can draw from Figs. 5.10 and 5.11 is that choosing a moderate value of 0.5 for the weighting factor leads to negligible reduction in energy saving gain (2.1 %), while resulting in significant battery life extension (33 %), comparing to the case of α = 1.

In order to evaluate the effectiveness of the proposed credit scheme to detect and isolate selfish players, we conduct a different simulation with a coalition of 50 mobile terminals and relay density of 50 % (25 sources and 25 relays); among them,

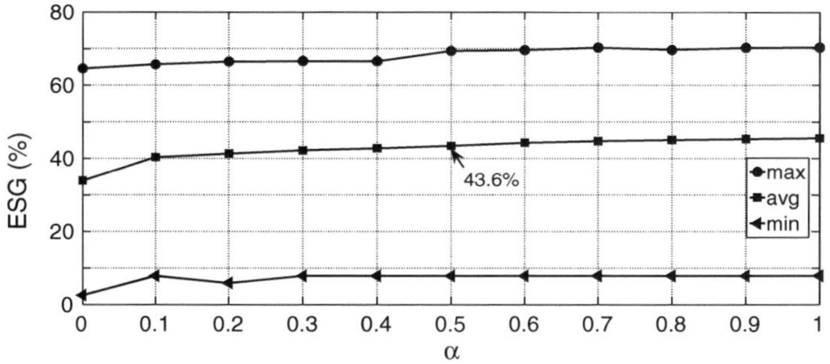

Fig. 5.11 Energy saving gain for different values of the weighting factor α

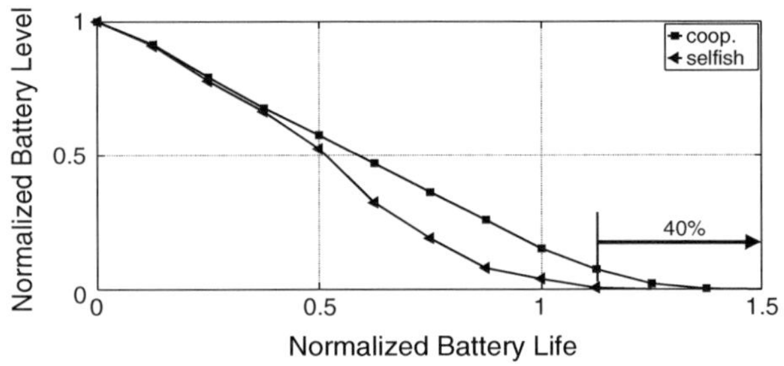

Fig. 5.12 Battery depletion rates of cooperative and selfish nodes [42]

5 nodes are selfish, while the rest 45 nodes are cooperative. The simulation starts with all mobile terminals having equal initial battery of 2 J and initial credit of 0.1. Similar to the previous simulation, every source node sends 10 packets per second with packet size of 1,024 bytes, and every 10 s the sources and relays change their roles, giving chance to every mobile terminal to act both as a relay and as a source equally likely. The simulation lasts until all mobile terminals run out of power. Figures 5.12 and 5.13 compare the average battery and credit levels of cooperative and selfish nodes, respectively. As seen from Fig. 5.12, on average, the battery of cooperative nodes last about 40 % more than the battery of selfish nodes. As shown by Fig. 5.13, all mobile terminals start with equal initial credit (0.1). The cooperative nodes increase their credit level, while the selfish ones lose their initial credit soon and are isolated from the coalition accordingly. As seen from Fig. 5.12, the battery level of selfish and cooperative nodes depletes with almost same pace in the beginning, when selfish nodes still have credit. However, soon after, selfish nodes are left without credit. This is the time when other nodes avoid cooperating with them, so their battery starts to deplete faster than the cooperative nodes.

Overall, the simulation results demonstrate that the energy saving gain depends not only on the total number of mobile terminals in the coalition but also on the percentage of the relay nodes. When the coalition size increases, both the communication overhead due to context exchange (for negotiation between mobile terminals) and the computation time for relay selection algorithm of Eq. (5.10) increase. Therefore, to keep the running time and the communication overhead of the cooperation at a practical level while ensuring a reasonable energy saving gain, a moderate cluster size of 30–50 nodes is recommended. Finally, as an instance, for

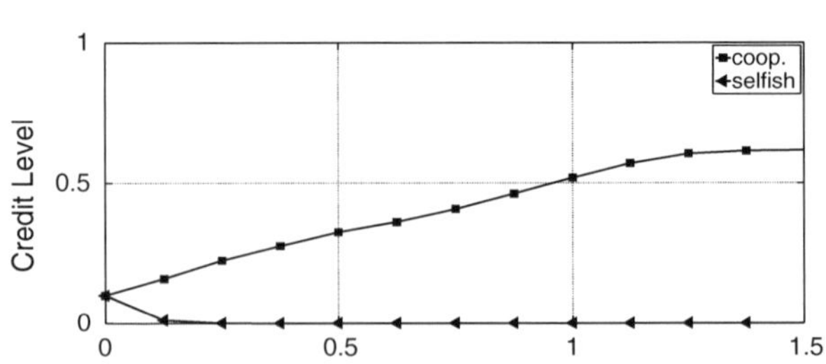

Fig. 5.13 Credit level of cooperative and selfish nodes [42]

a cluster composing of 25 relay nodes and 25 source nodes, cooperative communication can extend the battery lifetime of mobile terminals to more than double while successfully detecting and isolating selfish nodes from the coalition.

5.5 Conclusion

Cooperation has been identified as an effective technique to enhance the performance of wireless communication systems. In this chapter, we discussed the possibility of using this technique for improving the energy efficiency of mobile terminals. In particular, we applied coalitional game theory to model how a number of adjacent mobile terminals can join together and form a cooperative group to share their resources and cooperate in order to reduce their energy consumption. Spread of selfish behaviour is a crucial impediment hampering cooperation among mobile terminals. We discussed how cooperative game theory could be applied to solve this problem by offering credit to cooperative players while detecting selfish players and isolating them from cooperative groups. We simulated a scenario where mobile terminals adopted two-hop communications, exploiting their short range interfaces for out-of-band relaying. The simulations results validated that mobile terminals could extend their battery life time to more than double using the described short range cooperation technique. Moreover, we discussed how credit scheme could be used for detecting and isolating selfish nodes. We introduced credit account for each node which is maintained at a central trust entity embedded in the infrastructure side where the contribution of each mobile terminal is recorded and traced to detect whether the terminal is cooperative or not. We also validated the effectiveness of the scheme by simulations.

Acknowledgments This work is supported by the grant of the Fundação para a Ciência e a Tecnologia (FCT-Portugal), with the reference number: SFRH/BD/79909/2011.

References

1. Fitzek, F., Katz, M.D. (eds.): Cooperation in Wireless Networks: Principles and Applications: Real Egoistic Behavior is to Cooperate. Springer (2006)
2. Lin, Y., Hsu, Y.: Multihop cellular: a new architecture for wireless communications. In: Proceedings of INFOCOM, vol. 3, pp. 1273–1282 (2000)
3. Le, L., Hossain, E.: Multihop cellular networks: potential gains, research challenges, and a resource allocation framework. IEEE Commun. Mag. **45**(9), 66–73 (2007)
4. Salem, N.B., Buttyán, L., Hubaux, J.P., Jakobsson, M.: Node cooperation in hybrid ad hoc networks. IEEE Trans. Mob. Comput. **5**(4), 365–376 (2006)
5. 3G TR 25.924 version 1.0.0.0: 3rd Generation Partnership Project: Technical Specification Group Radio Access Network; Opportunity Driven Multiple Access (1999)
6. Aggélou, G.N., Tafazolli, R.: On the relaying capability of next-generation GSM cellular networks. IEEE Pers. Commun. **8**(1), 40–47 (2001)

7. Pabst, R., et al.: Relay-based deployment concepts for wireless and mobile broadband radio. IEEE Commun. Mag. **42**(9), 80–89 (2004)
8. Cover, T., Gamal, A.E.: Capacity theorems for the relay channel. IEEE Trans. Inf. Theor. **25** (5), 572–584 (1979)
9. Van der Meulen, E.C.: Three-terminal communication channels. Adv. Appl. Probab. **3**, 120–154 (1971)
10. Laneman, J.N., Tse, D.N.C., Wornell, W.: Cooperative diversity in wireless networks: efficient protocols and outage behavior. IEEE Trans. Inf. Theor. **50**(12), 3062–3080 (2004)
11. Laneman, J.N., Wornell, G.W.: Distributed space-time coded protocols for exploiting cooperative diversity in wireless networks. IEEE Trans. Inf. Theor. **49**(10), 2415–2425 (2003)
12. Sendonaris, A., Erkip, E., Azhang, B.: User cooperation diversity—Part I: system description. IEEE Trans. Commun. **51**(11), 1927–1938 (2003)
13. Sendonaris, A., Erkip, E., Azhang, B.: User cooperation diversity—Part II: implementation aspects and performance analysis. IEEE Trans. Commun. **51**(11), 1939–1948 (2003)
14. Hunter, T.E., Nosratinia, A.: Diversity through coded cooperation. IEEE Trans. Wireless Commun. **5**(2), 283–289 (2006)
15. Janani, M., Hedayat, A., Huntter, T.E., Nosratinia, A.: Coded cooperation in wireless communications: space-time transmission and iterative coding. IEEE Trans. Sig. Process. **52** (2), 362–371 (2004)
16. Sadek, A.K., Su, W., Liu, K.J.R.: Multinode cooperative communications in wireless networks. IEEE Trans. Sig. Process. **55**(1), 341–355 (2007)
17. Boyer, J., Falconer, D.D., Yanikomeroglu, H.: Multihop diversity in wireless relaying channels. IEEE Trans. Commun. **52**(10), 1820–1830 (2004)
18. Kramer, G., Gaspar, M., Gupta, P.: Cooperative strategies and capacity theorems for relay networks. IEEE Trans. Inf. Theor. **51**(9), 3037–3063 (2005)
19. Zheng, L., Tse, D.N.C.: Diversity and multiplexing: a fundamental tradeoff in multiple-antenna channels. IEEE Trans. Inf. Theor. **49**(5), 1073–1096 (2003)
20. Nosratinia, A., Hunter, T.E., Hedayat, A.: Cooperative communication in wireless networks. IEEE Commun. Mag. **42**(10), 74–80 (2004)
21. Nagpal, V.: Cooperative multiplexing in wireless relay networks. Ph.D. thesis, University of California, Berkeley (2012)
22. Yijia, F., Chao, W., Poor, H.V., Thompson, J.S.: Cooperative multiplexing: toward higher spectral efficiency in multiple-antenna relay networks. IEEE Trans. Inf. Theor. **55**(9), 3909–3926 (2009)
23. Shih, E., Bahl, P., Sinclair, M.J.: Wake on wireless: an event driven energy saving strategy for battery operated devices. In Proceedings 8th Annual International Conference on Mobile Computing and Networking, ACM, pp. 160–171 (2002)
24. Pering, T., Agarwal, Y., Gupta, R., Want, R.: CoolSpots: reducing the power consumption of wireless mobile devices with multiple radio interfaces. In: Proceedings of 4th International Conference on Mobile Systems, Applications and Services, ACM, pp. 220–232 (2006)
25. Yoo, J., Park, K.H.: A cooperative clustering protocol for energy saving of mobile devices with WLAN and Bluetooth interfaces. IEEE Trans. Mob. Comput. **10**(5), 491–504 (2011)
26. Gür, G., Alagöz, F.: Green wireless communications via cognitive dimension: an overview. IEEE Netw. **25**(2), 50–56 (2011)
27. C2POWER: Cognitive radio and cooperative strategies for power saving in multi-standard wireless devices. http://www.ict-c2power.eu/
28. GREEN-T: Green terminals for next generation wireless systems. http://greent.av.it.pt/
29. Radwan, A., Rodriguez, J.: Energy saving in multi-standard mobile terminals through short-range cooperation. EURASIP J. Wirel. Commun. Networking **2012**(159), 1–15 (2012)
30. Saghezchi, F.B., Radwan, A., Rodriguez, J.: Energy efficiency performance of WiFi/WiMedia relaying in hybrid ad-hoc networks. In: Proceedings of IEEE 3rd International Conference on Communications and Information Technology (ICCIT), pp. 285–289 (2013)

31. Saghezchi, F.B., Radwan, A., Alam, M., Rodriguez, J.: Cooperative strategies for power saving in multi-standard wireless devices, pp. 284–296. Springer, Berlin Heidelberg (2013). (The Future Internet)
32. Myerson, R.B.: Game Theory Analysis of Conflict. Harvard University Press, Cambridge (1991)
33. MacKenzie, A.B., DaSilva, L.A.: Game Theory for Wireless Engineers. Morgan and Claypool Publishers, San Refael (2006)
34. Felegyhazi, M., Hubaux, J.: Game theory in wireless networks: a tutorial. Technical report LCA-REPORT-2006-002, EPFL (2006)
35. Srivastava, V.: Using game theory to analyze wireless ad hoc networks. IEEE Commun. Surv. Tutorials **7**(4), 46–56 (2005)
36. Félegyázi, M., Hubaux, J., Buttyán, L.: Nash equilibrium of packet forwarding strategies in wireless ad hoc networks. IEEE Trans. Mob. Comput. **5**(5), 463–476 (2006)
37. Yang, J., Klein, A.G., Brown, D.R.III.: Natural cooperation in wireless networks. IEEE Sig. Process. Mag. **26**(5), 98–106 (2009)
38. Srinivasan, V., Nuggehalli, P., Chiasserini, C.F., Rao, R.R.: An analytical approach to the study of cooperation in wireless ad hoc networks. IEEE Trans. Wirel. Commun. **4**(2), 722–733 (2005)
39. Saad, W., Han, Z., Debbah, M., Hjørungnes, A., Basar, T.: Coalitional game theory for communication networks. IEEE Sig. Process. Mag. **26**(5), 77–97 (2009)
40. Saghezchi, F.B., Nascimento, A., Albano, M., Radwan, A., Rodriguez, J.: A novel relay selection game in cooperative wireless networks based on combinatorial optimizations. In: Proceedings of IEEE 73rd Vehicular Technology Conference (VTC Spring). Budapest (2011)
41. Saghezchi, F.B., Radwan, A., Nascimento, A., Rodriguez, J.: An incentive mechanism based on coalitional game for fair cooperation of mobile users in HANETs. In: Proceedings of IEEE 17th International Workshop on Computer Aided Modeling and Design of Communication Links and Networks (CAMAD). pp. 378–382 (2012)
42. Saghezchi, F.B., Radwan, A., Rodriguez, J., Dagiuklas, T.: Coalition formation game towards green mobile terminals in heterogeneous wireless networks. IEEE Wirel. Commun. Mag. **20** (5), 85–91 (2013)
43. Buttyán, L., Hubaux, J.P., Nuglets, A.: Virtual currency to stimulate cooperation in self organized mobile ad hoc networks. Technical report, no. DSC/2001 (2001)
44. Jakobsson, M., Hubaux, J.P., Buttyán, L.: A micro-payment scheme encouraging collaboration in multi-hop cellular networks, pp. 15–33. Springer, Berlin Heidelberg (2003). (Financial Cryptography)
45. Salem, N.B., Levente, B., Hubaux, J.P., Jakobsson, M.: A charging and rewarding scheme for packet forwarding in multi-hop cellular networks. In: Proceedings of MOBIHOC '03. Maryland, USA (2003)
46. Zhong, S., Chen, J., Yang, Y.R.: Sprite: a simple, cheat- proof, credit-based system for mobile ad hoc networks. In: Proceedings of IEEE INFOCOM '03, vol. 3. pp. 1987–1997 (2003)
47. Buchegger, S., Boudec, J.L.: Performance analysis of the CONFIDANT protocol. In: Proceedings of 3rd ACM International Symposium on Mobile Ad Hoc Networking and Computing, pp. 226–236. Lausanne (2002)
48. Michiardi, P., Molva, R.: CORE: a collaborative reputation mechanism to enforce node cooperation in mobile ad hoc networks. In: Proceedings of IFIP-Communication and Multimedia Security Conference (2002)
49. Bansal, S., Baker, M.: Observation-based cooperation enforcement in ad hoc networks. http://arxiv.org/pdf/cs/0307012v2 (2003)
50. He, Q., Wu, D., Khosla, P.: SORI: a secure and objective reputation-based incentive scheme for ad hoc networks. In: Proceedings of IEEE Wireless Communications and Networking Conference, pp. 825–830 (2004)
51. Rebahi, Y., Mujica, V., Simons, C., Sisalem, D.: SAFE: securing packet forwarding in ad hoc networks. In: Proceedings of the 5th Workshop on Applications and Services in Wireless Networks (ASWN). Paris (2005)

52. Jaramillo, J.J., Srikant, R.: DARWIN: distributed and adaptive reputation mechanism for wireless ad-hoc networks. In: Proceedings of the 13th Annual ACM International Conference on Mobile Computing and Networking (MOBICOM '07). Montréal (2007)
53. Hong, Y.W., Huang, W.J., Chiu, F.H., Kuo, C.C.: Cooperative communications in resource-constrained wireless networks. IEEE Sig. Process. Mag. **24**(3), 47–57 (2007)
54. Wei, H.Y., Gitlin, R.D.: Two-hop-relay architecture for next generation WWAN/WLAN integration. IEEE Wirel. Commun. Mag. **11**(2), 24–30 (2004)
55. Li, G.Y., et al.: Energy-efficient wireless communications: tutorial, survey, and open issues. IEEE Wirel. Commun. Mag. **18**(6), 28–35 (2011)
56. Shapley, L.S., Shubik, M.: The assignment game I: the core. Int. J. Game Theory **1**(1), 111–130 (1972)
57. Data sheet: Cisco Aironet 80.11 a/b/g wireless CardBus adapter. http://www.cisco.com/en/US/prod/collateral/wireless/ps6442/ps4555/ps5818/product_data_sheet09186a00801ebc29.pdf
58. IEEE 802.11g: Part 11, Amendment 4. http://standards.ieee.org/getieee802/download/802.11g-2003.pdf
59. ECMA-368: Standard: high rate ultra wideband PHY and MAC standard. 3rd ed., Dec. 2008. http://www.ecmainternational.org/publications/files/ECMA-ST/ECMA-368.pdf
60. Sandner, C., et al.: A WiMedia/MBOA-compliant CMOS RF transceiver for UWB. IEEE J. Solid-State Circuits **41**(12), 2787–2794 (2006)
61. Data sheet: Cisco Aironet 1200 Series Access Points. http://www.cisco.com/en/US/prod/collateral/wireless/ps5678/ps430/ps4076/product_data_sheet09186a00800937a6.pdf

Chapter 6
Cognitive Means Smart: Knowledge *Saves* Power

Dionysia Triantafyllopoulou, Klaus Moessner, Muhammad Alam, Ayman Radwan and Jonathan Rodriguez

Abstract The previous chapters mainly examined methods to save energy at the mobile handset, either by using short-range cooperation between mobile terminals, or by performing smart vertical handovers between heterogeneous radio access technologies. These techniques can be beneficial to mobile systems, but they have to be performed based on informed decisions; meaning that mobile devices need to be cognitive. Modern devices already collect significant amounts of information, but they have limited capability to exploit such context/information, and handover decisions are merely based on signal strength, or are network controlled and based on network load. In this chapter, we aim to go beyond the state-of-the-art by envisioning mobile terminals with the capability to make informed decisions based on a reservoir of context information made available through context providers; namely what is referred to as smart phones. We include a survey of the current state of the art for context extraction and management in context-aware systems; besides listing the current context extraction techniques and research efforts, we pinpoint the important properties of good context extraction techniques. Thereafter, we discuss how context information can be exploited in energy saving when performing network or node discovery mechanism, by instructing the mechanisms to scan for certain nodes/networks which are known to be in the vicinity. Finally, we discuss the range of context information that can be used to make informed decisions to save power.

D. Triantafyllopoulou (✉) · K. Moessner
University of Surrey, Guildford, UK
e-mail: d.triantafyllopoulou@surrey.ac.uk

M. Alam · A. Radwan · J. Rodriguez
Instituto de Telecomunicações, Campus Universitário de Santiago, 3810-193 Aveiro, Portugal

A. Radwan and J. Rodriguez (eds.), *Energy Efficient Smart Phones for 5G Networks*, Signals and Communication Technology,
DOI 10.1007/978-3-319-10314-3_6

6.1 Introduction

With every new communication system generation and every new air interface standard deployed, the radio environment becomes ever more complex and fragmented. Standard terminals meanwhile host more than a handful different air interfaces; making a choice about which air interface should be accessed is usually pre-determined by the user's subscription rather than based on informed choices which could help preserve energy, or use spectrum more efficiently. Cognitive Radio (CR) technologies have been discussed as a means to enable such informed choices. Mitola [1] and many more proponents of CR systems regarded the technology as one of the main enablers to overcome the spectrum scarcity issue, approaches including overlay-underlay, or opportunistic spectrum access have been widely discussed in research and some of the techniques have been taken into more modern spectrum management schemes, (e.g. some of the Self Organizing Network (SON) approaches do implement CR techniques).

However, there is more to cognitive radio than just more efficient spectrum usage. Mitola outlined in his thesis the notion of situation awareness and implied as well that not only spectrum access should be subject to learning mechanisms. He indicated that also the choice of coding or even encryption approach should be based on knowledge inferred from the current operational conditions.

The same applies to systems that aim to save power or preserve energy levels in mobile devices. The solutions in this book are built on this approach and identify, out of a wide range of operational conditions that determine the radio environment, those context parameters that can be exploited to achieve energy efficiency in wireless communication systems. As there are basic system functions, like network detection and attachment that cannot be omitted, the concentration was on parameters that influence the protocols design, and energy-efficient network and node discovery mechanisms.

6.2 Cognition in Wireless Communication Systems

6.2.1 Understanding the "Context" and Acting on It

Processing operational information and understanding context can be rather tedious. The scope and range of the information fragments that describe the operational situation can be rather narrow, which would allow rapid processing of the information and rapid reaction to changing circumstances. On the other hand, the definition of scope and range can be rather wide, which allows a more detailed description of the operational situation, however at the price of more extensive processing. Figure 6.1 illustrates this idea. In the actual context space (as defined by the ORACLE project [2]), different types of context information are captured, whereby non-relevant information will be filtered out using a rule based

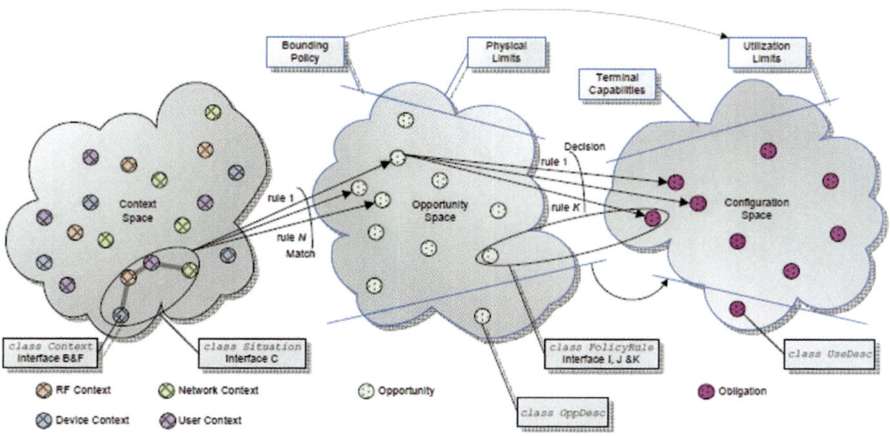

Fig. 6.1 From information to context to action [1]

mechanism. The resulting context information then describes the actual opportunity space, meaning it outlines the system view on the boundaries in which a reconfiguration decision has to be made. Depending on the rules that may be brought in through operational limitations (e.g. hardware restrictions), the range of possible configurations can be determined and actions can be taken.

To be able to implement such an approach and narrow context information to finally make decisions about instantiation of those system configurations that will help achieving the intended optimization goal, a robust filtering and decision making system is required. The following subsection explains how the proposed system exploits or treats context information.

6.2.2 Incomplete Information, Context, and Knowledge

As stated in the previous subsection, information that influences systems may be obtained from different sources. This implies that this information may be rather heterogeneous in terms of format, type as well as precision. A common notation is required to express the diverse information fragments in a homogeneous manner. As illustrated in Fig. 6.2, information fragments are collected, translated into a processable format, and then processed to generate knowledge.

Another issue is the relevance of information; not each information fragment may be relevant for a given problem, yet may be needed if another type of problem occurs. Only relevant information must be used for the generation of knowledge. Therefore proper "scoring" has to be introduced.

The third issue that affects context systems is the level of completeness of information. In case of gaps in the available information, the ability to estimate the required values decreases. Learning mechanisms can be used to achieve better

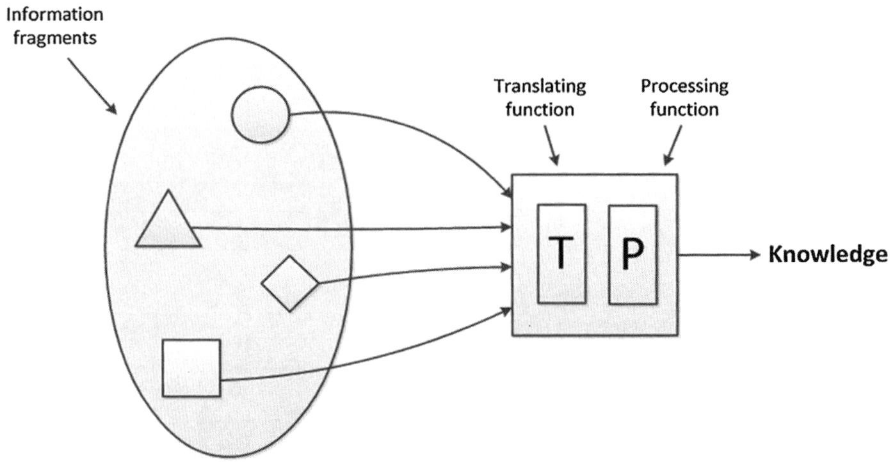

Fig. 6.2 Deriving knowledge from context

understanding of the surrounding world; they can be exploited to overcome the problems that incompleteness of information can generate (Fig. 6.3). Incomplete information introduces an uncertainty that needs to be reduced. However, collection and processing of information in the process of generating knowledge is costly. It is very important that its true value is assessed with precision.

Finally, context information must generate system knowledge to be really useful; the approach of how this is handled is described in the following sections.

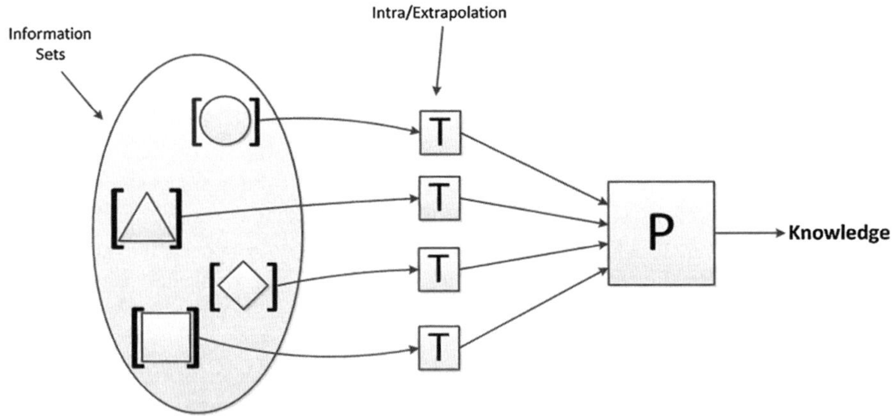

Fig. 6.3 Handling incomplete information

6.2.3 Exploiting Context

In this section, we aim to allow modern mobile communication terminals to take advantage of the availability of multiple wireless interfaces and the presence of neighbouring devices willing to establish co-operations in order to reduce their energy consumption. To this end, the current chapter is related to the design and validation of energy efficient network and node discovery mechanisms through the establishment of an architecture for context awareness support. More specifically, the knowledge of the available context in terms of neighbouring network and node availability is considered by the proposed architecture (earlier in Chap. 1) in two ways. First, in the case of vertical handovers, energy consumption minimization is achieved through the deployment of energy efficient network discovery and selection mechanisms that allow the mobile terminals to optimize the network discovery procedures, and finally, select the network type that satisfies their requirements. Similarly, the mobile terminals are allowed to adopt cooperative communication approaches, according to which they can identify neighbouring nodes that can act as relays, enabling the formation of appropriate clusters to allow transmission energy consumption minimization.

Figure 6.4 depicts the components and interfaces that constitute the context support architecture, categorizing the identified modules based on their logical location. Therefore, the different context awareness support functions reside either on the network or on the terminal side of the proposed architecture.

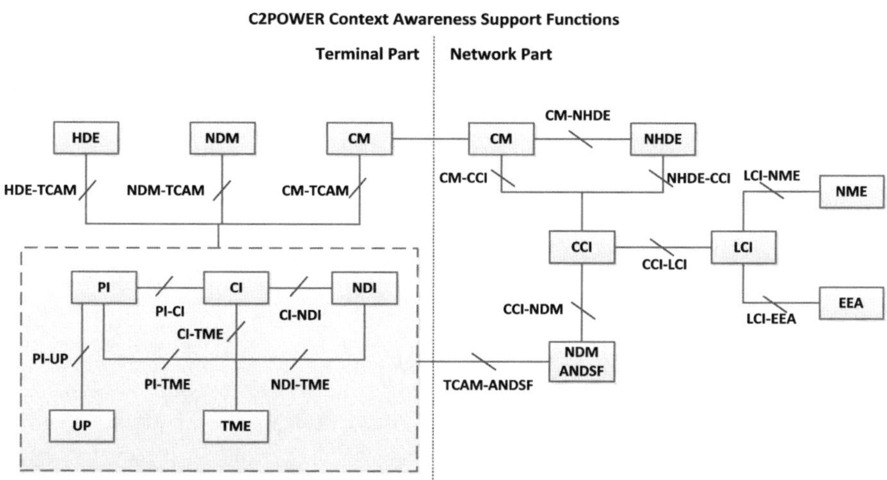

Fig. 6.4 Context support structure—system interfaces [3]

6.2.3.1 Context Awareness Support Functions on the Network Side

The modules that reside in the network part of the proposed architecture are the following:

- *Local Context Information* (*LCI*)
  It is responsible for collecting information on the energy consumption and energy requirements within the different networks that are considered in a particular efficiency policy.
- *Network Measurement Extractor* (*NME*)
  It is responsible for providing information and parameters regarding the network and cell load, as well as cell capacity, to the LCI.
- *Energy Efficiency Agent* (*EEA*)
  It collects cell statistics relevant to energy usage and general context information.
- *Central Context Information* (*CCI*)
  It collects, classifies and organizes the network and context information of the various LCIs that are attached to it and functions as a centralized source of information.
- *Network Discovery Module* (*NDM*)
  It collects and provides information on the unknown or non-registered networks that are available. In C2POWER, this is performed by the ANDSF.
- *Cooperation Mediator* (*CM*)
  It performs the negotiation process between the cooperation modules within individual terminals, which are willing to cooperate.
- *Network Handover Decision Engine* (*NHDE*)
  It is responsible for the approval of the handover decisions taken by the mobile terminals, based on the analysis of the context information provided by the CCI module and the policies that may apply.

6.2.3.2 Context Awareness Support Functions on the Terminal Side

The modules that reside in the terminal part of the proposed architecture are the following:

- *Terminal Context Aware Module* (*TCAM*)
  It contains all the functionalities that are required to extract, organize and understand the context regarding the environment in which the mobile terminal resides.
  - *Context Information* (*CI*)
    It is responsible for collecting and storing information on the radio environment, the capabilities of the mobile terminal, as well as the applications in use.

  - *Policy Information* (*PI*)
    It interprets information related to user settings and usage, and builds a set of policies based on the information provided by the CI module.
  - *User Preferences* (*UP*)
    This module is responsible for gathering the user defined information and the profiles that can be selected by the user.
  - *Terminal Measurement Extractor* (*TME*)
    It extracts raw context information and operation parameters from the terminal.
  - *Network/Node Discovery Information* (*NDI*)
    It is responsible for collecting and refining information about networks or nodes in the terminal's neighbourhood that can be used as access points or relays.

- *Handover Decision Engine* (*HDE*)
  It makes handover decisions based on the context information provided by the TCAM.
- *Network Discovery Module* (*NDM*)
  It uses information about terminal settings and available nodes in order to make targeted searches for available access networks.
- *Cooperation Module* (*CM*)
  It uses terminal context information to search for neighbouring nodes and negotiate with them in order to implement multi-hop collaboration policies.

The availability of information on the networking environment and the characteristics of the mobile terminals that are collected, processed and used by the various context awareness support functions are of fundamental importance in the envisioned architecture and the proposed solutions within this book. This information provides the different modules with the necessary awareness of the surrounding environment, on the network and terminal level, allowing them to efficiently adapt their operation towards the goal of energy consumption minimization.

Figure 6.5 depicts the four broad categories of network context information that are considered by the context awareness architecture and the information fragments that emanate from this categorization. These include (i) the security levels that can be provided, (ii) policies, based on the user-operator and operator-operator agreements, (iii) coverage of the network and the co-existence of different networks and (iv) the QoS, which describes the static and dynamic capabilities of the network.

Similarly, Fig. 6.6 depicts the five main categories of mobile terminal context information that are considered by the context awareness architecture. These include (i) the descriptions of the device capabilities, (ii) the information on the current location and velocity of the mobile terminal, (iii) the applications in use, (iv) the energy information, which includes the current battery level, the energy history and the energy consumption, and (v) the various parameters that constitute the user preference.

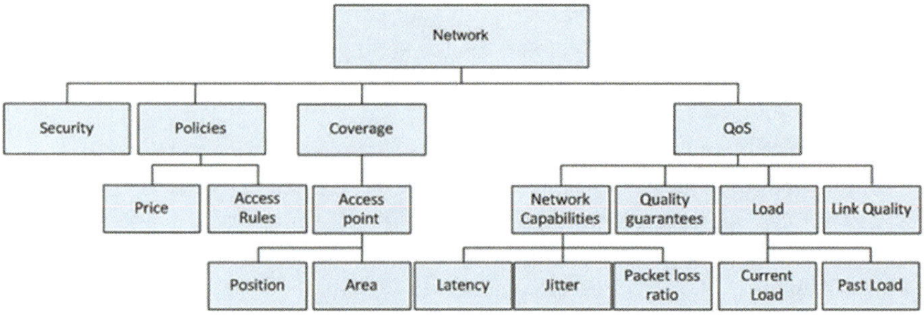

Fig. 6.5 Network context information [4]

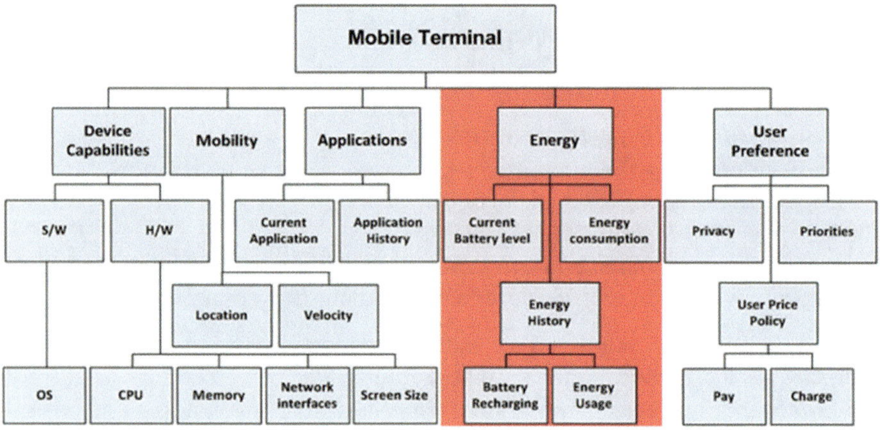

Fig. 6.6 Mobile terminal context information [4]

Figures 6.5 and 6.6 provide a high level description of the context information required for the correct functioning of the proposed solutions within the book. More detailed description can be obtained based on different solutions; an example of a detailed description of the types of network and mobile terminal context information that can be taken into consideration can be found in [4].

6.3 Exploiting Context to Save Energy

To demonstrate the energy saving potential when taking into consideration the available network and terminal context information, this subsection describes and evaluates the performance of two context aware algorithms that aim at reducing the energy consumption in modern heterogeneous networking environments.

First, we describe in detail an energy efficient network discovery mechanism, which utilizes information on the user location and the availability of neighbouring networks in order to determine the appropriate—in terms of energy efficiency—time to perform a network scanning and discover available networks that can be used as targets for handover. Based on simulations that compare the performance of the proposed algorithm with different periodic network discovery schemes, it is shown that the proposed network discovery algorithm can reduce the energy consumption, as well as lead to considerable improvement in the network detection delay with no compromise in the network detection rate.

Then, we present an energy efficient algorithm, which addresses the issues of node discovery and cooperative cluster formation in an energy efficient manner. More specifically, in a heterogeneous WiMAX—UWB environment mobile nodes exchange context lists that contain information that allows them to take efficient decisions regarding cluster formation. Extensive simulation results are provided to show the energy saving achieved with the use of the context based discovery process.

6.3.1 Network Context

In order to reduce the energy consumption and improve the provided QoS level, a user in a heterogeneous networking environment proactively scans for the available access networks to obtain up-to-date knowledge of the network topology in his/her vicinity and performs vertical handovers to less energy-consuming networks, when possible. Traditionally, in legacy networks, the procedure of network scanning is performed periodically, with a fixed pre-determined period, without considering any information on the user and system conditions and requirements. However, this often results in inefficient performance in terms of energy consumption due to the fact that context information, such as the availability of neighbouring networks, as well as the user's current position, speed and travelled distance since the last network scanning, are not taken into consideration. Therefore, unnecessary energy-consuming network scanning may be performed in case the user is moving very slowly or not moving at all, and the surrounding topology remains invariant. Moreover, in the case of high user speed, the network scanning may not be performed frequently enough resulting in increased probability of mis-detection of available less energy-consuming networks.

The recent literature on energy efficient network discovery mainly focuses on adaptively determining the most appropriate time for a user to perform a network scan, aiming to accurately detect available neighbouring networks without increasing the energy consumption [5–7]. More specifically, in [5], an energy efficient idle scanning strategy for local area networks is proposed. This method exploits the operating channels and Access Point (AP) density information provided by the Access Network Discovery and Selection Function (ANDSF) [8], defined by the 3rd Generation Partnership Program (3GPP). In [6], the authors propose a system that considers information on the user mobility and the AP density information to determine the Wireless Local Area Network (WLAN) [9] scanning period. However, the AP density information used is not explicitly provided by the network, but is calculated by the previously performed network scanning. In [7], the issue of vertical handover between integrated IEEE 802.16e [10] and WLAN networks is addressed. The proposed scheme allows a mobile user to decide whether to attempt AP discovery and how to set the 802.11 active scanning intervals. However, this scheme is designed to use standard IEEE 802.16e signaling and does not consider the possibility of vertical handover between IEEE 802.16e/WLAN and 3GPP networks.

To address the problem of network scanning period adaptation and improve the energy efficiency of the network scanning procedure, we propose a novel ANDSF-assisted network discovery algorithm that exploits information on the user location and on the location of available networks, in order to decide when to perform a network scanning. The aim of the proposed algorithm is to avoid unnecessary energy-consuming network scanning and mis-detection of available networks that can be used as targets of handover. The performance of the system that employs the proposed algorithm is compared against a system that performs network scanning with a fixed period, without taking into consideration the user or the network context information.

6.3.1.1 System Model

The system model, as shown in Fig. 6.7, consists of a large geographic area covered by a number of neighbouring Long Term Evolution (LTE) [11] cells (called eNodeB according to 3GPP), and a number of randomly distributed WLAN APs [9]. The User Equipments (UEs), while moving randomly all over this area, perform eNodeB selection and inter-eNodeB handovers based on the measurement of the received signal strength. Moreover, they regularly perform network scanning in order to detect the presence of a WLAN AP that they can connect to. When an available WLAN AP is discovered by a UE, the UE performs a vertical handover to the WLAN for better service and energy-saving transmission purposes. When the UE moves out of the coverage of the WLAN AP, the UE performs a WLAN-LTE handover. An ANDSF server is deployed at the Evolved Packet Core (EPC) acting as a database to record the network context, e.g., the WLAN AP location information. A UE is assumed to be able to estimate its current location using mature localization techniques such as Assisted Global Positioning System (AGPS) [12]. Moreover, a UE can estimate its movement information, e.g., its travelled distance, through its embedded accelerometer that is prevalent in current mobile phones. This operation consumes much less energy compared to GPS-like location estimation operations [13], and is often utilized by various phone applications.

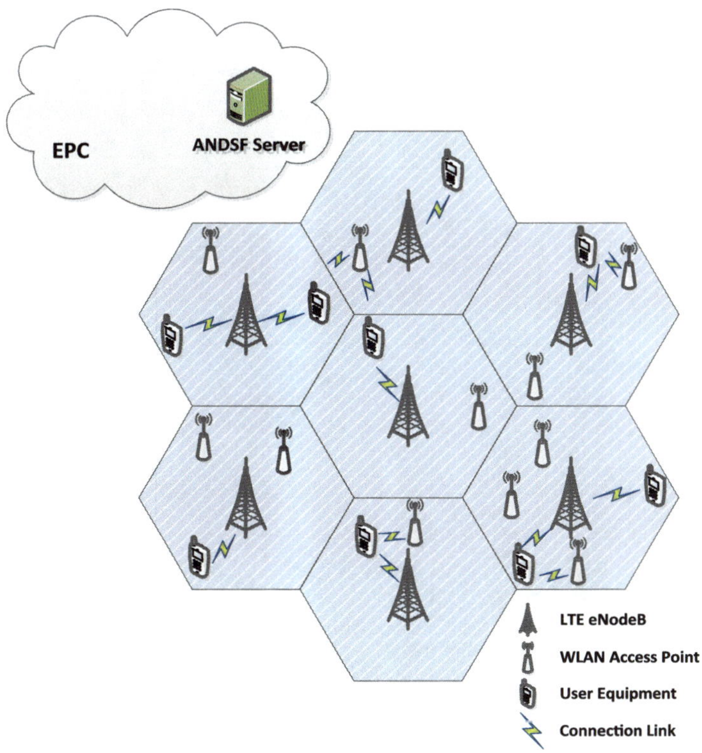

Fig. 6.7 System model

6.3.1.2 Algorithm Description

To improve the system performance in terms of energy efficiency and avoid unnecessary queries to the ANDSF and the subsequent network scanning, the proposed algorithm introduces a new UE context parameter that is taken into consideration in the decision for a network scanning initiation: the *query distance* d_q. The query distance is set as the distance between the UE and the closest WLAN AP discovered in the previous ANDSF query, as illustrated in Fig. 6.8. According to the proposed algorithm, when the UE has travelled a distance q larger than d_q since the last ANDSF query, the UE will estimate its current location and send a new query with its location information to the ANDSF server. The ANDSF will check the network context and respond to the UE's query with a list of the available networks in the UE's coverage area, if any. The UE will use this information to perform the network scanning and possibly also perform a vertical handover. If there is no WLAN AP accessible by the UE at the current location, the ANDSF will check the WLAN AP map and set a new query distance d_q to the UE. The UE will reset q and use its embedded accelerometer to estimate its travelled distance. While more complex mobility prediction algorithms may be employed to further improve

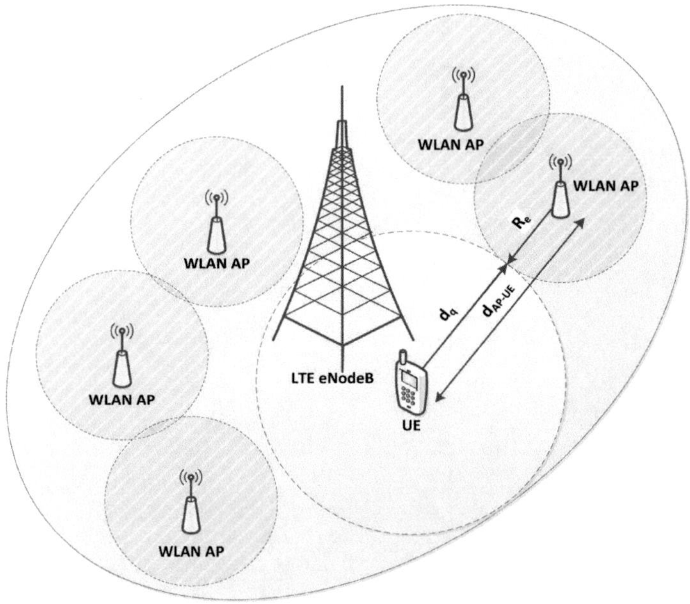

Fig. 6.8 Illustration of the query distance d_q

the performance, the algorithm proposed in this paper achieves a good balance in the tradeoff between algorithm complexity and effectiveness, as shown in the next subsection, and is completely compliant with the 3GPP specifications. The pseudocode of the proposed algorithm is shown in Fig. 6.9.

6.3.1.3 Performance Evaluation

To evaluate the performance of the proposed network discovery algorithm, a simulation model was constructed in C++. The performance of the proposed algorithm was compared against an algorithm that performs periodic network scanning with a fixed period ranging between 1 and 6 s, without taking into consideration the user or network context information. In the simulation scenario considered, the total energy consumption per user, the average network detection delay and the average network detection rate were measured versus the WLAN AP density, i.e. the number of available WLAN APs per km^2, and the user speed. The simulation parameters used are summarized in Table 6.1.

Figure 6.10a depicts the total energy consumption per user versus an increasing number of freely accessible APs per km^2, ranging from very low to increased AP density [16]. All users are travelling at 30 km/h speed. As it can be seen, the total energy consumption in the systems, which perform periodic network scanning, increases with the increase in the network scanning period. This is resulting from

ANDSF-Assisted Network Discovery Algorithm:

d • Travelled distance since the last ANDSF query;
d_q• Query distance;

IF $d > d_q$
Query ANDSF;
IF available WLANs within the UE coverage exist
Perform network scanning;
Perform vertical handover;
ELSE
d_q• Distance to the closest AP estimated by the ANDSF, based on the UE location and APmap;
d • Null;
END
ELSE
Update d;
END

Fig. 6.9 Pseudocode of the ANDSF-assisted network discovery algorithm

Table 6.1 Network discovery algorithm simulation model parameters

Parameter	Value
Maximum LTE power in the uplink	250 mW
LTE cell radius	1 km
Path loss model in the LTE network	$PL = 128.1 + 37.6 \cdot \log\ d$, d in km [14]
LTE uplink power control	Open loop with fractional path-loss compensation
Maximum WLAN power in the uplink	100 mW
WLAN cell radius	50 m
Energy consumption per WLAN scan	11 mW [5]
Operational power	100 mW
Mobility model	Random walk model [15]
Number of LTE cells	7
Number of users	500
Simulation time	500 s

the fact that a reduction in the network scanning frequency leads to greater opportunities to discover WLAN network access points that can serve as a target for handover and reduce the transmission energy consumption. This figure also depicts the ANDSF-assisted network discovery algorithm that outperforms in terms of energy consumption all systems that perform periodic scanning. This result is based on the fact that the use of context information for the network discovery allows the initiation of the network discovery procedure only when the user has travelled a distance which is long enough to increase the possibility of detecting an available

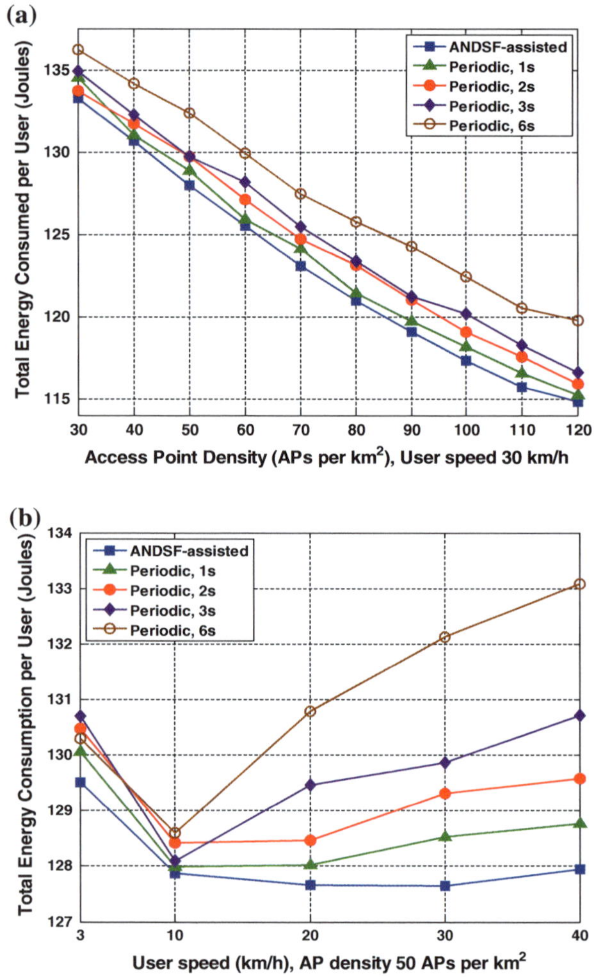

Fig. 6.10 Total energy consumption per user versus **a** the AP density and **b** the user speed

network. Thus, on the one hand, the user avoids unnecessary energy consuming network scanning when there are no WLANs in his/her proximity, while, on the other hand, the UE initiates network scanning frequently enough to allow quick discovery of available networks. The total energy consumed per user is relatively high for all schemes because, according to the simulation model, all the users are assumed to continuously transmit data during the entire simulation time. Similarly, Fig. 6.10b depicts the total energy consumption per user versus increasing user speed that ranges from pedestrian to vehicular. The AP density is 50 APs/km^2. The system that employs the proposed network discovery algorithm outperforms the systems that perform periodic scanning in terms of total energy consumption. This is a result of the fact that the use of both the user and the network context information for the network discovery allows the initiation of the network discovery

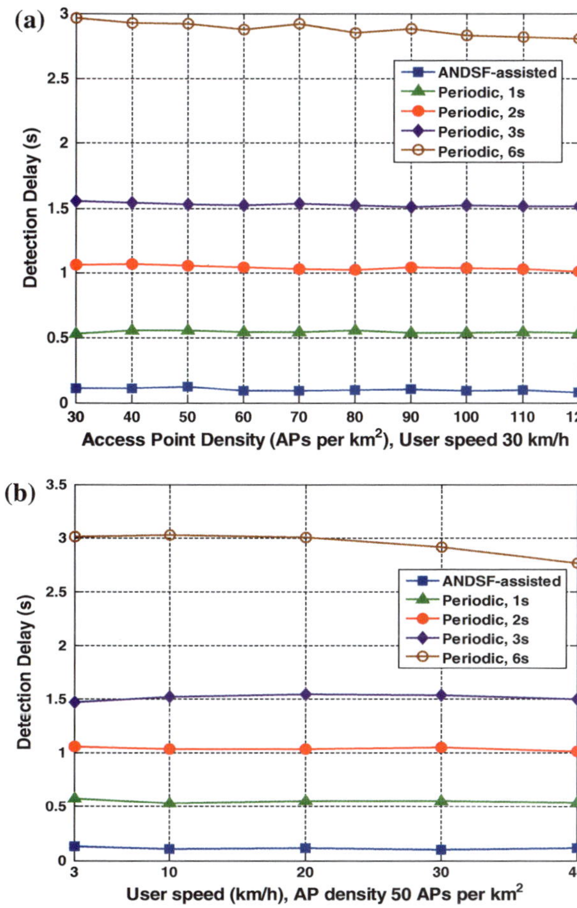

Fig. 6.11 Average network detection delay per user versus **a** the AP density and **b** the user speed

procedure once it is considered necessary, based on the travelled distance by the user. It is interesting to see that the total energy consumption of all schemes first decreases, as the user speed increases since mobility brings the users the increased possibility to access a WLAN. With the further increase in user speed, the duration a user can stay in a WLAN is reduced, resulting in high transmission energy consumption.

The average network detection delay, which is defined as the interval between the time a user first enters the coverage area of a WLAN and the time that the user discovers this network through network scanning, versus the AP density and the user speed is depicted in Fig. 6.11a, b, respectively. As it can be seen, the system that employs the proposed network discovery algorithm takes advantage of the user location and the available networks, to achieve significant reduction in the network detection delay compared to all the systems that perform network scanning periodically. Note that the fading channel is not considered here, and thus, the detection

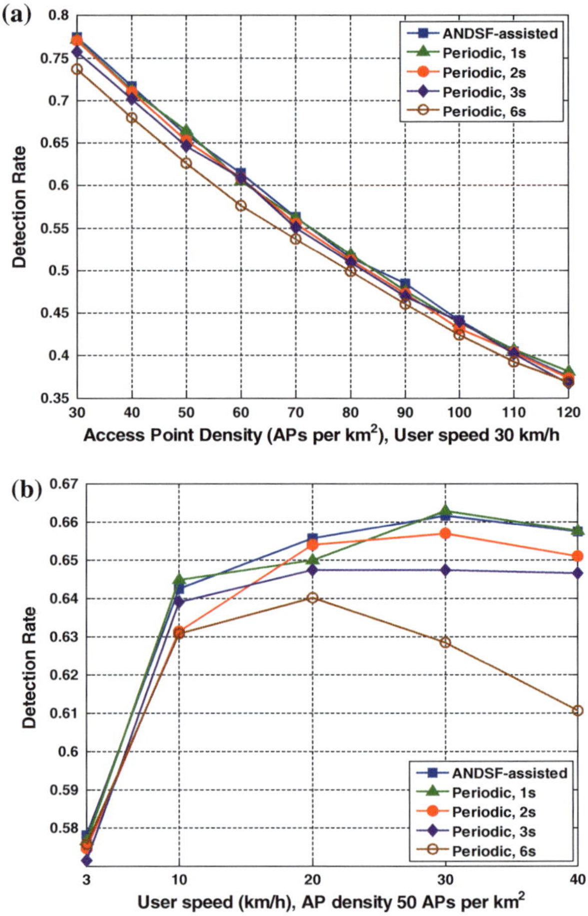

Fig. 6.12 Average network detection rate per user versus **a** the AP density and **b** the user speed

delay is evaluated based on the ideal coverage area of a WLAN hotspot. The detection delay in a real-world environment may be increased depending on the real WLAN coverage area.

Figure 6.12 depicts the average network detection rate, defined as the ratio of the number of the networks a user has successfully detected over the number of networks the user has entered the coverage area of. The system that employs the proposed network discovery algorithm has a similar, and sometimes slightly improved, detection rate compared to the systems that perform periodic scanning. Thus, it can be seen that the proposed algorithm can guarantee improvement in the total energy consumption and the network detection delay with no loss in the network detection rate.

6.3.2 Terminal Context

Energy is a critical resource in battery-powered mobile devices. The problem of trimming down the *energy consumption of mobile devices can be addressed by acting on different parts of the protocol* stack, from physical to application layer. In this book, we look at cooperative communication as one potential solution. The node discovery process is the initial process that a node has to perform in order to start communicating with its neighbours. The knowledge on the available neighbours is important for Medium Access Control (MAC) protocols, as well as for the routing, cooperation initiation, mobility control, and topology control procedures. Node discovery, from an energy point of view, should be as fast as possible, as nodes spend a considerable amount of energy gathering information about their neighbours. The energy consumption of every node during the discovery process depends on a number of activities, e.g. the amount of time spent in sensing beacons (Packets) or beacons transmitted to search for neighbouring nodes in the range, the balanced active and sleep mode for the node discovery process, etc. The primary goal of an energy efficient node discovery process is to maximize the probability of node discovery in a specified time.

Most of the node discovery protocols and mechanisms are based on the conventional scanning which is expensive in terms of energy due to spending long periods in the "listening" state, and performing broadcast communication. To conserve energy, MAC protocols with proper sleep/wake duty cycles have been proposed for wireless networks, providing a balance in sleep and wake time of mobile terminals. Although suitable duty cycles contribute to energy conservation, keeping a node in sleep mode weighs down a network or a part of network. Furthermore, for a solitary node to discover another node or network, even an efficient duty cycle cannot contribute effectively because the node does not know the suitable point in time when to turn on its interface to find a cooperative node. On the other hand, sometimes a node is in the vicinity of a cluster or a node but lacks the knowledge about the cluster's or node's attributes to cooperate with. Keeping in mind these problems, we propose a node discovery algorithm, to tackle both the node discovery and the cooperative cluster formation in an energy efficient way.

6.3.2.1 System Model

The system model is shown in Fig. 6.13 and involves Mobile Terminals (MTs) that can communicate through a long range network, e.g., Worldwide Interoperability for Microwave Access (WiMAX), LTE, etc., and a short range network, e.g., UWB, Bluetooth, etc. Without loss of generality, we consider WiMAX and UWB, but the approach has general validity. The model covers cluster discovery and cooperation instantiation for a mobile terminal that is searching for a network or device. The cluster is based on the centralized approach, where the cluster-head is the

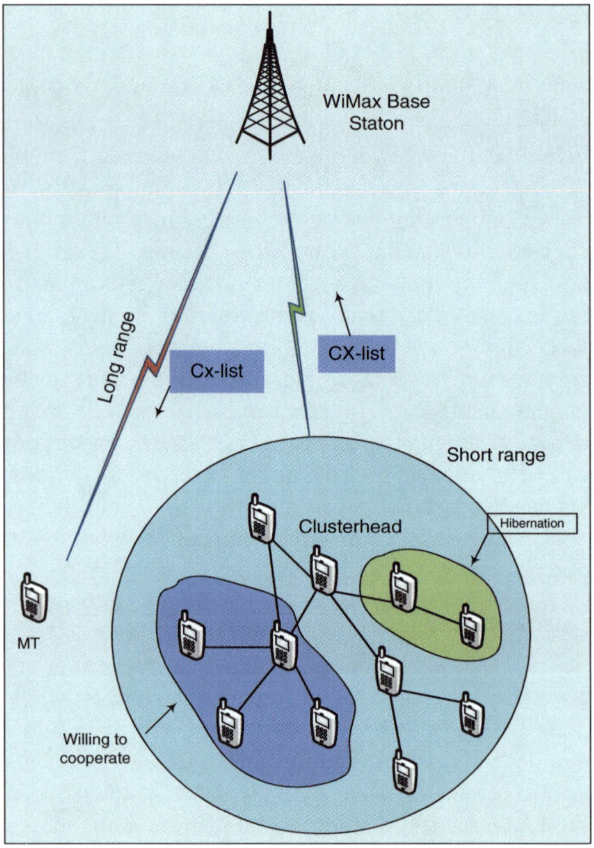

Fig. 6.13 System model for node discovery

centralized coordinating entity. Our approach uses context information about the cluster, e.g. location of cluster, battery levels of member devices, and willingness to cooperate.

Furthermore, inside the cluster we define two groups of member devices: (i) group consisting of devices that are considering switching to the hibernation state to save energy; since once a device is in hibernation mode it should not be considered as candidate for cooperation, and (ii) a group consisting of the devices in active mode and the devices that are interested in cooperating with the external devices, depending on their battery levels and willingness to cooperate. As shown in Fig. 6.13, a cluster connected with the Base Station (BS), and the clusterhead of each cluster (in case multiple clusters exist) sends its cluster information to the BS so that, when a device looks for short range connectivity, the BS can provide information about the nearby cooperative clusters. The information is then utilized by the mobile terminal to connect to the devices in an energy efficient way.

For the exchange of the Cx-list inside the cluster, and with the BS we consider the three following use-cases:

Case 1 In the first case, the Cx-list is extracted periodically only from the clusterhead that does not update the information except when some serious topology change is triggered.

Case 2 In the second case, the clusterhead collects the Cx-list and periodically or on demand, sends it to the BS as well as to other members.

Case 3 In the third case, all member devices in the cluster pack the information into a Cx-list that is sent to the clusterhead and also to the member devices in an ad hoc fashion. In this case, the cost of sending the Cx-list is shared among all devices. A Cx-list exchanged between all devices allows them to know the characteristics of each other and also information about external devices that want to be part of this cluster.

6.3.2.2 Algorithm Description

Both the discovering MT and the clusterhead are connected to the long-range BS (WiMAX in our case). Once the battery level (BL) of the MT reaches a defined low threshold (BLth), it triggers the BS for a Cx-list of nearby co-located short range networks. The Cx-list is the list of context information that the clusterhead sends to the BS for future connectivity with nearby devices. The Cx-list contains useful information, e.g., members of the cluster willing to communicate, Battery levels, current transmission power, Cluster ID, etc. The Cx-list constructed by the clusterhead floats via the long-range network to the discovering MT device. The Cx-list facilitates the searching MT to know important characteristics of the clusterhead or its members, e.g., battery level, channel information, etc. The context information is used for three different goals:

- Creating a list of possible co-located MTs that are willing to cooperate.
- Facilitating the communication. The exchanging of reserved channels, location etc., is used to direct the scanning, without having to check the presence of MTs by scanning all available channels.
- Setting up a rendezvous with other MTs. The timing information can be used to know in advance when the superframe starts. The timing information will be used to shorten the active mode (transmission and reception) phase, starting just a little before the probable start of the superframe, to take into account the delay in receiving this synchronization information from the long-range network.

The flowchart of the proposed node discovery algorithm is shown in Fig. 6.14.

6.3.2.3 Performance Evaluation

The proposed node discovery mechanism is simulated using OMNeT++4.0 [17], an open source simulator. We installed the Numbat WiMAX/Mobile IPv6 framework for OMNeT++ [18] for long range simulation results. For energy consumption

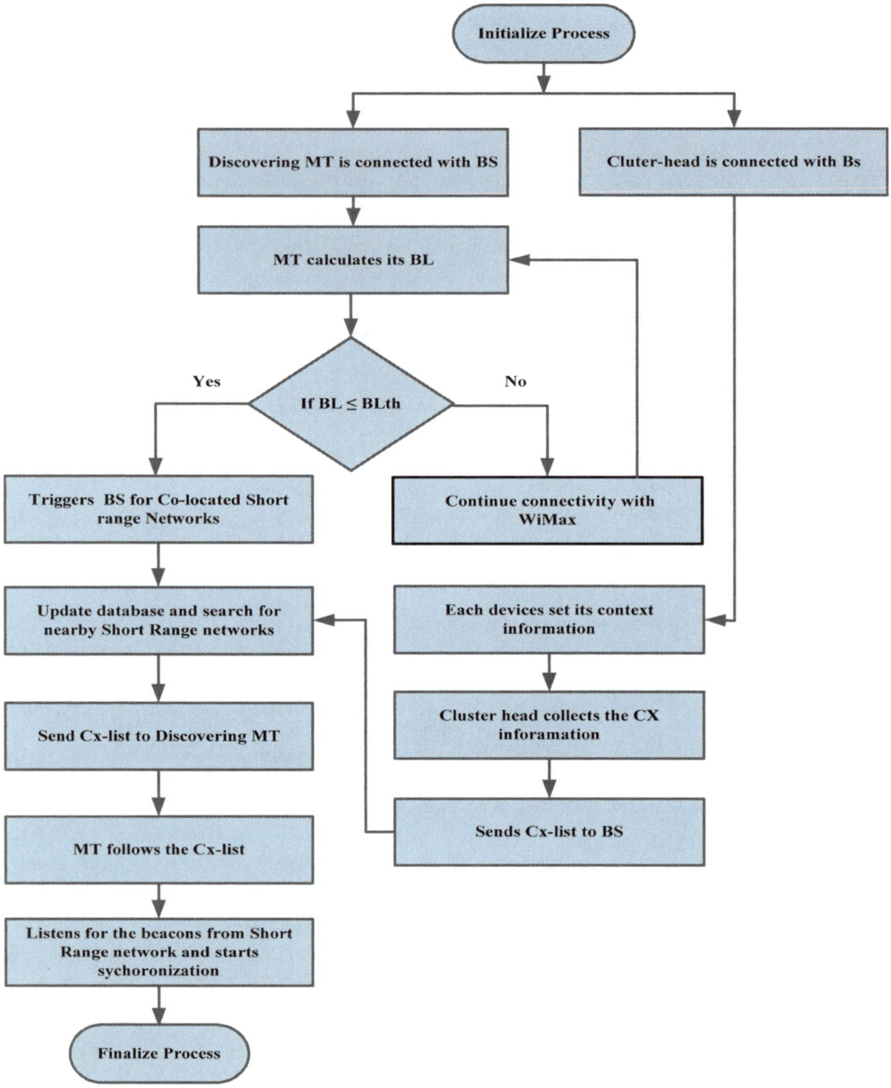

Fig. 6.14 Flowchart of the proposed node discovery algorithm

calculation, we used the Energy Framework of OMNeT++, which is a collection of modules that allow flexible and extensible modeling of battery consumption of wireless devices [19]. From the existing modules available in OMNeT++, we used the *SimpleBattery* module, which is a simple linear battery model that provides a common interface to all battery using devices. The battery status of each device is recorded as a time series with the events and results being published in vectors.

Table 6.2 Node discovery algorithm simulation model parameters

Parameter	Value
Simulation area	500 m × 500 m
Short range technology	UWB
Long range technology	WiMAX
Battery module	Omnet++ 4.0 Simple battery
Nodes in the cluster	10
Mobility Models	Constant speed Linear mobility Rectangular (Defined areas)
Node speed	2 m/s
Node update interval	1 s
Number of nodes	5
Cx-list size	32, 64, 128, 256, 512 bytes
Initial node battery	7 mAh
Voltage	3.3 V
Maximum transmission power (UWB)	1 Mw
Backoff time	0.0003 s
SIFS	0.00019 s
Time from RX to TX mode	0.00018 s
Time from RX to Sleep mode	0.000031 s
Time TX to RX mode	0.00012 s
Time TX to Sleep mode	0.000032 s
Time Sleep to RX mode	0.000103 s
Time Sleep to TX mode	0.000203 s

The simulation scenario consists of a cluster of 10 nodes and a searching node. The searching node moves with different motilities and tries to find the cluster by using both context based discovery and conventional scanning. The parameters of the simulation setup are listed in Table 6.2.

In the proposed scenario, if MT-A does not find any cooperative node or cluster, and continues its connection with the long range technology, its battery will be drained out very swiftly. Figures 6.15 and 6.16 verify the need for short range technologies for energy saving and also validate our energy saving mechanism by using WiMAX as the long range technology and UWB for short range in the node discovery process. We have generated a series of traffic with different data payload sizes and checked their energy consumption in the UWB and WiMAX networks. Figure 6.15 shows that the long range communication is expensive in terms of energy consumption. The energy consumption of the MT also depends on the data rates, payload size and the distance between the communicating devices. Since

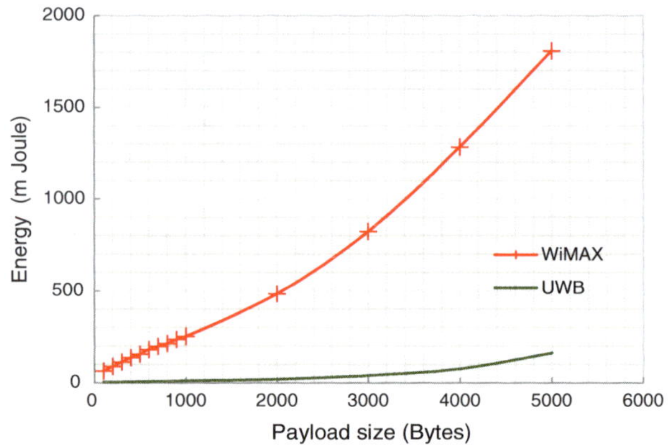

Fig. 6.15 Energy consumption of different payload sizes over UWB and WiMAX

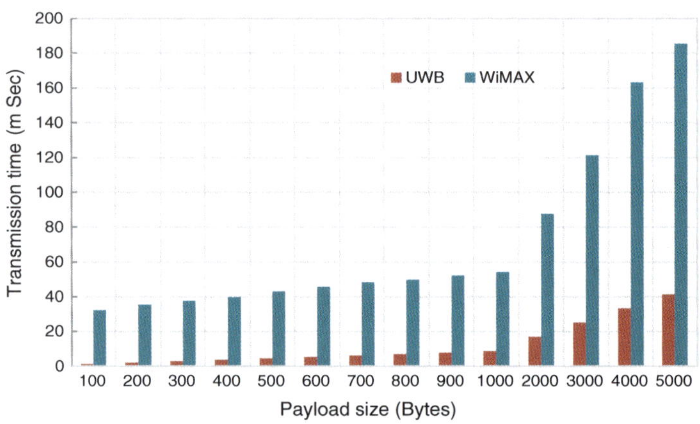

Fig. 6.16 Transmission time UWB and WiMAX with data payload size

UWB offers high data rates, the transmission time is lower compared to WiMAX and the result in Fig. 6.16 shows that the transmission time is proportional to the data payload size.

Turning on the short range interface at the exact location once the discovery node is in the coverage area of the short range cooperative cluster is very important, as the node has to consider the communication delay of the context list. We performed a series of simulations to check the transmission time when the beacon packets success rate is zero and when its success rate is almost 100 % during the discovery process. We can see in Fig. 6.17 that the beacon packets success rate is 0 when the node is not in the coverage area, but once it starts to reach the boundaries of the cluster the success rate goes increases and reaches its peak value in 5 s. For optimal connectivity, the discovering node has to switch on its short range interface

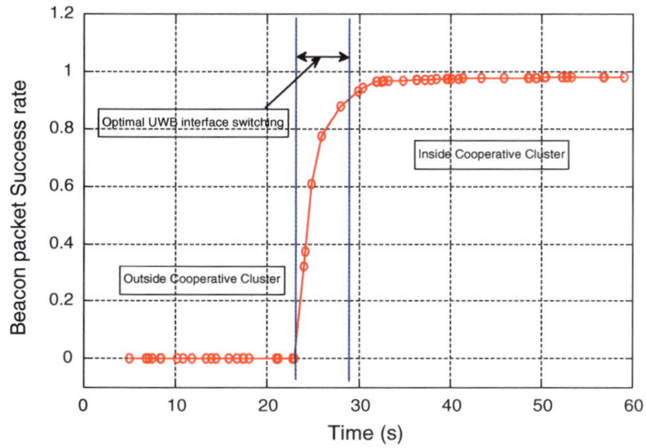

Fig. 6.17 Beacon packet success rate

and receive the fresh context list in 5 s. The coverage area of the UWB in this case is exactly 10 m.

In order to facilitate the searching to discover a cooperative node or cluster, the clusterhead of each cluster gathers the information into a context list about all the member nodes and sends it to the WiMAX BS. The context information flows over the WiMAX connection and is received by the device. Figure 6.18 shows the energy consumption of a context list with different sizes over UWB and WiMAX. The size of the Cx-list ranges from 32 to 512 bytes in order to check the energy consumption variation and to validate our mechanism. The results show that the energy consumption for different sizes of the Cx-list of WiMAX is higher than UWB, but negligible compared to the overall energy gain of the mechanism.

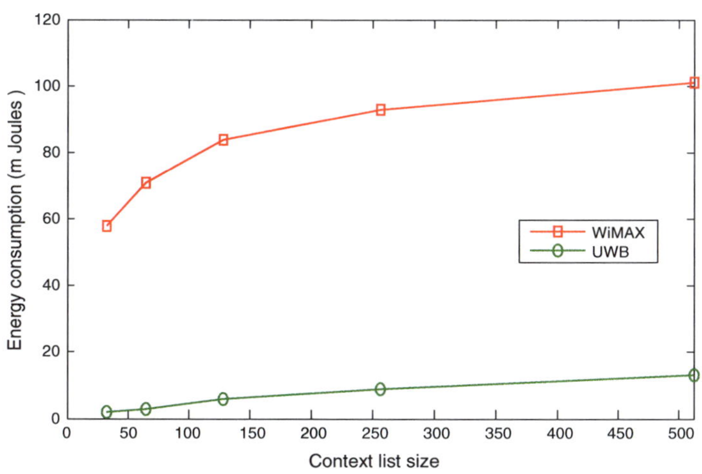

Fig. 6.18 Energy consumption comparisons of Cx-list via WiMAX and UWB

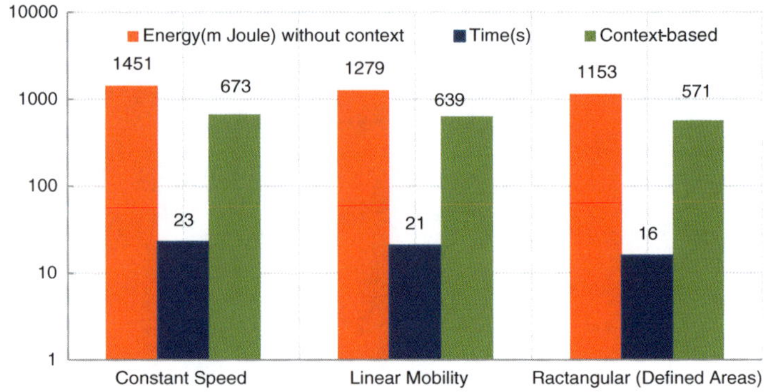

Fig. 6.19 Energy cost of node discovery and time to discover node with and without context information at different mobility models

The time to discover the cluster varies with the different mobility models. This is the time considered in the context based discovery to keep the short range interface off and save energy. The energy consumption of the node discovery with different mobility models is shown in Fig. 6.19. We have used Constant speed, Linear and Rectangular (defined areas) mobility models. In all simulations, the node transmission power has been kept constant to 1 mW and the speed to 2 m/s (a suitable speed for a pedestrian mobile user). Figure 6.20 shows the energy consumption of the node when it scans blindly without any context information to find a cooperative cluster and with context based scanning. A number of experiments have been performed to record the average time when the node is in the range of short range technology. Despite the reasonable duty cycle of sleep and active mode of the MAC, the energy consumption of the node in the discovery process is high. Figure 6.20 shows that the context based discovery process saves a considerable amount of energy.

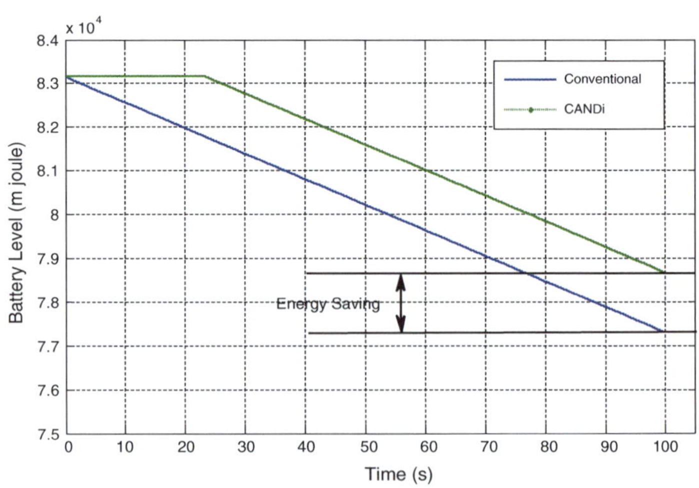

Fig. 6.20 Energy consumption of scanning process

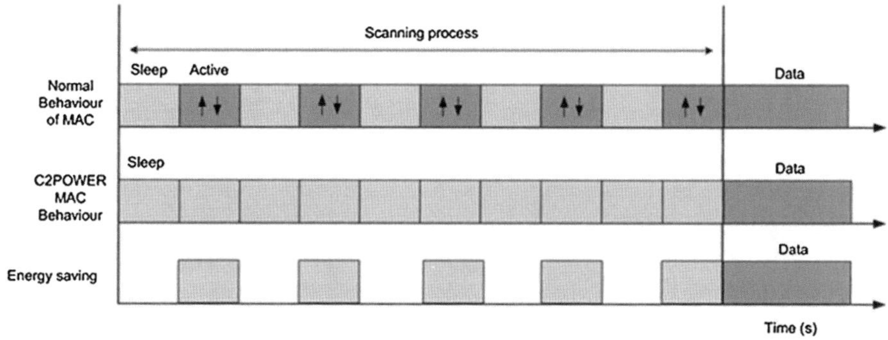

Fig. 6.21 Energy consumption of scanning of proposed algorithm versus conventional periodical scanning

With the conventional scanning process, the MAC layer periodically switches between the sleep and the active modes. Active mode means both sending and listening states, while the sleep mode considers both the idle and sleep modes. In the absence of prior knowledge of nearby nodes or clusters, the node scanning process is quite expensive in terms of energy. On the other hand, energy efficient 5G-enabled nodes have to use context based scanning to switch to the active state and start communicating once they reach the coverage area of the cluster. As shown in Fig. 6.21, the terminals save 50 % energy in the short-range discovery process compared to conventional discovery process based on blind scanning.

6.4 Conclusion

Achieving context awareness is challenging and can come at a cost in terms of complexity. This chapter provides an insight into smart communications, and details how we can attain, process and exploit information towards energy saving. In the first instance we identify the type of context and their categorization in order to provide a footprint towards the design of the context-aware architecture. The architecture then is involved to define context-aware functional blocks that are present both on the network and mobile side, that are responsible for keeping up–to-date knowledge on the dynamics of the environment, and to make this available for energy efficient protocols/algorithms. Furthermore, this chapter provides two approaches that are pivotal towards reducing energy consumption in mobile handsets, specifically towards scenarios that involve vertical handover events, and cooperation; aligned with the two key scenarios envisioned in Chap. 1. In line with this vision, the chapter details the design and validation of energy efficient network and node discovery mechanisms based on using the context awareness support. In both cases, the knowledge of the available context in terms of neighboring networks and node availability will effectively influence when to switch on/off the radio interface on the user handsets in order to save energy; which in contrast to today's technology, node and network scanning is "blind" and "continuous" representing a key consumer of energy in legacy mobile handset.

References

1. Mitola, J.: Cognitive radio for flexible mobile multimedia communications. Mobile Netw. Appl. **6**(5), 435–441 (2001)
2. IST-2004 027965 ORACLE project, opportunistic radio communications in unlicensed environments
3. ICT-248577 C2POWER consortium, deliverable D3.4, Context aware modules specification and implementation—final version", Apr (2012)
4. ICT-248577 C2POWER consortium, deliverable D3.1, identification of context parameters, Oct (2010)
5. Doppler, K., Ribeiro C.B., Kneckt, J.: On efficient discovery of next generation local area networks. In: Proceedings of *IEEE WCNC*, pp. 269–274, (2011)
6. Kim, K.H., Min, A.W., Gupta, D., Mohapatra, P., Singh, J.P.: Improving energy efficiency of Wi-Fi sensing on smartphones. In: Proceedings of *IEEE INFOCOM* pp. 2930–2938 (2011)
7. Choi, Y., Choi, S.: Service charge and energy-aware vertical handoff in integrated IEEE 802.16e/802.11 networks. In: Proceedings of *IEEE INFOCOM*, pp. 589–597 (2007)
8. 3GPP TS 24.302, v11.0.0 (2011-09): Access to the 3GPP evolved packet core (EPC) via non 3GPP access networks; Stage 3; (Release 11)
9. IEEE standard for information technology—telecommunications and information exchange between systems—local and metropolitan area networks—specific requirements Part 11: Wireless LAN medium access control (MAC) and physical layer (phy) specifications, IEEE Std. 802.11-2007, June (2007)
10. IEEE standard for local and metropolitan area networks part 16: air interface for broadband wireless access systems, IEEE Std. 802.16-2009, May (2009)
11. 3GPP TS 36.912, v10.0.0 (2011-03): Feasibility study for further advancements for E-UTRA (LTE-Advanced); (Release 10)
12. Assisted GPS: a low-infrastructure approach. GPS World. March 1, 2002. http://www.gpsworld.com/gps/assisted-gps-a-low-infrastructure-approach-734
13. Kjasrgaard, M.B.: Location-based services on mobile phones: minimizing power consumption. IEEE Pervasive Comput. **11**(1), 67–73 (2012)
14. Cai, Y., Lu, S., Zhang, L., Wang, C., Skov, P., He, Z., Niu, K.: Reduced feedback schemes for LTE MBMS. In: Proceedings of *IEEE VTC-Spring*, pp. 1–5 (2009)
15. Hong, H., Gerla, M., Pei, G., Chiang, C.C.: A group mobility model for ad hoc wireless networks. In: Proceedings of ACM MSWiM, pp. 53–60, (1999)
16. Schwartz, R., Johansson, M.: Carrier WiFi offload, building a business case for carrier WiFi offload. Wireless 20/20, Mar 2012. http://www.wireless2020.com/docs/CarrierWiFiOffload WhitePaper03202012.pdf
17. http://www.omnetpp.org/
18. http://klub.com.pl/numbat/
19. http://www.omnetpp.org/component/content/article/9-software/3647-energyframework

Chapter 7
Smart Phones for 5G Network

Du Yang and Jonathan Rodriguez

Abstract This chapter reinforces the notion of smart phones based on the context platform previously proposed in Chap. 6, playing a major role in next generation networks. The future is heading towards location based services and today nearly every new phone contains a GPS chip. Furthermore, the penetration of the so called smart phone is very high, with almost all subscribers on a data plan. This enables many new location-based applications that are very flat in their operational structure. We explore new opportunities for context information that not only supports ubiquitous mobile network access, but beyond that allows efficient use of radio access technology. Based on the availability of geo-aided positioning made available through the context platform, this chapter explores how we can use positioning context for enhancing network performance in terms of novel approaches such as community-based sequential paging for LTE-A cellular network, and location aided scheduling for fractional frequency reused LTE-A relay network.

7.1 Introduction

In previous chapters, we have focused on reducing energy consumption at the end-user side by exploiting several features provided by smart phones including multi-RAT, cooperative, cognitive sensing, and context-aware information. In this chapter, we will shift our attention to the network side, and explore the promising improvements of smart phones deployment on network performances.

We consider smart phone as a mobile device that supports multi-RAT, has various sensors embedded, and is capable of learning its environment and of pro-

---

D. Yang (✉)
Instituto de Telecomunicações, Aveiro, Portugal
e-mail: duyang@av.it.pt

J. Rodriguez
Instituto de Telecomunicações, Campus Universitário de Santiago,
3810-193 Aveiro, Portugal

A. Radwan and J. Rodriguez (eds.), *Energy Efficient Smart Phones for 5G Networks*, Signals and Communication Technology,
DOI 10.1007/978-3-319-10314-3_7

viding rich user-centric context-information. All the above-mentioned features could be exploited in communication network design and management. As an initial study, we will focus on utilizing location information, which can be obtained from the GPS chip embedded in almost every smart phone. A cellular network using Long-Term-Evolution Advanced (LTE-A) technology is considered in this chapter. Two location-based methods are separately proposed in Sects. 7.2 and 7.3. One of them aims to reduce the paging signaling overhead, and the other aims to improve the user satisfaction through advanced scheduling. The proposed methods are demonstrated with sufficient simulation results. Conclusion, future directions and challenges are summarized in Sect. 7.4.

7.2 Community-Based Sequential Paging for LTE-A Cellular Network

Mobility management plays an important role in cellular networks in effectively delivering services to Mobile Terminals (MTs). It involves two basic operations: location update and paging. When a MT moves from one area to another, the MT reports this new area's unique id to the network through a location update procedure. The network keeps a tracking record for every user. When an incoming call to an MT arrives, the network identifies the current serving cell of that MT within the last reported area through a paging procedure.

Activating the location update at the cell-level will cause a significant signaling overhead, as the cell size of today's networks is becoming smaller and smaller. In Long-Term-Evolution Advanced (LTE-A) networks, cells are grouped into Tracking Areas (TA), and TAs are further grouped into user-dependent Tracking Area List (TAL). Location update is activated when a MT enters a new TA, which is not included in its current TAL [1]. However, this inevitably increases paging traffic since paging messages are sent by every cell in a TA. Various paging schemes are proposed to reduce the paging messages overhead. One of them is called sequential paging [2, 3], whose basic idea is to first estimate the probability of the MT to reside in those cells, and then sequentially activate paging procedure, starting from cells having high probability to those having lower probability, until the MT is found. The accuracy of the location estimation is the key to success for any sequential paging scheme.

Most of the sequential paging schemes proposed in the literature exploit a single user's location history. For example, the sequential paging proposed in [2] records the id of the last interacted cell, which refers to the cell where a certain MT was paged, made calls or performed location update. Upon call arrivals, paging signaling is sent firstly to that last interacted cell, and then to the surrounding cells. Another example is the Mobility-Pattern based Scheme (MPBS) proposed in [4], which exploits the daily movement pattern of a single user (e.g. home-office-home). Here, upon call arrivals, paging signaling is sent first to those cells where the MT in question is mostly likely to reside according to his/her regular life pattern. This scheme becomes less effective if the users do not follow their daily movement pattern during holidays.

In this Section, we are interested in how to reduce the paging messages for the given LTE-A location update scheme. We exploit a group of users' location tracking history to identify the movement correlation between them, and design a community-based sequential paging scheme. Human movement is strongly affected by the need for humans to socialize and/or cooperate with each other, in one form or another, resulting in some degree of clustering [5] and correlation in their movements. For instance, a person may go to different leisure places on Sundays, which don't follow any regular pattern, and thus cannot be captured by the MPBS method. However, that person probably goes to these places with the same fellow community such as family or close friends. It was demonstrated in [6] that by using additional traces from another user with high correlation, the accuracy of the location estimation improves at least by one order of magnitude with respect to the case of using only his/her own mobility data. In our proposed algorithm, upon call arrivals, paging signaling is sent first to the MT's last interacted cell, and also to the cells where the other users in his highly correlated community are likely to be. We compare our new scheme against the scheme exploiting only the knowledge of the last interacted cell. Our simulation results demonstrate that the proposed community-based paging algorithm is capable of reducing paging messages by 20 % in high mobility scenario.

7.2.1 Preliminary: LTE-A Mobility Management

In this section, we briefly introduce the current mobility management methods deployed in LTE-A cellular system including (1) the organization of cells; (2) location update schemes; and (3) conventional paging schemes.

7.2.1.1 Location Organization

Figure 7.1 shows the mobility management architecture of LTE-A. The Mobility Management Entity (MME) is connected to a group of evolved Node-Bs (eNodeBs) and is responsible for MT tracking and paging procedures. The radio coverage area of an eNodeB (or an eNodeB's sector) forms a cell. Every cell has a unique cell identifier. The cells are grouped into TAs, which have a unique TA Identity (TAI), e.g. TA_1 contains $CELL_1$ and $CELL_2$ in Fig. 7.1. Cells and TAs do not overlap with each other. Several TAs are further grouped into a Tracking Area List (TAL), e.g. TAL_1 contains TA_1 and TA_2. TALs are allowed to overlap with each other. TAL is assigned on a per-user basis, which means TALs for different UEs could have different size. The central TA of a TAL is where the MT currently resides. The knowledge of TAL is synchronized at both MME and MT.

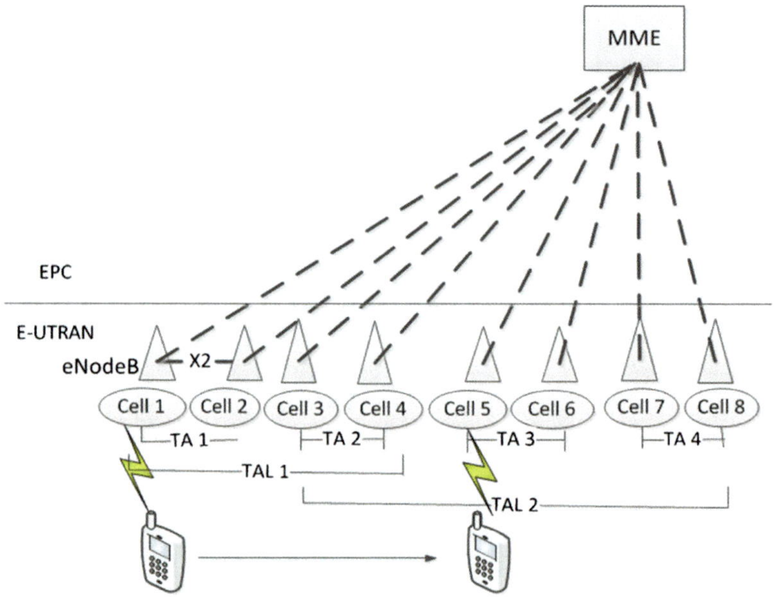

Fig. 7.1 Mobility management architecture of LTE-A

7.2.1.2 Location Update

When an MT does not have packets to be received and/or transmitted for an extended period of time, the eNodeB removes that MT's location information from its database. The MT is then in the power-conservation state. In this state, the location of the MT is only known at the MME, which records its last interacted cell, the corresponding TA and its current assigned TAL. Supposing that the MT Id is n, the recorded location information at the MME can be represented as:

$$MT_n\colon \{CELL(n), TA(n), TAL(n)\} \tag{7.1}$$

During the idle mode, the MT wakes up periodically to listen to system information on Down-Link (DL) broadcast channel. The mobility is fully controlled by the MT, since the network is not aware of the MT existence continuously. The MT must perform signal quality measurements with respect to the serving and neighboring eNodeBs. Based on the measured signal quality, the MT selects a new serving eNodeB when it moves away from the current one. The MT does not always report the change of serving eNB. Only when the TAI advertised by this new serving eNodeB is not included in the MT's TAL, the MT will perform a location update to the MME indicating its present location. For example, as shown in Fig. 7.1 MT_1 is initially located in $CELL_1$ TA_1 and its TAL containing TA_1 and TA_2. This MT received an incoming call when it was in $CELL_1$. After this call ended, MT_1 turned into power saving mode, the eNodeB of $CELL_1$ removed its information, and the MME keeps the following location record:

$$MT_1 : \{CELL_1, TA_1, \{TA_1, TA_2\}\} \tag{7.2}$$

Then, this MT started to move. On its way towards Cell 5, it received TAI information broadcast by the eNodeB and compared the received TAI with its TAL list. There is no location update until the MT moved into $CELL_5$, which is not included in its TAL. After the location update, the recorded information at the MME becomes

$$MT_1 : \{CELL_5, TA_3, \{TA_2, TA_3, TA_4\}\} \tag{7.3}$$

7.2.1.3 Conventional Paging Schemes

When an incoming call to the MT occurs, the MT can be found by sending paging messages over all cells in this MT's TAL if the above-mentioned location updates perform successfully. However, this so-called blanket paging scheme causes heavy paging traffic because the number of cells in a TAL is relatively large. To reduce the paging traffic, a Cell-TA-TAL (CTT) sequential paging method is proposed in [1], which is detailed as follows;

- CTT (Cell-TA-TAL) Scheme: when an incoming call arrives, the MME first asks the last interacted cell to page the MT. If it fails, the other cells in the last interacted TA page the MT. If it fails again, the eNodeBs of the remaining TAs in the TAL page the MT. If no positive reply received, this MT will be considered as unreachable.

This scheme assumes that the MT has a higher probability of remaining in the last interacted cell or TA. This is a reasonable assumption for high Call-to-Mobility Ratio (CMR) scenario, which means the user has a relatively high call arrival rate over its moving rate. For example, the user remains in his/her office for hours, and has incoming/outgoing calls roughly 20 min. In the case of low CMR scenario, this scheme is not capable of reducing paging traffic anymore.

7.2.2 Movement Correlation Vector and Its Acquisition

We introduce a time-variable real number $c_{n,m}(t,\ t + \Delta t)$, $\left(0 \le c_{n,m} \le 1\right)$ to characterize the movement correlation between MT_n and MT_m. At a given time, the value of $c_{n,m}$ is equal to $c_{m,n}$. For N users, if their mutual movement correlation coefficients are higher than a threshold value (e.g. 0.7), these N users are considered to be a community, which have a high probability of moving together during time interval

(t, $t + \Delta t$). Hence, for any user n, a time-variable vector $\mathbf{c}(n)$ can be formulated characterizing its movement correlation with other users in the same community.

$$\mathbf{c}(n) = \left[\ldots, c_{n,m}, \ldots\right], \quad n, m \in [1, N], (\text{given } t = t_0) \tag{7.4}$$

This movement correlation vector can be obtained through off-line analysis. By recording the MT's movement history of a group of socially connected users for a certain period of time (e.g. 1 month), it is possible to derive movement correlation between them using data mining techniques. However, since real movement traces have been rarely made available for public use, we are not able to derive real movement correlation vectors for a real scenario. Instead, we assume in this section that movement correlation vectors are already available.

7.2.3 Community-Based Sequential Paging Scheme

In the proposed community-based paging scheme, the MME is required to record one more item, $T(n)$ for MT_n, which is the time elapsed since the last location update. We assume that the knowledge of the movement correlation vector obtained offline is already available at the network side. The databases at MME for MT_n are extended as

$$MT_n\colon \{CELL(n), TA(n), TAL(n), T(n), \mathbf{c}(n)\} \tag{7.5}$$

When an incoming call for user n arrives, the MME searches over the other (N-1) users within this MT_n's community. We introduce a notation Φ_{CELL_i} to represent a group of users. If $m \in \Phi_{CELL_i}$, then $CELL(m) = CELL_i$, $T(m) < T(n)$, which means that the last interaction between MT_m and the network occurred in $CELL_i$, and this happened after the interaction between MT_n. In other words, we only consider those users in the community having a more recent location updates than the target user n. Moreover, we only consider the cells included in user n's TAL, $CELL_i \in TAL(n)$, since user n must be located within this TAL using the above-mentioned location update scheme. $|\Phi_{CELL_i}|$ represents the cardinality of the set. The probability of user n to reside at $CELL_i$ is formulated as

$$P_{CELL_i} = \frac{\sum\limits_{m \in \Phi_{CELL_i}} c_{n,m}\left(1 - \frac{T(m)}{T(n)}\right)}{|\Phi_{CELL_i}|}, \tag{7.6}$$

$$\begin{aligned} \Phi_{CELL_i} = \{m\colon CELL(m) = CELL_i, T(m) < T(n), m \neq n\} \\ \forall CELL_i \in TAL(n) \end{aligned} \tag{7.7}$$

The community-based estimated cell is defined as the one having the highest probability, which is formulated as

$$\hat{i} = \arg\max_{i} P_{CELL_i} \tag{7.8}$$

Similarly, we could derive the community-based estimated TA as follows:

$$P_{TA_j} = \frac{\sum_{m \in \Phi_{TA_j}} c_{n,m}\left(1 - \frac{T(m)}{T(n)}\right)}{|\Phi_{TA_j}|} \tag{7.9}$$

$$\Phi_{TA_j} = \{m: TA(m) = TA_j, T(m) < T(n), m \neq n\}\, \forall TA_i \in TAL(n) \tag{7.10}$$

$$\hat{j} = \arg\max_{j} P_{TA_i} \tag{7.11}$$

where Φ_{TA_j} is a set containing users from the community who reside in TA_j, and having more recent location updates than the target user n.

The proposed **Community-based CTT Scheme** is detailed as follows:

When an incoming call for MT_n arrives, the MME

- first asks the last interacted cell *CELL*(n) and **the community-based estimated cell** $CELL_{\hat{i}}$ to page the MT;
- If it fails, the other cells in the last interacted TA *TA*(n), and other cells in **the community-based estimated TA** $TA_{\hat{j}}$ page the MT;
- If it fails again, the other TAs in the TAL *TAL*(n) page the MT.

It is possible that the community-based estimated cell is the same one as the last interacted cell. In this case, the proposed C-CTT paging scheme performs the same as the conventional CTT paging scheme.

7.2.4 Simulation Results and Discussion

We consider an 8 × 8 Km^2 area as illustrated in Fig. 7.2 which is divided into 64 cells. Every cell covers an 1 × 1 Km^2 area. A TA is a 4 × 2 Km^2 area consisting of 8 cells. A TAL is all the TAs surrounding the current TA where the MT is currently residing. For example, if MT_n currently resides in TA_1, its TAL includes 4 TAs ($\{TA_1, TA_2, TA_5, TA_6\}$) corresponding to 32 cells. While if MT_n moves into TA_2. The new TAL assigned to it includes 6 TAs ($\{TA_1, TA_2, TA_3, TA_5, TA_6, TA_7\}$) corresponding to 48 cells.

In order to simulate the movement correlation between users, we consider a single community consisting of N number of users. The total simulation time is set to 120 min with a time resolution of 1 min. The number of incoming calls during

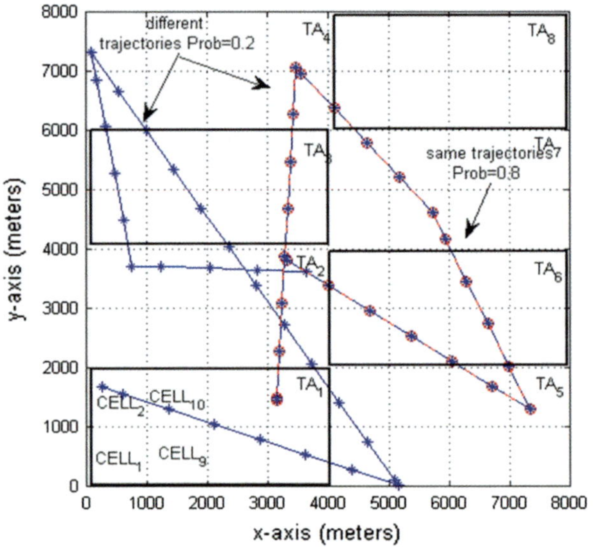

Fig. 7.2 An illustration of the simulation scenario having 64 cells, 8 TAs and the movement trajectory of two users (*Blue Star*: MT-1; *Red Circled*: MT-2)

this period obeys Poisson distribution with mean value λ_{call}. These incoming calls occur at any time during the simulation period with equal probability, and are independent between users. The mobile users move freely within this area. We employed the most commonly used random-walk mobility model to generate the MT moving trajectory. The moving speed varies from 3 Km/h (pedestrian) to 48 Km/h (vehicular). We measure only the number of paging messages used for UE_1. The movement correlation between MT_1 and other users is set equally to $c_{1,m} \in [0.1, 1], \forall m \neq 1$. To simplify our simulation, the movement correlation coefficients between other users are set to 1, meaning $c_{n,m} = 1$, $\forall n \neq 1$, $m \neq 1$, $n \neq m$. In other words, all the users except MT_1 always moves together, while UE_1 moves with the community with a probability $c_{1,m}$. This movement correction vector remains unchanged during the simulation. For example, Fig. 7.2 shows the trajectories of a community having two users. The movement correlation vector for $MT_1$is $\mathbf{c}(1) = [1, 0.8]$, and for MT_2 is $\mathbf{c}(2) = [0.8, 1]$. The blue star curve represents MT_1's trajectory, while the red circled curve represents MT-2's moving traces. These two trajectories could be the same with probability 0.8, or completely different with probability 0.2.

We first evaluate how the movement pattern (characterized by speed v and pausing time T_{pause}), and the incoming call pattern (characterized by mean value λ_{call}) influence the call-to-mobility-ratio (CMR) ρ. As illustrated in Fig. 7.3, the CMR decreases exponentially when the movement speed increases from 3 Km/h to 10 Km/h, and then slowly decreases when the movement speed increases further. As we expected, having longer pausing time increases the CMR value even at high mobility scenario. The CMR also increases when the number of incoming calls increases. Since we are more interested in the low CMR scenario, we will employ the non-stop case having $T_{pause} = 0$ and $\lambda_{call} = 10$ for the rest of the simulation.

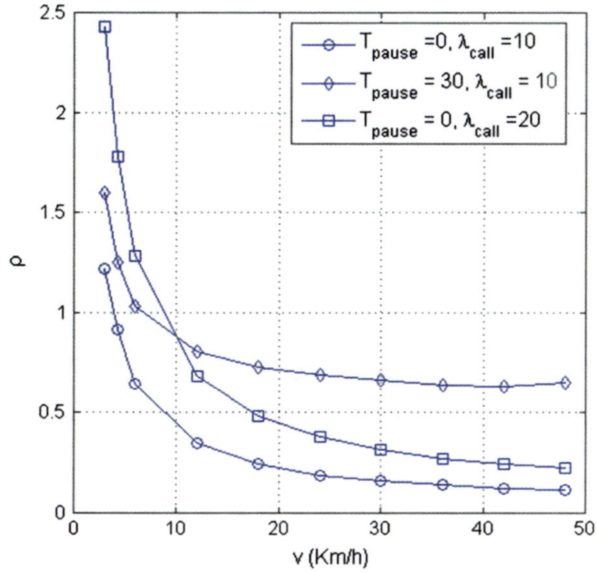

Fig. 7.3 CMR versus movement speed at different T_{pause} and λ_{call}

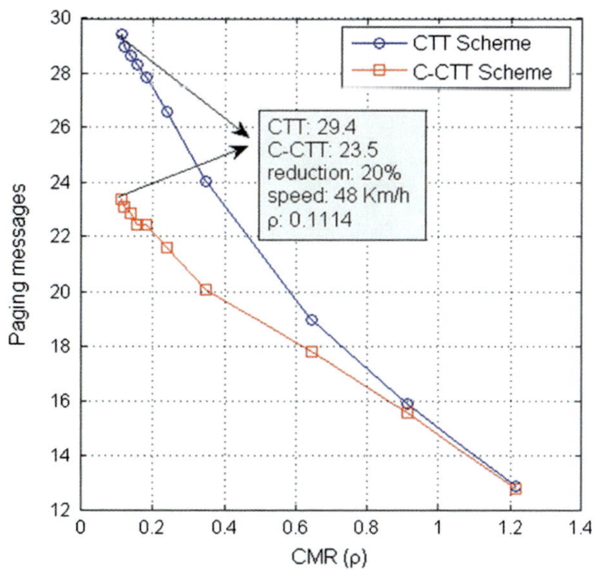

Fig. 7.4 Comparison of paging messages signaling overhead using CTT scheme and the proposed C-CTT scheme

Figure 7.4 compares the required paging messages using the conventional CTT scheme and the proposed community-based CTT scheme in the scenario of having $T_{pause} = 0$, $\lambda_{call} = 10$ and $v \in [3, 48]$ Km/h. We assumed that there are four users in the community, and our target user UE_1 always moves together with the other users. In other words, the collocation coefficient is $c_{m,n} = 1$ for all users. It demonstrates

Fig. 7.5 Comparison of paging messages signaling overhead versus community size using CTT scheme and the proposed C-CTT scheme

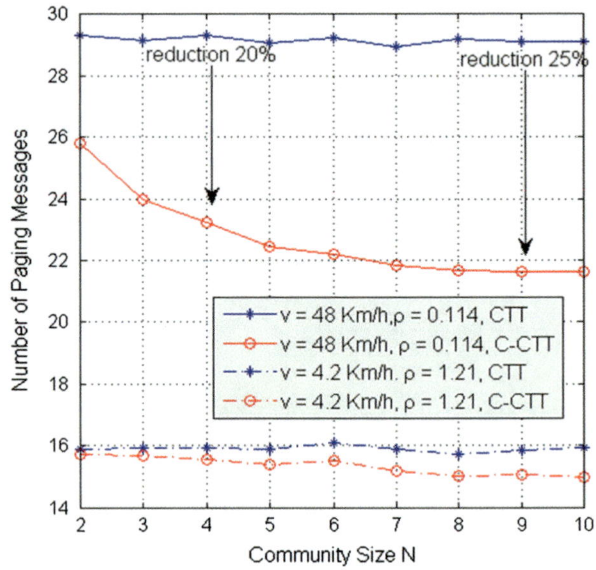

that the proposed paging algorithm is capable of significantly reducing the paging message overhead, especially at low CMR values. For example, when the CMR value $\rho = 0.114$ corresponding to $v = 48$ Km/h, the proposed algorithm is capable of saving 20 % in paging messages overhead. This gain diminishes when the CMR value reaches to $\rho = 1.2$. To sum up, the proposed algorithm is capable of significantly reducing the paging messages signaling overhead, especially in low CMR scenario.

In Fig. 7.5, we evaluate the influence of the community size. It demonstrates that the proposed algorithm provides higher gains over the conventional paging algorithm when the size of community increases. For example, when the community size increases from 4 to 9 users, the required paging messages decrease by 5 % for CMR $\rho = 0.114$. The reason is that the proposed algorithm uses not only the target user's historical location record, but also the historical location records of other users in the same community to estimate the target user's current location. Since the incoming calls occur independently for each user, the proposed algorithm in essence increases the actual CMR of the target users. Because the location information updates are more often than before, it becomes easier to accurately estimate the target users' current locations. As a result, the paging messages overhead can be reduced.

We further investigate the robustness of the proposed algorithm. We assume that the target user has a probability of moving with the community less than 1. Figure 7.6 shows the required number of paging messages when this probability increases from 0.1 to 1. The gain of using community-based sequential paging scheme decreases from 20 % overhead reduction to 6 % as the probability decreases from 1 to 0.1, and becomes closer to the conventional CTT paging scheme.

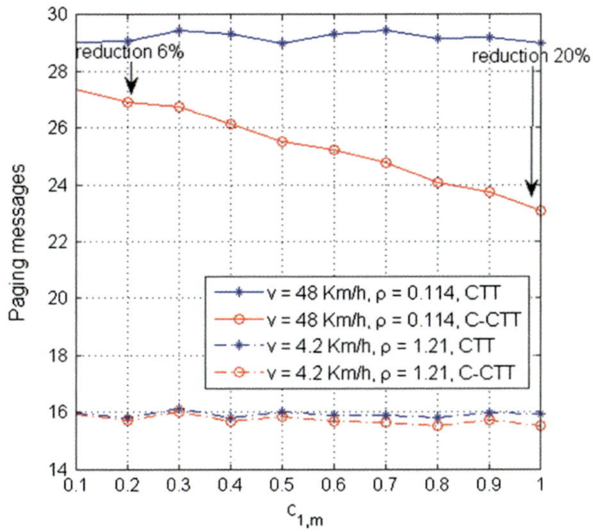

Fig. 7.6 Comparison of paging messages signaling overhead versus probability using CTT scheme and the proposed C-CTT scheme

7.3 Location-Aided Round Robin Scheduling for Fractional Frequency Reused LTE Relay Network

Multi-hop cellular network integrates numbers of relay nodes into the conventional single-hop cellular network, which was first proposed by Lin and Hsu [7] in 2000. These relay nodes are capable of improving their own cells performances in terms of cell-edge throughput, fairness, energy saving and etc. Because of these benefits, multi-hop cellular network has been considered by the standardization community such as 3GPP and IEEE802. The research in decode-and-forward type of relaying technology was started in the Rel-9 3GPP LTE standard and enhanced in Rel-10 and Rel-11.Two types of Relay Nodes (RNs) have been defined in 3GPP LTE-Advanced [8]. Specifically, a Type-I RN located outside the coverage area of its serving Base Station (BS), which is called its donor eNodeB. A Type-I RN has its own cell id, and is non-transparent to Mobile Terminals (MTs). On the other hand, a Type-II RN located within the coverage of its donor eNodeB, shares the same cell id with its donor eNodeB, and is transparent to MTs. are interested in the Type-II relay network.

MTs in the Type-II relay network suffers inter-cell interference not only from the surrounding BSs, but also from the RNs in other cells, which may locate closer to the MT than the interfering BSs. The cell-edge throughput will not be improved if the interference issue is not properly controlled. One solution is to employ Fractional Frequency Reuse (FFR), where the three different frequency bands are assigned to a certain RN, to its donor eNodeB, and to its adjacent interfering RNs, separately [9]. As a result, the RNs using the same frequency band are no longer in any adjacent cells, so that the resultant co-channel interference is reduced.

One consequence of employing FFR is that the allocated bandwidth at the BSs and the RNs are only a fractional of the entire available bandwidth, which may not satisfy the heavy traffic demand occasionally occur in busy hours. In this scenario, one solution is to allow the BSs and RNs to temporarily use the same frequency band. We propose a location-aided round robin [10] scheduling algorithm to achieve this goal. By exploiting the location information of the MTs, the proposed algorithm is capable of minimizing the resultant co-channel interference.

7.3.1 System Model

The network layout considered is illustrated in Fig. 7.7. Without RNs, directional antennas are employed at each eNodeB, which separates the entire site into three cells. For example, the center site with an eNodeB id of 5 consists of three cells having id 13, 14, and 15 separately. On top of such a single-hop network, one low-power RN is employed in each cell covering the cell-edge area using omnidirectional antenna. We considered an FDD LTE-A downlink transmission scenario. The RNs operate in half-duplex mode. More explicitly, in the even number Transmission Time Interval (TTI), the eNodeBs communicate to the MTs and the RNs. In the odd number of TTIs, the eNodeBs communicates to the MTs only, and the RNs transmit

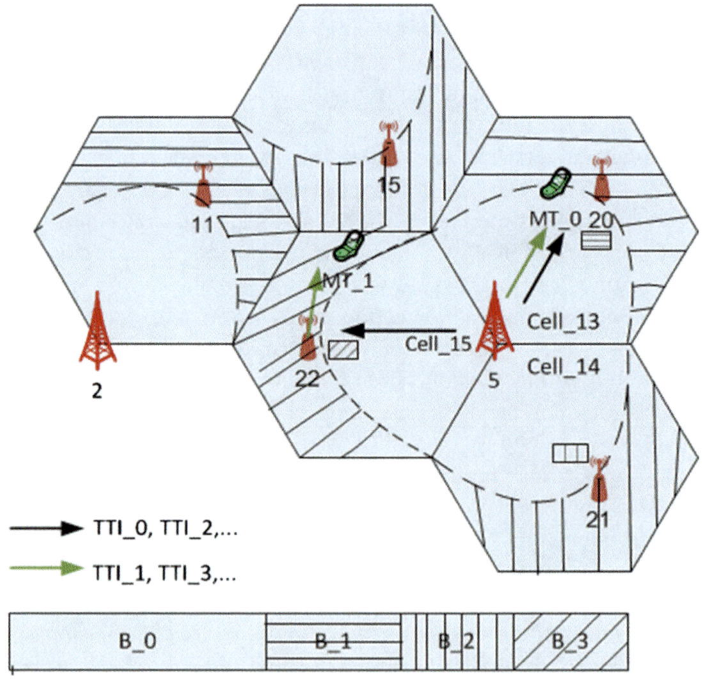

Fig. 7.7 The layout of a relay-aided LTE-A cellar network

information to their associated MTs. The scheduling of all MTs is centrally controlled by the eNodeBs.

For the original single-hop network, the entire available frequency band is reused in every cell. Hence, a cell-edge MT MT_1 located in $cell_{15}$ shown in Fig. 7.7 suffers intra-site interference from $cell_{13}$ and $cell_{14}$, as well as inter-site interference from other eNodeBs. If the frequency reuse factor remains 1 per cell, the same MT will additionally suffer co-channels interference from RNs in all other cells. To reduce the co-channel interference, especially the strong interference from RNs in adjacent cells, the entire available bandwidth B is divided into four proportions. More explicitly, donor eNodeBs uses frequency band B_0 to communicate with its associated MTs located in the center of the entire site. The frequency band B_1, B_2 and B_3 are used by RN_{20}, RN_{21} and RN_{22} separately to communicate with the donor eNodeB at even TTIs, and with MTs located at the cell-edge at odd TTIs.

RNs are transparent to MTs since all control information are handled by the eNodeBs. Suppose that there is a call request for MT_1, the diagram shown in Fig. 7.8 represents the process of a call set-up. Moreover, eNodeBs and RNs only transmit reference signal within their allocated frequency band. As a result, MT_1 located in $cell_{15}$ can measure the channel quality of frequencies in B_0 only for those channels connected to eNodeBs, and the channel quality of frequencies in B_3 only for those channels connected to RNs. Depending on the feedback Channel Quality Information (CQI) CQI (B_0, *eNodeB*) and CQI (B_3, *RN*), the eNodeB schedules this user in some Resource Blocks (RB) in B_3, and informs this user via control channels. Based on this RB mapping, the MT receives data information at T_5 via an RN. Since this MT only receives data and reference signal in frequency band B_3 at odd number of TTIs, it becomes aware that it is connected via an RN. If the CQIs of frequency band B_3 is much higher than those CQIs of B_0, it indicates that the connection via RNs is stable, and this user can only feedback CQIs of B_3 to reduce feedback overhead and save energy. Otherwise, it can feedback all detected CQIs.

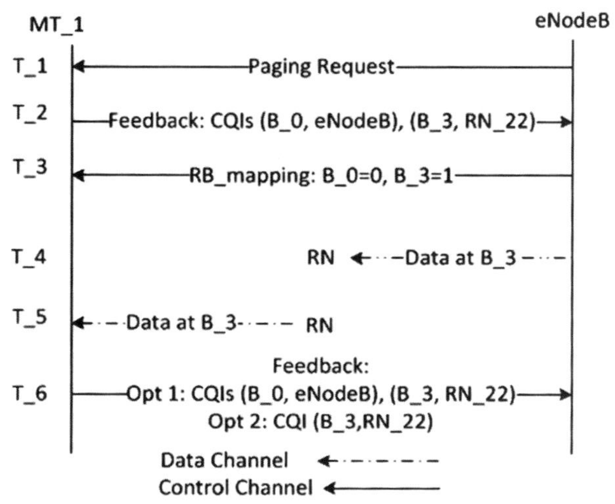

Fig. 7.8 A brief illustration of the signaling process of setting up a call between the eNodeB and a cell-edge MT

7.3.2 Target Scenario and Problem Formulation

The FFR relay network illustrated in the previous section can reduce the co-channel interference. However, the resultant frequency reuse factor reduces below 1 per cell, which may cause some problems such as the scenario shown in Fig. 7.9. In this scenario, round robin scheduler is employed at the donor eNodeB in $cell_{15}$. There are P MTs connected directly to the eNodeB in frequency band B_0 having their QoS requirements just satisfied, and K edge MTs connected to the network via RN_{22} using frequency band B_3. The service qualities of these K users are superior to their requirements. A paging request arrives at eNodeB asking to communicate with a new user $MT_{(P+1)}$ which is located near the eNodeB. If FFR is not applied, the scheduler could reduce the allocated bandwidth to the K edge MTs, and allocate the redundant resource in frequency band B_3 to $MT_{(P+1)}$. However, with fixed FFR, this request will be denied or delayed since the frequency band B_0 has been used up. To overcome the drawbacks of FFR, one possible solution is to allow the direct communication temporarily to use some frequency resources in B_3. In fact, the reused frequency resource could be chosen from B_2 and B_1 as well, which may introduce less interference because of the directional antenna. Nevertheless, the principle is the same regardless the specific frequency band. So we will continue using B_3 in the following sections.

Although satisfying the QoS requirement is our ultimate objective, we decided to use overall achievable throughput as performance metric for simplicity. From now on, we use RBs to represent different frequency bands, since RB is the basic unit for scheduling in LTE-A system. One RB occupies 180 kHz bandwidth and one TTI (1 ms) time duration, which consist of $12 \times 14 = 168$ modulated symbol assuming normal cycle prefix. If a user MT_1 can achieve spectral efficiency of η_1 (bits/symbol), the achievable throughput of MT_1 using one RB is $168\eta_1$ Kbit/s. We further simplify the problem by assuming that each user is located with only one RB. More explicitly, our target problem is rephrased as follows.

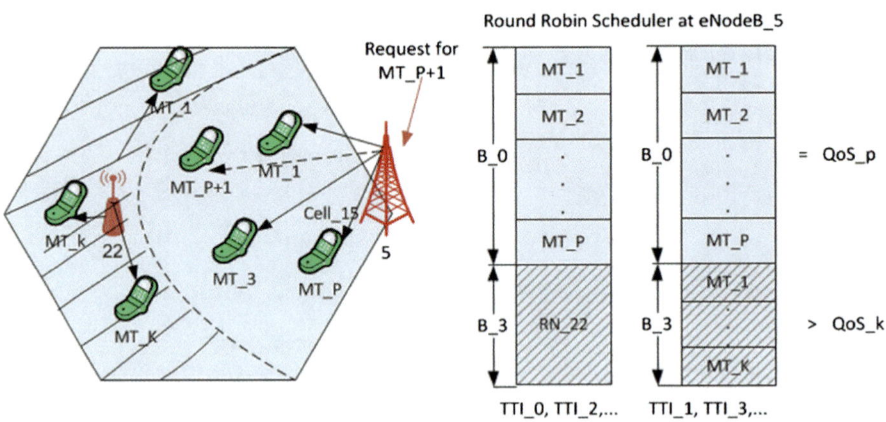

Fig. 7.9 Target Scenario

- Assumption 1: The frequency band B_0 consists of P RBs, which are equally shared by P MTs denoted as MT_p^c (1 RB/MT). The frequency band B_3 consists of K RBs, which are also equally shared by K cell-edge MTs denoted as MT_k^e (1 RB/MT). The superscription $(\cdot)^c$ and $(\cdot)^e$ represent "center" and "edge" separately.
- Assumption 2: All MTs employ the opt2 feedback scheme shown in Fig. 7.8 for energy saving purpose.
- Assumption 3: The location information of all MTs are available. A database recording the long-term averaged path-loss and shadowing for every location in $cell_{15}$ is also available.
- Target 1: Choose one MT p out of $\{1, 2, \ldots, P + 1\}$ cell-center MTs, choose another cell-edge MT k out of $\{1, 2, \ldots, K\}$ cell-edge MTs. The chosen MT_p and MT_k will employ the same RB k originally scheduled to MT_k. As a result, cell-center $MT_{(P+1)}$ can be served.
- Target 2: Minimize the throughput loss at cell-edge while at the same time maximizing the achievable throughput incensement for center MTs.

Before proceeding to the next subsection, we want to justify the validity of Assumption 3. The location information of a MT has become more available with the popularity of GPS-aided smart phones. Moreover, both 3GPP and IEEE802 standardization bodies are including positioning information into the future communication system [11]. Furthermore, user-behavior studies show that people location usually follows a particular pattern every day, and it is highly predictable [12]. As a result, it is highly likely that in future communication systems, the location knowledge of a large proportional users and the corresponding long-term channel quality at these locations will be available at the eNodeBs.

7.3.3 Location-Aided Round Robin Scheduling

The first target can be easily satisfied even by randomly choosing one RB m currently used by MT_k from frequency band B_3, and allocating this RB to the new terminal $MT_{(P+1)}$. However, this cannot satisfy the second target, since the location of MT_k and $MT_{(P+1)}$ is random. If MT_k is close to the eNodeB, it will receive high interference. An intuitive solution is that the chosen MT_k and $MT_{(P+1)}$ should be located as far as possible from each other, so that the resultant co-channel interference is minimized. Since the center MTs only sends the CQI (B_0, $eNodeB$) as feedback, the instantaneous interference signal strength caused by the RN in frequency band B_3 for the center MTs is not available. Similarly, the instantaneous interference signal strength caused by eNodeB in frequency band B_3 for the cell-edge users is not available either. However, the location information of every MT and their corresponding long-term channel quality including pathloss and shadowing at these locations are available. As a result, we can compare the long-term channel degradation between all K edge users and the donor eNodeBs, and choose the one having the highest degradation, so as to minimize the potential co-channel interference from eNodeBs. Similarly, we compare the long-term channel degradation between all P center MTs and the RN, then also choose the one having the highest channel degradation, so as to minimize the potential co-channel interference from RNs. These two chosen users will share the same RB. Other users are scheduled normally using the round robin algorithm. More precisely, the proposed algorithm is summarized in the following steps:

- For each MT_k^e with coordinates (x_k^e, y_k^e), searching out the database so as to determine the channel degradation value between this MT and the donor eNodeB $P_{eNodebB-MT_k^e}$.
- Choose the cell-edge user $l^e = \arg\max_k\{P_{eNodeB-MT_1^e}, \ldots, P_{eNodeB-MT_K^e}\}$;
- For each MT_p^c with coordinates (x_p^c, y_p^c), searching out the database so as to determine the channel degradation value between this MT and the RN $P_{RN-MT_p^c}$.
- Choose the cell-center user $m^c = \arg\max_p\{P_{RN-MT_1^c}, \ldots, P_{RN-MT_p^c}\}$;
- Assign RB l in frequency band B_3 to center user m.
- Schedule other users using the normal round robin algorithm.

7.3.4 Simulation Results

In this subsection, we first demonstrate the essential of using FFR, by comparing the achievable spectral efficiency between relay aided network and single-hop cellular network, with and without FFRs. Second, we demonstrate that the proposed scheduling algorithm statistically improves the achievable throughput of the entire cell; it has better performance than the random scheduling one.

7.3.4.1 FFR-Relay Versus Non-FFR Relay Versus Single-Hop Network

We consider an $1200 \times 1000\ m^2$ area with a spatial resolution of $10 \times 10\ m^2$, where 7 eNodeBs are located with a separation distance of 500 m. The network layout is illustrated in Fig. 7.7. More simulation parameters are summarized in Table 7.1. Single antenna is employed at BSs, RNs and MTs. Only considering the pathloss and shadowing, we calculate the achievable spectral efficiency η at different locations. Then we compare the situation using one relay per cell aided network to the single-hop network case. Their spectral efficiencies are denoted as η_r and η_s respectively. Figure 7.10 represents the difference $\eta_r - \eta_s$ in the considered area. It shows that only a small area has efficiency improvement, some area has even poorer performance than single-hop network. The benefits of relaying are diminished because of the co-channel interference. While Fig. 7.11 compares the spectral efficiency between relay aided network with FFR η_{FFR_r} and single-hop network η_s, which demonstrates that FFR is essential for achieving cell-edge throughput for a network having close located eNodeBs.

Table 7.1 Simulation parameters

Transmit power at eNodeB	P^{BS}_{max}	40	Watts
Transmit power at RNs	P^{RN}_{max}	1	Watts
Distance between RN and eNodeB	$d_{BS\text{-}RN}$	150	meter
Pathloss between eNodeB and MTs	M2135 [13] NLoS		
Pathloss between RNs and MTs	M2135 NLoS		
Pathloss between eNodeB and MTs	M2135 LoS		
Shadowing	Zero mean, 8 dB deviation		

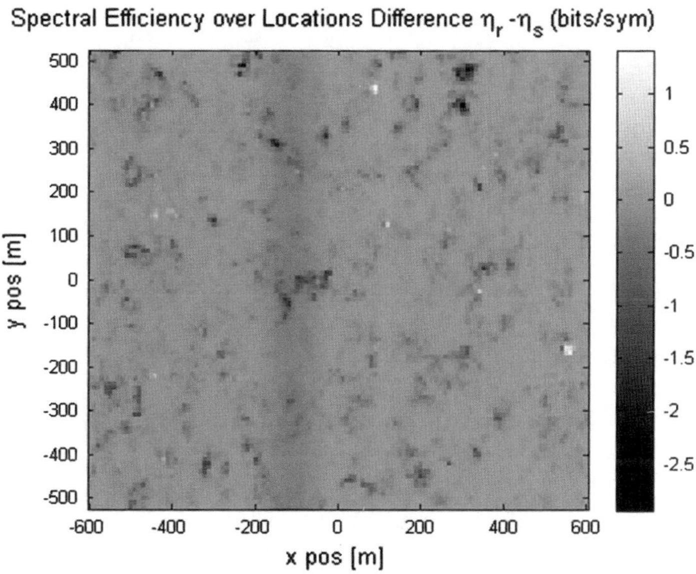

Fig. 7.10 The spectral efficiency comparison between relay aided network without FFR and single-hop network

7.3.4.2 Performance of the Proposed Algorithm

The relay aided network with FFR employed in the previous subsection is employed. We consider a transmission scenario in one of the central cell $cell_{15}$. Using the problem formulated for Fig. 7.9, we set up our simulation with the following parameters $P = 9$, $K = 4$. The MTs locations are randomly generated and the simulation results are averaged over 5,000 different locations.

Figures 7.12 and 7.13 demonstrate the experimental cumulative density function of the total throughput of cell-edge users, and cell-control users, respectively. Three scenarios are compared: (1) there are 9 cell-central users, and 4 cell-edge users. One orthogonal RB is assigned to each of them; (2) One RB in the cell-edge frequency

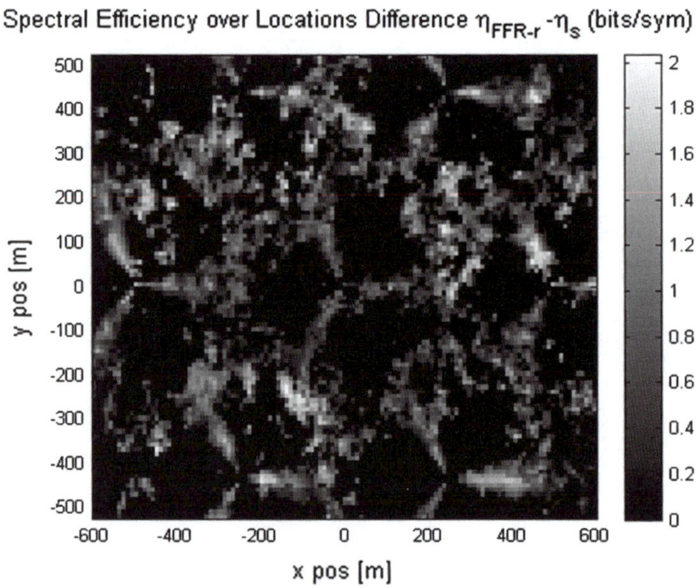

Fig. 7.11 The spectral efficiency comparison between relay aided network with FFR and single-hop network

band is randomly chosen and allocated to the 10-th cell-central MT; and (3) One cell-edge MT is chosen to share its RB with a chosen cell-central MT, using the proposed location-aided round robin scheduling algorithm.

Basically, if the CDF curve shifts to the right direction, it indicates higher throughput is achieved. As a result, the two experimental CDF demonstrate that

Fig. 7.12 Experimental Cumulative Density Function of the total throughput of four edge MTs

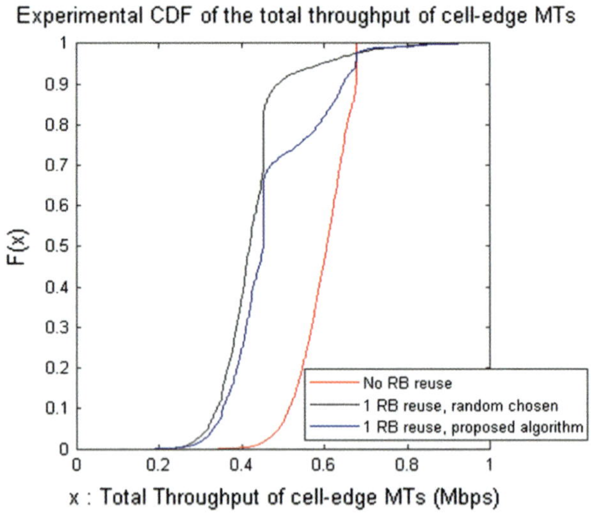

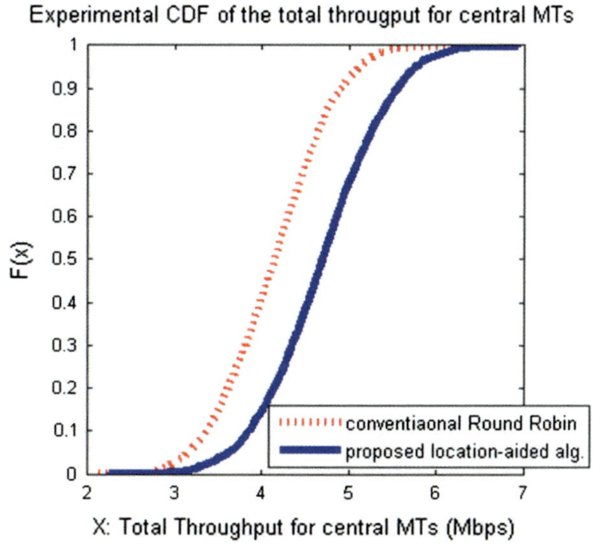

Fig. 7.13 Experimental Cumulative Density Function of the total throughput of central MTs

first, the achievable total throughput for the central MTs are increased by the reuse of an RB. The proposed algorithm achieves higher total throughput than the random selection method. Moreover, the achievable total throughput at the cell-edge is reduced because of the co-channel interference of reusing one RB at the cell center. Using the proposed algorithm, it is capable to compensate the degradation compared to the random selection approach.

7.4 Conclusion

The booming market in the smart phone industry has been driven by advances in device capability that can support a plethora of interesting applications with rich content. A specific market case is the use of GPS information for navigation services. With the increasing development and popularity of the smart phone, it will set the stage for even more meaningful knowledge that can be measured, collected, or derived at the user handset. In this chapter, we demonstrated how a smart phone can be used by the network operator to manage network resources more effectively in an era where spectral and energy resources are costly.

In the first instance, we exploit the context pertaining to movement correlation between users, and propose a community-based sequential paging scheme, and evaluate its paging messages signaling overhead. Our simulation results demonstrate that given the same location update scheme employed in the LTE-A standard, the proposed algorithm is capable of reducing by 20 % the paging messages in the best scenario having high movement rate. The proposed algorithm is especially beneficial for a group of user having high movement correlation and high movement rate.

Secondly, we demonstrate how location can be used to improve the operator's cell capacity. The 3GPP Type-II relay aided cellular network is capable of enhancing the throughput at the cell-edge. However, to achieve the promising gain, the co-channel interference has to be carefully controlled. Frequency fractional reuse is an efficient way of reducing the co-channel interference. To overcome the drawback of FFR, we propose a location-aided round robin scheduling algorithm, which exploit the MTs' location information to allow some frequency band to be temporarily reused within a cell area, so as to improve center area's throughput.

In summary, this user-centric context-aware information may not be accurate, stable and may need to be protected for privacy. Hence, privacy security and robustness are major challenges when using this information for communication system enhancement. Nevertheless, by appropriately using the knowledge provided by smart phones, it can be a promising solution towards more efficient communication network in terms of throughput, complexity, energy consumption, and user satisfaction.

References

1. Liou, R.-H., Lin, Y.-B., Tsai, S.-C.: An investigation on LTE mobility management. IEEE Trans. Mob. Comput. **12**(1), 166–176 (2013)
2. Wang, J., Zhang, H., Toril, M., Wille, V.: Trial results of intelligent paging in GERAN. IEEE Commun. Lett. **11**(10), 829–831 (2007)
3. Xiao, Y., Chen, H., Du, X.: Performance analysis of blanket paging, sequential probability paging, and pipeline probability paging for wireless systems. IEEE Trans. Veh. Technol. **56**(5), 2745–2755 (2007)
4. Ma, W., Fang, Y.: Mobility management strategy based on user mobility patterns in wireless networks. IEEE Trans. Veh. Technol. **56**(1), 322–330 (2007)
5. Newman, M. E. J., Park, J.: Why social networks are different from other types of networks. Phys. Rev. **68**(3), 036122 (2003)
6. De Domenico, M., Lima, A., Musolesi, M.: Interdependence and predictability of human mobility and social interactions. In: Proceedings of the Nokia Mobile Data Challenge Workshop, 2012
7. Lin Y., Hsu, Y.: Multihop cellular: a new architecture for wireless communications. In IEEE Annual International Conference of Computer and Communications (INFOCOM), vol. 00, no. c, pp. 1273–1282 (2000)
8. 3GPP TR 36.814 V9.0.0 (2010-03): Further Advancements for E-UTRA Physical Layer Aspects (2010)
9. Krishnan, N., Yates, R., Mandayam, N.: Bandwidth sharing for relaying in cellular systems. IEEE Trans. Wireless Commun. **11**(1), 117–129 (2012)
10. Hahne, E.L.: Round-robin scheduling for max-min fairness in data networks. IEEE J. Sel. Areas Commun. **9**(7), 1024–1039 (1991)
11. 3GPP TS36.355 V10.1.0: LTE Positioning Protocols (LPP). pp. 1–114 (2011)
12. Song, C., Qu, Z., Blumm, N., Barabási, A.-L.: Limits of predictability in human mobility. Science (New York) **327**(5968), 1018–1021 (2010)
13. Series,M.: Guidelines for Evaluation of Radio Interface Technologies for IMT-Advanced (2009)

Chapter 8
Showcasing 5G Handsets

Ayman Radwan, Jonathan Rodriguez, Marios Raspopoulos, Peter Trapps, Laurent Dussopt, Alexandre Giry, Frederic Fraysse, Sami Aissa and Olivier Perrin

Abstract Applications are increasingly becoming more sophisticated and rich in content, requiring more capable devices with higher energy requirements. This provides the impetus for an overhaul of the design process to engineer future emerging handsets to include energy efficiency as an optimization metric, complementing legacy design goals that include multi-mode operation, miniaturization and flexibility. Throughout this book, multiple innovative solutions have been discussed to limit the energy consumption of mobile handsets which exploit not only the available communication infrastructure in legacy and future emerging mobile networks, but also exploit a context-aware architecture that can provide rich context. Until now, the aforementioned solutions have been tested both analytically and in software through system level simulation studies. However, to attain a real practical insight, this chapter takes a step further and implements the proposed algorithms/protocols on demonstrative hardware platforms to showcase the applicability of the scenarios under real operating conditions. The chapter presents three showcases for demonstrating short-range cooperation and vertical handovers for power saving, as well as energy efficient RF design. The concept pertaining to the first showcase demonstrates how nodes/handsets belonging to a cluster can cooperate to optimize their connection to the radio network in order to save battery power. The second showcase demonstrates how a flexible multi-standard handset can select the radio network which has the lowest energy demand while providing the required quality of

A. Radwan (✉) · J. Rodriguez
Instituto de Telecomunicações, Campus Universitário de Santiago,
3810-193 Aveiro, Portugal
e-mail: aradwan@av.it.pt

M. Raspopoulos
Sigint Solutions Ltd, Nicosia, Cyprus

P. Trapps
Veebeam, Cambridge, UK

L. Dussopt · A. Giry · F. Fraysse
CEA-LETI, MINATEC Campus, Grenoble, France

S. Aissa · O. Perrin
Cassidian, Elancourt, France

A. Radwan and J. Rodriguez (eds.), *Energy Efficient Smart Phones for 5G Networks*, Signals and Communication Technology,
DOI 10.1007/978-3-319-10314-3_8

service. This showcase requires a mobility platform which provisions seamless connectivity in heterogeneous radio technologies. The third showcase is a horizontal one, in that it demonstrates the benefits of the RF hardware building blocks previously proposed within the book (which includes the envelope-tracking Power Amplifier, Doherty Power Amplifier, Tunable Matching Network integrated circuit and a miniature multi-band antenna) for energy saving, where the showcase scenario would apply to both the cooperation and handover cases.

8.1 Introduction

This chapter presents practical validation to some of the solutions presented throughout the book. The chapter illustrates three demonstrative showcases verifying the energy savings achieved through using short range cooperative communications and smart cognitive handover mechanisms, as well as the energy efficiency of a reconfigurable radio transceiver which constitutes newly designed energy efficient power amplifiers, miniature antenna and matching network.

The scope of the first showcase is to demonstrate how nodes belonging to a cluster can cooperate to optimize their connection to a RAT in order to save the battery power of mobile handsets. To facilitate the measurement and analysis, we have designed a 'test bed' consisting of a number of energy saving enabled Nodes, that in reality reflect typical handsets, whose power and energy efficiency can be measured within a heterogeneous networking environment. Two separate scenarios have been selected, a simple 2-Node arrangement that can be configured to operate in both wired and wireless environments, and a more complicated 4-Node wired setting which delivers stability and reproducibility for more complex scenarios. In both cases, the power consumption of the Device Under Test (DUT) is measured and logged by the nodes whilst transmitting/receiving data across the network incorporating the nodes and the Wi-Fi AP, with the DUT in different physical locations. The 2-Node wireless test-bed was set to test the performance of the proposed cooperative solutions integrated with the fabricated energy efficient matching network and antenna to further validate the concept in a wireless mode. This arrangement offers insight into the issues to be addressed, and the realistic boundaries to be drawn when considering how a real product would integrate the TMN into a multi-radio system.

The scope of the second showcase was to test how a flexible device could select the RAT which has the lowest energy demand while providing the required quality of service. This required a mobility platform which provisions seamless connectivity in the presence of heterogeneous RATs. Therefore in order to evaluate the energy efficient Handover algorithms (Vertical Handover and Macro to Femto Handover) and demonstrate the benefit in battery savings in a multi-technology terminal, an experimental setup was put in place that integrates various functional entities, including a context aware module (which provides location-specific information about the user, power consumption of each adapter of the handset, remaining battery power), Handover decision algorithms (discussed in Chap. 5), IP mobility implementation (which executes the handovers to ensure session continuity) and finally the energy evaluation module (which evaluates and demonstrates energy efficiency of the tested handset). The investigations presented in this chapter have been carried out for a typical use-case where an end-user roams between multiple locations (i.e. home, shopping centers, office, etc.) and handovers are performed according to the smart handover decision algorithm proposed.

Finally, a showcase was developed to verify the benefits of the hardware technology blocks developed in Chaps. 2 and 3. The showcase demonstrates an energy-efficient reconfigurable radio transceiver, that is common to both the cooperative and handover scenario and can be considered a horizontal showcase rather than a separate and third scenario. The demonstrator comprises an envelope-tracking Power Amplifier, Doherty Power Amplifier, Tunable Matching Network integrated circuit and a miniature multi-band antenna. The objective of the proposed set-up is to demonstrate experimentally the performances of the hardware prototypes and the energy efficiency improvement obtained with respect to a standard architecture (i.e. standard PA and no TMN). A figure of merit, referred to as the Power Added Efficiency, is defined to characterize the global power budget of the demonstrator from the input RF power to the radiated RF power taking into account the power consumption of each block.

8.2 Energy Efficient Short Range Communications Showcase

In this subsection, energy savings through cooperative short range communications will be showcased. For measurement and analysis, a testbed consisting of a number of energy efficient enabled nodes was designed. The nodes represent multi-standard mobile handsets, which posses two technology radios. In this showcase, the nodes carry WiFi [1, 2] and Wireless Ultra-Wideband radios [3].

The testbed allows the power and energy efficiency of the nodes to be measured in a heterogeneous networking environment. Two separate scenarios have been selected, a 2-Node arrangement that can be configured to operate in either wired or wireless environments and a more complicated 4-Node wired test platform which delivers stability and reproducibility for more complex scenarios. It has to be noted here that although those algorithms are proposed for wireless mobile handsets, wired testing is a regular custom for measurements and testing, for repeatability and fair comparison. The wired environment provides a controlled environment, where exact tests and scenarios can be reproduced more than once; hence frequent tests and comparisons between different technologies can be carried out. The 2-Node wireless platform has been used to undertake measurements in a typical indoor office environment to demonstrate nodes integrated with the TMN-antenna module previously designed and have been discussed in Chap. 2, as well as provide baseline measurements that can be used to align the wired platform as closely as possible with wireless operation. To this end, the 2-Node arrangement is reproduced in the cabled RF environment, which allows the power consumption of the DUT to be measured in a controlled radio environment and for initial set-up and calibration of the wired environment attenuators.

In both setups, the power consumption of the DUT is measured and logged by the nodes whilst transmitting/receiving data across the network incorporating the nodes and the WiFi AP, with the DUT in different physical locations. The data is processed to allow the energy efficiency in different configurations to be determined.

8.2.1 Testbed Architecture

The main testbed architecture is shown in Fig. 8.1. Each Node consists of a DUT, an RF attenuator board, a power measurement block and the telemetry controller. Each node is equipped with WiFi and ultra-wideband (WUWB) air interfaces. The WiFi is considered the infrastructure based technology of medium range access.

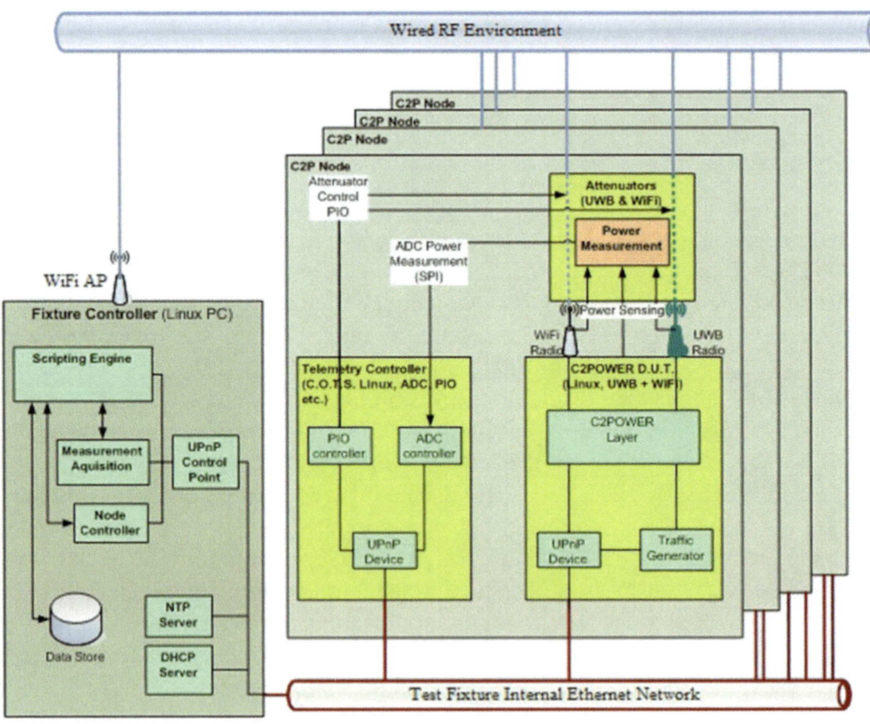

Fig. 8.1 Short range cooperation testbed architecture [5]

The UWB represents the short range cooperative technology. The WUWB is connected to the node through a Wireless USB (WUSB); hence in the rest of the chapter WUSB and WUWB are used interchangeably and they both mean the Ultra-wideband radio interface connected to the node under testing. The testbed has the capability to measure the power consumption of the three entities that comprise the DUT (CPU, WiFi radio and UWB radio module), whilst simultaneously running TCP/IP traffic across the network. A brief description of the different components of the testbed is provided below.

The node representing a mobile handset is pictured in Fig. 8.2 and can be seen in schematic form in Fig. 8.1. It consists of three major modules; the DUT, a Telemetry Controller (TC) and an Attenuation and Power Measurement Board (APM). Both the DUT and the TC have individual Ethernet connections to the testbed internal network. Each node has a layer that implements the defined algorithms and is controllable via another UPnP service interface. Control via the UPnP interface makes it possible for the TFC or other external controller to enable or disable the implemented algorithms. Each node possesses both Wi-Fi and UWB (WUSB) radio modules.

The DUT has an architecture broadly similar to that of a modern mobile handset. It consists of a main 32-bit 400 MHz RISC CPU running embedded Linux with

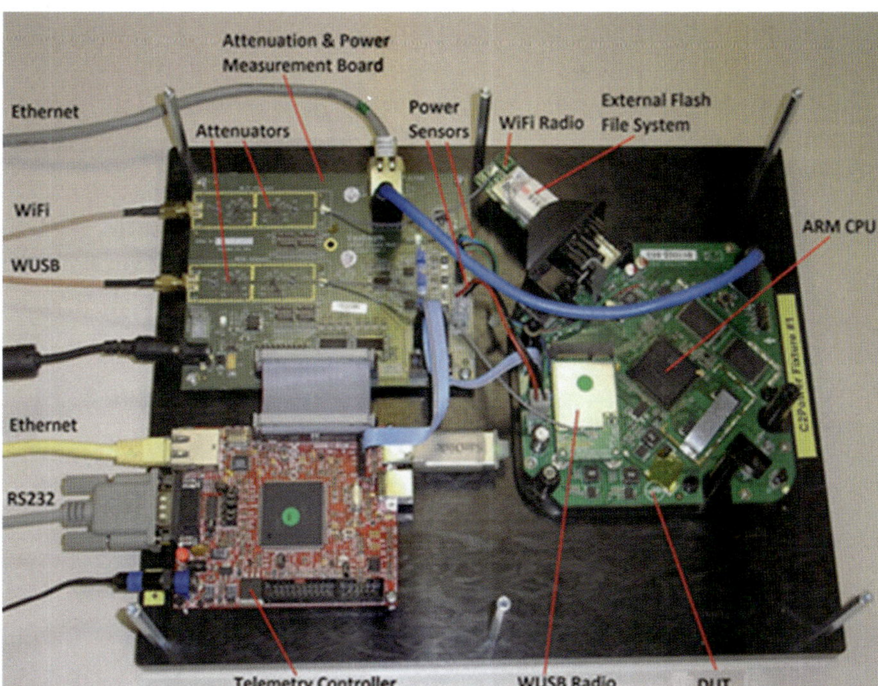

Fig. 8.2 Testing node representing the mobile handset [5]

associated UWB and Wi-Fi radio modules and an external flash file system in a USB thumb drive for convenient updating. A traffic generator acts in the role of an application on the handset and receives commands via a service interface in the UPnP device. It has the facility to report throughput, error rates, etc., when requested via the same UPnP interface.

The Telemetry Controller (TC) is a general purpose COTS (Component Of The Shelf) development board with a 32 bit CPU, 64 MB RAM, and 512 MB NAND flash running embedded Linux and has a wide selection of interfaces, including Ethernet, PIO and SPI. There are three main software modules in the TC. They are a UPnP device similar to that in the DUT, a PIO controller and an ADC controller.

The PIO controller sets the attenuators on the APMB in response to commands delivered over the Ethernet via a UPnP service interface. The ADC controller receives instructions to gather voltage and current readings from the power sensors on the APMB. The ADC controller can be configured over UPnP to buffer a number of readings for delivery to the TFC in response to later requests. Each measurement is time stamped by a local clock synchronized with the NTP server on the TFC, which ensures that all readings from all nodes are synchronized across the testbed. In the case of this demonstrator, an Ethernet interface is used.

The RF attenuator and Power Measurement Board (APMB) has two RF attenuators inserted between the antenna connector of each radio and the wired radio environment. The variable attenuator device (HMC425LP3) is broadband 6-bit GaAs digital attenuators from Hittie, covering a range of 2.2 GHz to 8.0 GHz with a typical insertion loss of 3.8 dB [4].

Each one is capable of inserting a wide enough range of attenuation approximately the same as 1 m → 20 m of separation in a line of site over-the-air arrangement. The maximum attenuation is sufficient to break the link in both the Wi-Fi and UWB cases. So it is possible to emulate the bulk of real world arrangements of nodes.

Power Measurement—Three voltage/current measurement sensors are connected to the DUT to measure the power consumption of: the main CPU and ancillary hardware, and each of the Wi-Fi and UWB radio modules. An ADC under the control of the TC via an SPI bus digitizes and captures the measurements at up to 10 K samples per second. In practice the sampling rate is limited by the software latency on the TC.

The RF connections of the mid-range (WiFi) and short-range (UWB) radios are to be routed into the antenna ports of the TMN which provides an ideal opportunity to test its performance in a typical indoor office environment. Bear in mind that the architecture and resource profile of a DUT perfectly resembles that of a typical mobile handset. This arrangement offers insight into the issues to be addressed, and the realistic boundaries to be drawn when considering how a real product would integrate the TMN into a multi-radio system.

The detailed description of the architecture and its components has been avoided to concentrate the attention of the reader on the experiment setup and results, which are more critical for the validation of the achieved energy savings of the proposed solutions.

8.2.2 Integration of the TMN-Antenna Module

The nodes included in the testbed incorporate both a 2.4 GHz WiFi module and a WiMedia compliant UWB (3.1–7.9 GHz) radio module. These radios aim to be representative of short and long/medium range radios found in most modern handsets. To determine the suitability of the TMN Antenna for integration into the energy saving enabled nodes, a number of parameters needed to be considered. The selection criteria are presented below.

The TMN module has been integrated on the 2-Node wireless test Bed and operates under the control of the DUT. The two node scenario was run firstly in a wired environment and the results logged and graphed and then repeated wirelessly with the TMN in place and the results compared.

TMN Operating Band—The TMN was designed to meet the WiMAX standard requirement with an actual (measured) upper limit of about 3,690 MHz. The lowest UWB band is just within this limit. Since the UWB radio modules operated on WiMedia Bandgroup3 (6,336–7,920 MHz), which is well above the operating range of the TMN-Antenna, the UWB modules were re-configured to restrict them to operating in BG1, TFC5 (3,168–3,696 MHz).

Radio Selection Switch—In order for the TMN-Antenna module to correctly re-tune itself to be compatible with the currently selected radio, it must be provided with a signal indicating which radio has been selected for data transmission. Additional firmware was developed to control two GPIO lines from the DUT connected to the RF switch on the Antenna board and coordinate the switchover with the selected radio.

Board Mounts and RF connections—The TMN Antennas are mounted on top of the Nodes. They are horizontally polarized on the intermediate node and vertically polarized on the Originator Node. Vertical polarization for the Originator Node gave the best performance/range for the UWB radio and did not adversely impact the WiFi performance. The TMN DSP board has been mounted above the TC close to the Antenna board to ensure the routing of the RF connections is kept short.

DC Supply for TMN DSP Board—The boards are powered from the +5 V supply on the DUT. A set of power measurements was taken with just the Antenna board and then after fitting the DSP board in order to verify the additional power consumption of the DSP board.

Power Handling Capacity—The output power of the WiFi module is about +17 dBm (max) and the power of the UWB radio is around −14 dBm. This falls well within the capability of the TMN-antenna board, which is designed to handle 1 W of input power.

8.2.3 Experiment Setup

In this showcase, two different experimental configurations have been set and are described in the following three subsections.

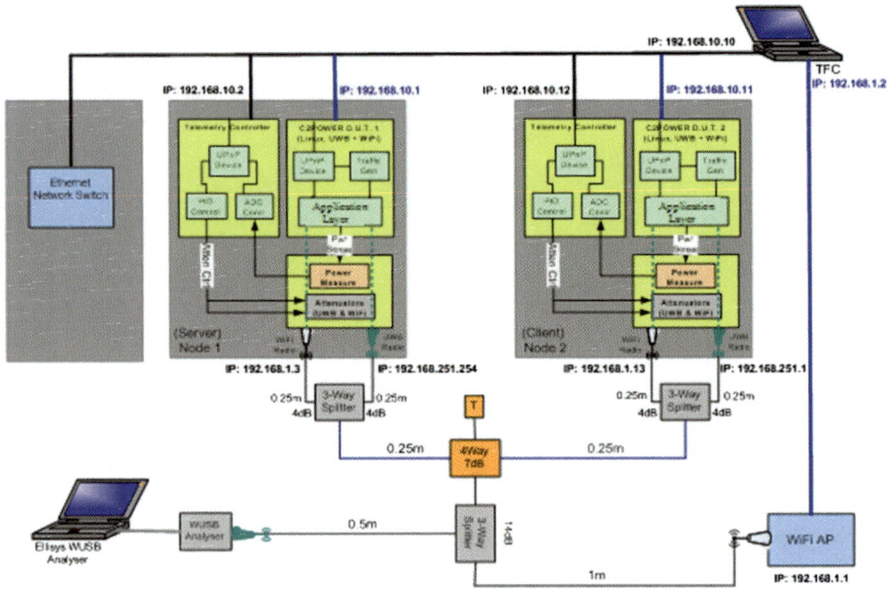

Fig. 8.3 2-node wired configuration demo [5]

8.2.3.1 2-Node Wired and Wireless Demo

To start with, Fig. 8.3 shows the test-bed configuration for a 2-Node Wired test platform, which uses the on-board attenuators on the TC to change the effective range between the nodes and AP. The attenuators on the TC are configured to recreate (from the RF perspective of each Node) the relative positioning of the nodes. These can be controlled using Linux shell scripts on each TC and from the TFC using the uPnP protocol. A typical test-bed configuration is adopted for the 2-node wireless Demo case, where the TMN Wireless Antenna module is mounted on top of the tested nodes.

For the wireless case, the wired attenuators for the WUSB radio on the TC are bypassed in this configuration, as the data transmissions take place over-the-air and will be attenuated by normal propagation in the office space. It should be considered that the WUSB radio has a maximum range of about 10 m. Since WiFi has a far greater range than the confines of a small office (tens of meters in an indoor environment) a fixed amount of attenuation (attenuators on the TC) is required to simulate the AP being near to the edge of range. The TFC is configured to act as a data traffic server, as well as providing a conduct for the separate wired uPnP command and control messages used to interrogate the Nodes.

A Netgear dual band (2.4/5 GHz) 802.11a/b/g Access Point [6] was used to create a typical real world WiFi infrastructure, configured for 802.11g (OFDM) operation in the 2.4 GHz band. This standard is a commonly used standard and provides an on-air maximum physical layer bit rate of 54 Mbit/s or about 23 Mbit/s max useable TCP throughput. The AP and Nodes have been configured with static IP addresses for convenient test automation.

A WUSB Analyzer is used to monitor the on-Air UWB traffic and to verify that in the case of the 'RELAY' scenario, that traffic is being routed over the UWB radio via the Intermediate Node.

The DUT power consumption is measured and recorded on the Originator and Intermediate Nodes whilst the nodes are transferring data from the AP. The RF cabling and splitters provide a closed RF radio environment that allows the test scenario to operate in a repeatable and consistent manner free from outside interference.

The two Nodes in this configuration are referred to as 'Originator' and 'Intermediate'. The test scenario involves a data transfer of 200 MB from the TFC (acting as server) to the Originator Node, over a TCP connection. The power consumption (derived from separate I and V measurements) of the CPU, WUSB and WiFi radio modules on each node is measured, displayed and recorded whilst the Originator Node is performing the data transfer across the network from the AP.

The measurements are undertaken using two scenarios, these being:

- DIRECT—where the data is transferred from the Originator to the AP using just the WiFi Radio
- RELAY—where the data is transferred via the Intermediate Node using the WUSB Radio between the Nodes and then from the Intermediate Node to the AP using the WiFi radio.

A schematic representation of the experimental setup is shown in Fig. 8.4. The experimental setup can represent typical scenarios where lots of users are on their mobile handsets in a static or low mobility environment, e.g. airports, coffee-shops, etc. The figure shows that the originator node can be far from the AP in the WiFi space; hence cannot be transmitting at the maximum available data rates; this would result in higher energy consumption due to the longer transmission periods at maximum power transmission. The Originator node then opts to use Relay mode, whereas data is transmitted from the Originator to the Intermediate node using WUSB Radio at high data rates, due to the close range between mobile handsets, then the Intermediate will transfer the data to the AP using its WiFi radio at higher data rates, since it is closer to AP than the Originator. The results are compared between the two scenarios to evaluate the energy savings worth of such Relay configuration, which are represented in the results subsection.

8.2.3.2 4-Node Wired Demo

Figure 8.5 shows the test-bed configuration of the 4-Node Wired test platform that uses the on-board attenuators on the TC to change the effective range between the nodes themselves and between the nods and the AP. These can be controlled using Linux shell scripts on each TC and from the TFC using the uPnP protocol to change the relative position and movement of the nodes. This is an evolution of the two

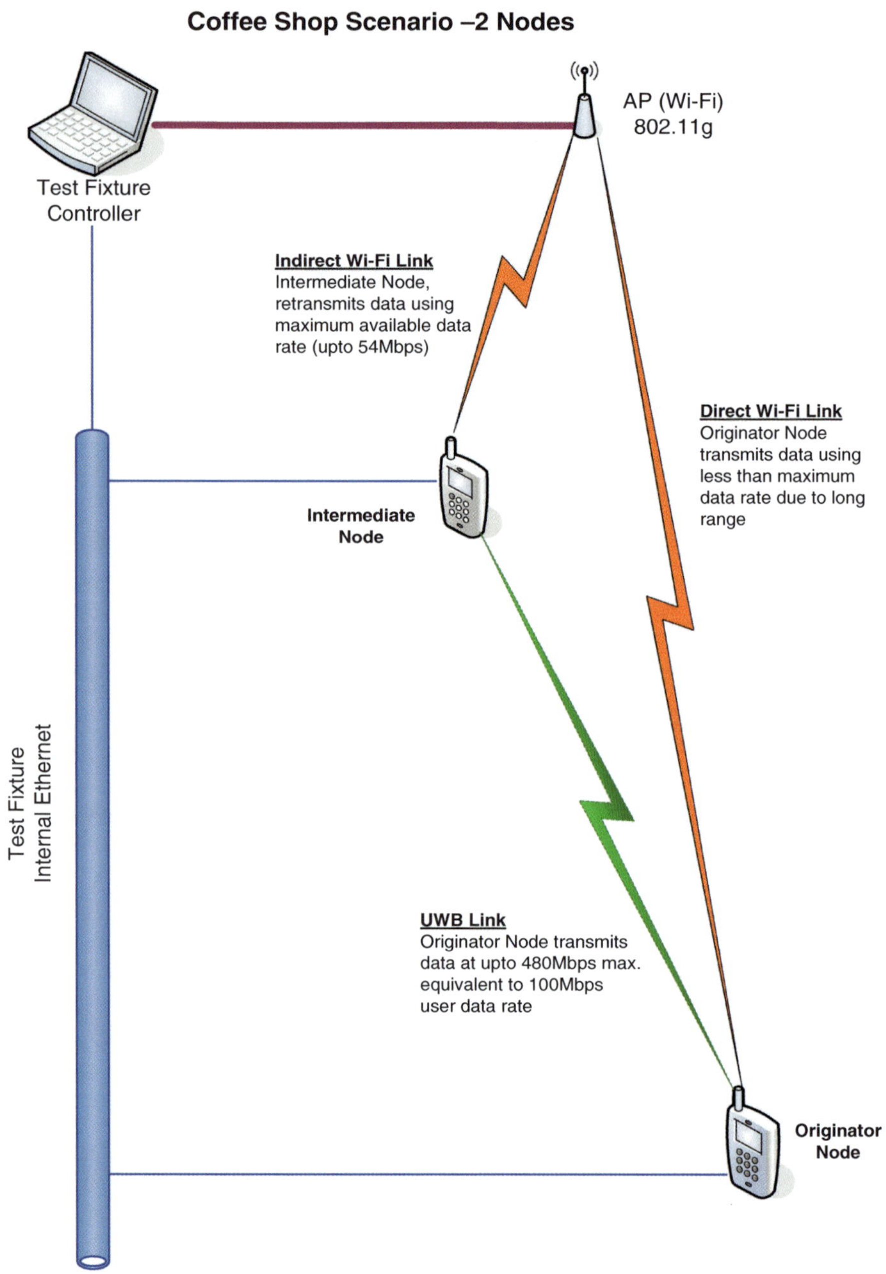

Fig. 8.4 Schematic representation of the real scenario of the 2-node setup [5]

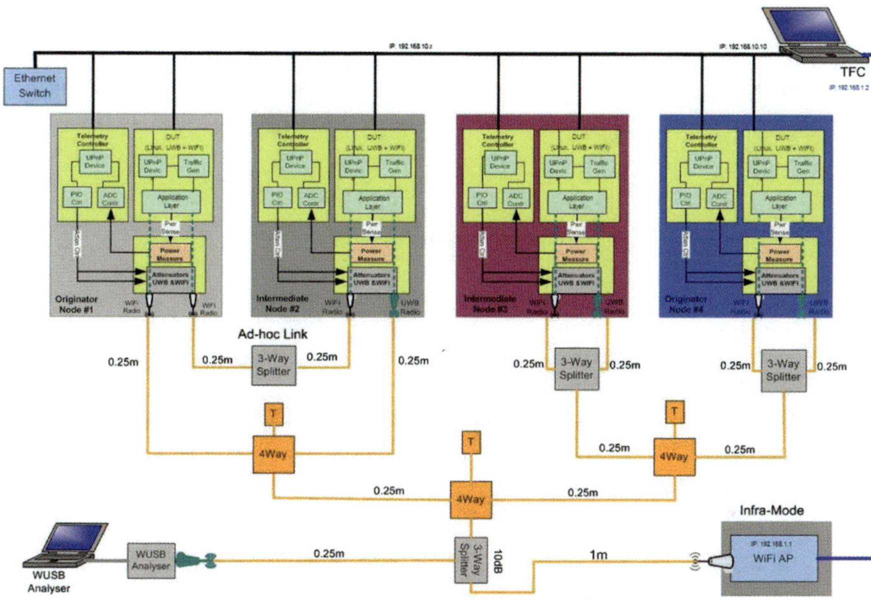

Fig. 8.5 4-node wired configuration demo [5]

node wired system, to incorporate two additional nodes. The RF cabling and splitters provide a closed RF radio environment that allows the test scenario to operate in a repeatable and consistent manner free from outside interference.

The attenuators on the TC are configured to recreate (from the RF perspective of each Node) the relative positioning of the nodes, along the lines of an open or closed area with low mobility scenario. The DUT power consumption is measured and recorded on the Originator and Intermediate Nodes whilst performing the traffic tests in a similar manner to that done for the 2-Node tests.

Figure 8.6 shows the physical equivalent configuration for the test scenario. There are two Intermediate Nodes and two Originator Nodes. This scenario simulates a handset transmitting data and increasing its range from the wireless AP, whilst its energy consumption and that of its neighbors is measured.

This configuration is based on the simple 2-node wired configuration, but expanded to four nodes to simulate the 'Cluster Head' scenario. The RELAY path in this instance consists of a node (Intermediate #1) connected to the AP via standard 'Infrastructure' mode and a second Head node (Intermediate #2) connected to this via Ad hoc mode used to extend the distance and allow a greater throughput in the RELAY path. Originator Node #4 (the main node of interest) is connected to Intermediate Node #2 using its short range WUSB radio. The second Originator Node #3 is also connected to Intermediate #2, but just performs a passive IDLE function in this scenario.

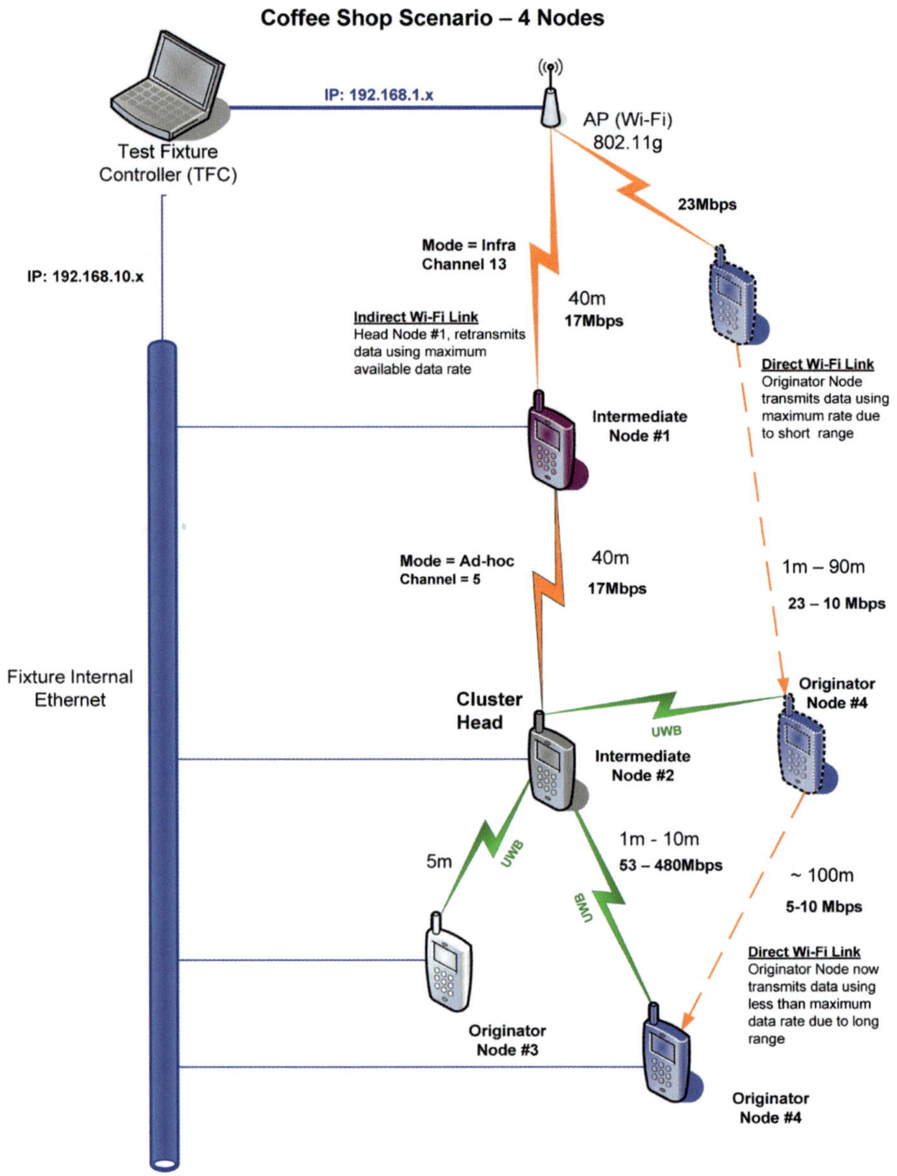

Fig. 8.6 Physical representation of the 4-Node test configuration [5]

Originator Node #4 initiates its data transfer from the TFC (acting as a server) using its WiFi radio (Direct Scenario), whilst the RF path attenuation is increased to simulate the Originator Node moving away from the WiFi AP. The traffic throughput and power consumption are logged throughout the time of the data transfer. Once the node has moved to its farthest point (signified by a low traffic rate) a decision is made to route the data via the two Intermediate Nodes using the WUSB radio on the Originator Node #4.

8.2.4 Measurements and Results

8.2.4.1 Measurement Capture and Results Processing

The test scenarios are standardized on a data transfer of 200 MB from the TFC to the Originator Node. The power consumption of the CPU, WUSB and WiFi radio modules on each node is measured. The power consumption was compared for the time of the file transfer. The Telemetry Controller generates a set of time-stamped log files of the measured power consumption (W, I, V) of the CPU, WiFi and WUSB radio modules in each DUT, plus an aggregate of the three (RCVR). The Originator DUT generates a tabularised log of data throughput, time-stamped at 1 s intervals. Following, we show a set of results that support and validate the energy savings achieved using short range cooperation.

The measurements on the radio module were used to select the most effective mode of operation for the WiFi radio. The test involves transmitting a 200 MB file for 200 secs and measuring the power consumption inside an RF screened room. Briefly, the different power saving modes of the WiFi radio are:

- Constant Awake Mode **(CAM)**—Keeps the radio powered up continuously.
- Fast Power Save Mode **(Fast_PSP)** switches between PSP and CAM based on network traffic.
- Power Save Mode **(MAX_PSP)**—The Wireless Adapter must wake up periodically.

Based on measurements results of power consumption in the different saving modes, fast Power Save Mode (Fast_PSP) was selected as a practical compromise for the wireless testing.

8.2.4.2 2-Node Wired Configuration

This section describes and presents the results of the tests conducted on the 2-Node wired test (Fig. 8.3) using the cabled RF environment for both the DIRECT and RELAY scenarios (Fig. 8.4) at various emulated ranges from the AP and between nodes.

The energy consumption figures of the Originator and Intermediate Nodes for each distance from the AP were summarized for both the DIRECT and RELAY scenarios.

An attenuation mapping table was developed based on a free space loss model, but adjusted to provide a similar traffic throughput to the indoor office environment measured from the tests of the 2-node wireless configuration. This takes into

Table 8.1 2-node wired energy consumption (originator) 200 MB file transfer

2-node wired—cabled RF environment				
Energy consumption (originator node)—200 MB file transfer				
Range (m)	Direct scenario	Relay scenario	Energy saved	Energy saved
	Energy (J)	Energy (J)	(J)	Ratio (%)
1	461.68	421.03	40.64	8.80
2	575.90	424.70	151.21	26.26
3	584.22	421.78	162.44	27.80
4	617.49	430.56	186.93	30.27
5	942.40	424.81	517.59	54.92
6	965.76	436.58	529.19	54.79
7	994.65	425.46	569.19	57.22
8	1015.96	472.51	543.45	53.49
9	1063.83	562.00	501.83	47.17
10	1366.44	604.43	762.01	55.77

account the Tx Power, the Rx sensitivity of the radios and the losses in the attenuator paths and cabling environment. The aim was to make the cabled RF environment provide a representative model of the wireless environment.

The following graphs and tables show the energy consumption and the energy saved with respect to the Originator and Intermediate Nodes for both the DIRECT and RELAY scenarios for the duration of the 200 MB file transfer.

Table 8.1 and Fig. 8.7 show the energy consumption and the energy saved in the 'Cabled Environment' with respect to the Originator Node for both the DIRECT and RELAY scenarios over the duration of the 200 MB file transfer. The energy saving is really appreciated for the duration of the file transfer and can reach up to 55 %, at the longest distance of the Originator from the AP.

The previous results show highly appreciated energy savings from the point of view of the Originator node, but for a more comprehensive view, the energy consumption of the Intermediate node has to be taken into consideration. The total energy consumption of both the Originator and Intermediate nodes is shown in Table 8.2 and Fig. 8.8. It can be observed that energy savings can be achieved for the overall network, reaching up to almost 50 %, even with taking the consumption of the Relay into consideration.

The previous results clearly demonstrate the gains in energy savings achieved from using short-range cooperative relaying.

8.2.4.3 2-Node Wireless Configuration

This section describes and presents the results of the test conducted using the 2-Node test bed configured with the TMN Antenna. This configuration test is

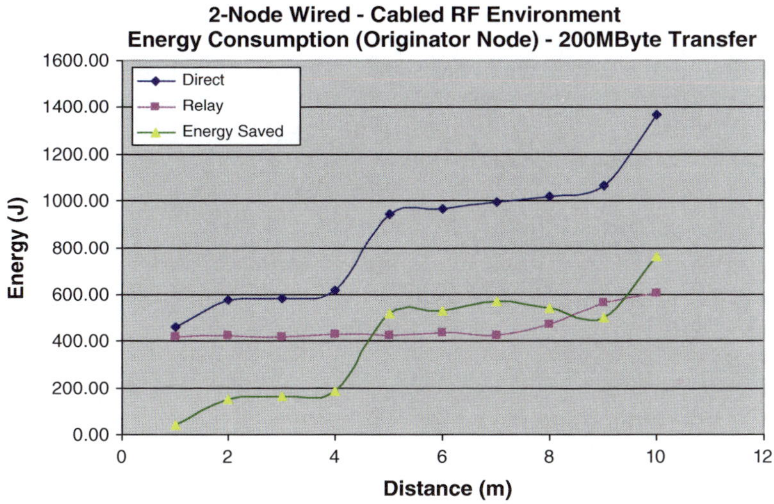

Fig. 8.7 2-node wired energy consumption (originator)—200 MB File Transfer

Table 8.2 2-node wired total energy consumption (originator + intermediate) 200 MB file transfer

2-node wired—cabled RF environment				
Energy consumption (originator + intermediate nodes)				
200 MB file transfer				
Range (m)	Direct scenario	Relay scenario	Energy saved	Energy saved
	Energy (J)	Energy (J)	(J)	Ratio (%)
1	858.19	896.11	−37.92	−4.42
2	1068.82	904.33	164.49	15.39
3	1067.23	885.41	181.82	17.04
4	1108.00	883.43	224.57	20.27
5	1715.53	861.06	854.47	49.81
6	1752.75	922.37	830.38	47.38
7	1830.68	893.92	936.76	51.17
8	1865.79	987.61	878.18	47.07
9	1945.45	1167.71	777.74	39.98
10	2485.31	1244.06	1241.25	49.94

crucial for the validation of the results obtained from the RF-cabled environment tests. This as well provides validity to the results obtained from the more complicated setup of the 4-node wired configuration. Additionally, it provides validity to the matching network and antenna module designed in Chap. 2.

Table 8.3 and Fig. 8.9 show the energy consumption and the energy saved with respect to the Originator Node for both the DIRECT and RELAY scenarios over the

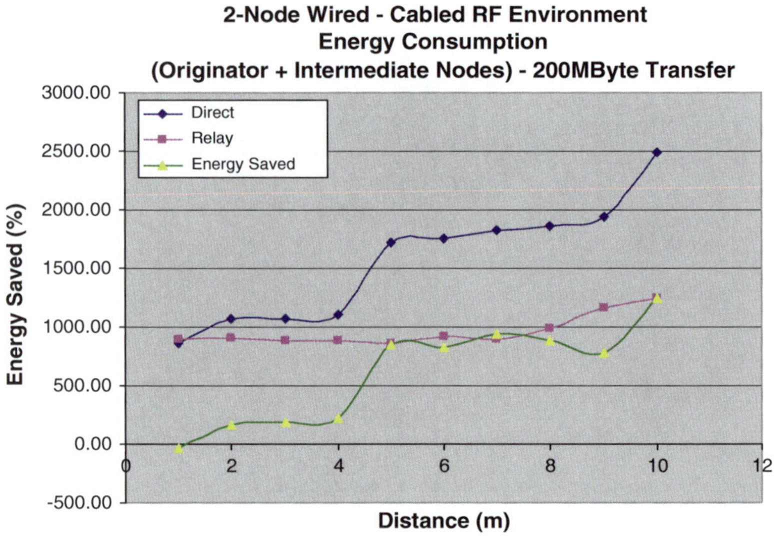

Fig. 8.8 2-node wired total energy consumption (originator + intermediate) 200 MB transfer

duration of the 200 MB file transfer. The energy saving achieved ranges from 14 % up to 68 %, which increases with the increase of the distance from the originator to the AP.

For fair comparison, Table 8.4 and Fig. 8.10 show the total energy consumption (summation) of both the Originator and Intermediate Nodes for the duration of the file transfer. From this perspective there is an overall advantage to using the RELAY path due to the faster throughput of this link. This will depend however on

Table 8.3 2-node wireless energy consumption (originator) 200 MB file transfer

2-node wireless (TMN-antenna)—open office environment				
Energy consumption (originator node)				
200 MB file transfer				
Range (m)	Direct scenario	Relay scenario	Energy saved	Energy saved
	Energy (J)	Energy (J)	(J)	(%)
1	587.93	505.87	82.07	13.96
2	916.18	568.79	347.39	37.92
3	1134.23	603.25	530.98	46.81
4	1826.25	828.95	997.29	54.61
5	1158.15	556.11	602.04	51.98
6	1671.08	812.13	858.95	51.40
7	1825.35	982.61	842.74	46.17
8	1929.37	605.06	1324.31	68.64

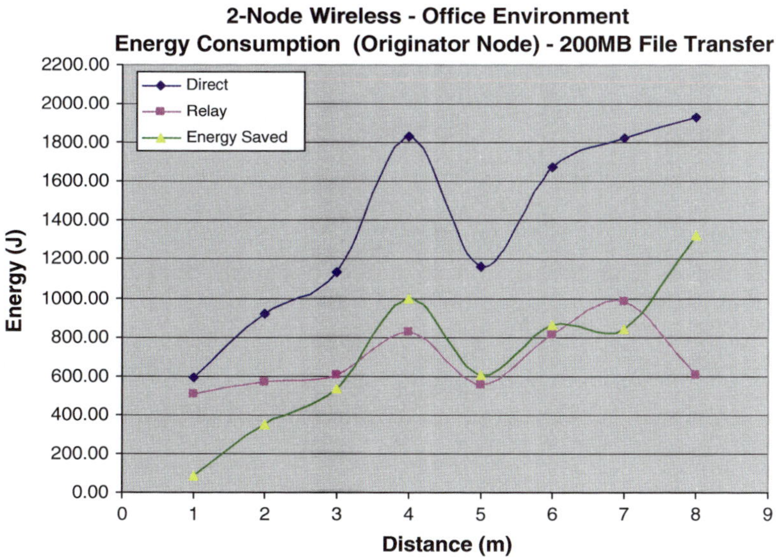

Fig. 8.9 2-node wireless energy consumption (originator) 200 MB file transfer

Table 8.4 2-node wireless energy consumption (originator) 200 MB file transfer

2-node wireless (TMN-antenna)—open office environment				
Energy consumption (originator node)				
200 MB file transfer				
Range (m)	Direct scenario	Relay scenario	Energy saved	Energy saved
	Energy (J)	Energy (J)	(J)	(%)
1	1142.82	1088.23	54.60	4.78
2	1689.75	1192.94	496.82	29.40
3	2036.66	1262.22	774.44	38.02
4	3062.00	1774.84	1287.16	42.04
5	2084.64	1189.31	895.33	42.95
6	2990.17	1723.61	1266.55	42.95
7	3264.73	2079.39	1185.34	36.31
8	3448.62	1270.82	2177.80	63.15

the propagation conditions, building environment and any interference impacting throughput. The energy savings considering the whole system can reach up to 63 %.

Comparing the results from the wired and wireless configurations, it can be seen that energy savings can be achieved using short-range cooperative relaying in both cases. Both tests follow a similar trend. It has to be noted that the distance in the wireless setup was limited to 8 m due to distance limitation in the testing space.

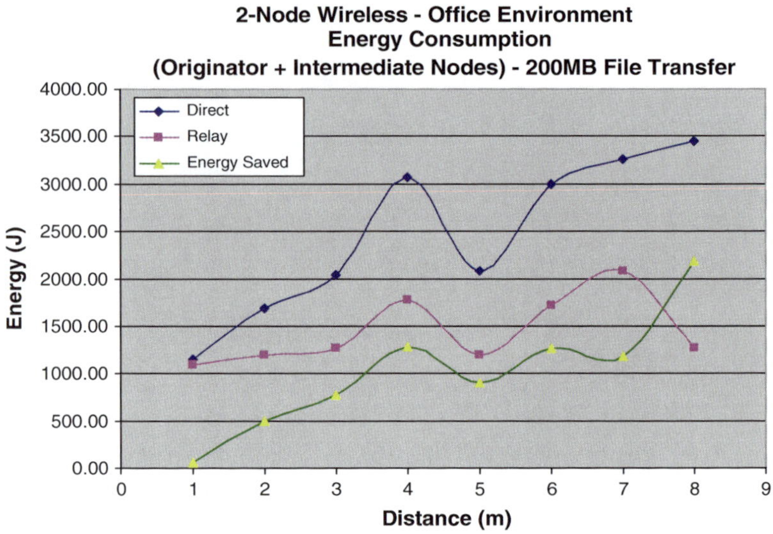

Fig. 8.10 2-node wireless energy consumption (originator + intermediate) 200 MB transfer

The energy saving ratio increases with the increase of the distance between the Originator and the AP. The energy saving ratios in the wireless test are a bit higher, due to the uncontrollable interference caused by other devices in an open space environment. As the distance increases, the data rates achieved in the Direct scenario decrease hence increasing the interference. When using the Relay mode, the distance between the Intermediate node and the AP is much shorter, hence higher data rates are reached, decreasing the transmission time and decreasing the expose to interference. This results in higher energy savings in the wireless configuration compared to the wired configuration. It also has to be noted that the open office environment benefits from multi-path at mid range which results in a slightly higher energy saving than was expected. This is likely to vary with different room layout, furnishings etc.

8.2.4.4 4-Node Wired Configuration

This section describes and presents the results of the tests conducted on the 4-Node wired test for the test-bed configuration shown in Fig. 8.5 for the DIRECT and RELAY scenarios as described earlier in the chapter. This configuration represents a more complicated setup, where the data is transferred between the Originator node to the Cluster head using WUSB radios, then the data is relayed through another relay to the AP using WiFi radios. This configuration basically represents a 3-hop relaying scenario, with two intermediate nodes as shown in Fig. 8.6.

For this scenario, a different perspective will be used to present the data, that of effective range where a node may be physically close to another but suffer a low data rate due to interference, obstructed line of sight or other propagation issue. The power consumption is measured for 8 different TCP/IP throughput figures, from maximum to minimum over the attenuated cabled environment. This was achieved by increasing the WiFi attenuation between the Originator Node #4 and the AP (DIRECT path) or by increasing the WUSB attenuation between the Originator #4 and Intermediate Node #2 (RELAY path). The power consumption is measured at a number of TCP data rates, aligned to the discrete data rates of the WiFi and WUSB radios, as their rate adapt to the increasing attenuation.

The energy consumption is measured for the Originator and the Intermediate nodes for transferring a 200-MB file.

The total radio energy (WiFi and WUSB) of the Originator Node consumed by the DUT for the duration of the 200 MB file transfer is shown in Table 8.5 and Fig. 8.11.

Considering the radio energy consumed, clearly transmitting via the RELAY path via the two intermediate nodes offers a distinct advantage. There is a substantial energy saving for the RELAY scenario for all data rates due to the lower power consumption of the WUSB radio on the originator node. Extrapolating from the curves yields a saving of between 20 and 25 % depending on the data rate.

Table 8.6 and Fig. 8.12 show the total energy consumption (summation) of both the Originator and the two Intermediate Nodes for the duration of the file transfer.

From the perspective of the total system energy consumed there is a penalty, as the two Intermediate Nodes will consume additional energy relaying the data from the Originator. This will vary depending on the propagation conditions, building environment and any interference impacting throughput. Based on the results in the figure, a penalty of between 14 and 33 % occurs depending on the data rate. This makes us conclude that situations involving a relay via 2 WiFi hops are only beneficial, if their data rates are relatively high, or the intermediate nodes are able to

Table 8.5 4-node wired total radio energy (originator) 200 MB file transfer

4-node wired—cabled environment			
Total radio energy (originator node)			
200 MB file transfer			
WiFi data rate	WUSB data rate	Direct scenario	Relay scenario
Mbits/s	Mbits/s	Energy (J)	Energy (J)
23.30	16.20	108.80	105.38
21.80	17.70	114.80	104.81
18.20	17.60	148.75	104.54
14.60	17.60	190.39	109.79
11.00	18.00	245.53	105.92
10.60	17.60	232.19	114.98
8.62	14.30	293.09	141.79
5.52	9.78	459.94	202.18

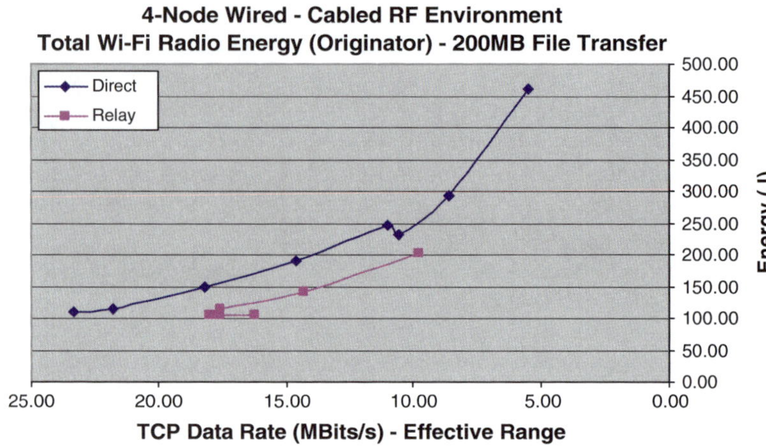

Fig. 8.11 4-node wired total radio energy (originator) versus range 200 MB file transfer

Table 8.6 4-node wired energy consumption (originator + intermediate) 200 MB transfer

4-node wired—cabled environment			
Energy consumption (originator + intermediates)—200 MB file transfer			
Direct TCP rate	Relay TCP rate	Direct scenario	Relay scenario
Mbits/s	Mbits/s	Energy (J/Mbit)	Energy (J/Mbit)
23.30	16.20	1327.55	2095.28
21.80	17.70	1422.67	1913.55
18.20	17.60	1587.52	1918.51
14.60	17.60	1596.23	1942.65
11.00	18.00	2783.82	1850.69
10.60	17.60	2823.86	1921.69
8.62	14.30	3536.75	2336.39
5.52	9.78	5461.74	3378.72

act as store and forward agents. Then the UWB nodes are able to save power by sending their traffic at high data rates. Otherwise a stream is set up between the AP and the Originator node, which is effectively throttled to the data rate of the slowest link in the relay chain and can prevent the full benefits being realized.

8.2.5 Showcase Wrap up

Summing up the first showcase of the chapter, we can deduce that using short rang cooperative relaying is definitely a source of energy savings for mobile handsets. The Originator (source) node always save energy, which can reach up to more than

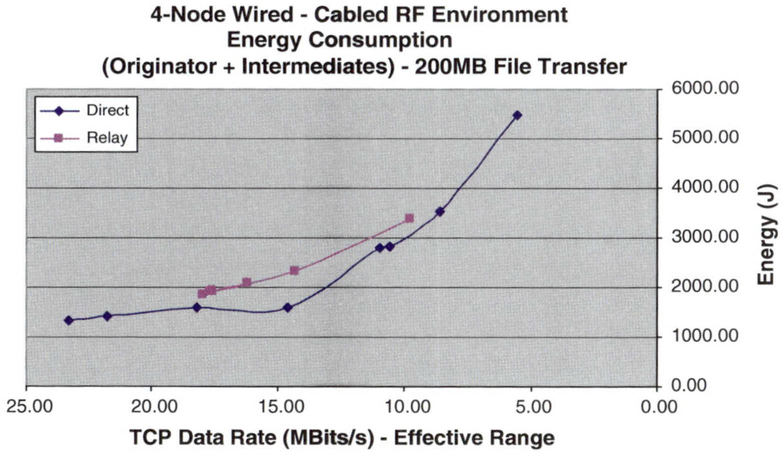

Fig. 8.12 4-node wired energy consumption (originator + intermediate) 200 MB transfer

50 % savings, whereas the Intermediate (relay) node has to pay the price by increasing its energy consumption for helping other mobile handsets in need. This cooperation between different mobile handsets is essential for the successful energy savings of handsets using cooperation. Methods to guarantee the cooperation of users are required, such as rewarding cooperative nodes. Such methods have been previously discussed in Chap. 5 and business models based on such cooperative behavior is presented in the following chapter.

Based on the above results in the subsection, the cabled RF environment provides a useful baseline against which to compare the results of wireless tests. The power saving figures of the two-node testbed either 'Wireless' or 'Cabled' show similar power savings and throughput behavior providing confidence that the wired environment was a realistic emulation. This enabled the four node test bed to be developed by providing values with which to configure the attenuators on the TC to represent equivalent distances between Nodes and to the AP.

Considering the results of the two-node wireless test bed, the energy saving varied from 14 to 68 %, with the increase in the distance between Originator and AP. If we look at the energy consumed by just the radios, the saving varied from 31 to 76 %. This indicates that applications which disconnect when idle and handset operating systems that can shut down radios aggressively will benefit the most. The results of the two-node Cabled environment, gave similar results to that of the wireless open office, although a bit pessimistic. The wired environment facilitates testing over a greater 'range' and reproducibility than that available within the confines of an office environment and based on the results can be justified as representative of a wireless environment when incorporating high numbers of nodes.

The energy consumption increased with distance, as the WiFi radio had to increase its transmit power in an attempt to keep throughput up in the face of MAC rate adaptation and the longer transmission time of slower data rates thus favoring relaying and even multi-hop relaying as range to the AP increases.

External RF interference from other WiFi AP in the vicinity caused variation in available bandwidth, such that the throughput varied by typically ±5MBits/s from one run to the next. This inconsistency and the unpredictable nature of its impact was one of the original motivators for developing the wired platform.

Inspecting the results of measurements of the 2- and 4-Node scenarios demonstrates that significant energy savings are achievable when using a mix of high-speed short-range with the lower speed longer range radio technologies. However, the savings experienced by individual users will vary significantly depending on their location and the type of activity in which they are engaged. High density environments where nodes are slow-moving or static, such as a busy airport lounges, public transport or coffee shops with users operating tablets, smart phones, and laptops, offer rich opportunities for relaying and will show the greatest benefits. Fast-moving mobile or rural situations will offer little chance to form worthwhile relays due to rapidly changing data rates and so cannot expect to derive much benefit. Situations involving a relay via 2 WiFi hops as demonstrated in the 4-Node scenario are only beneficial, if their data rates are relatively high, or the intermediate nodes are able to act as store and forward agents. Then the UWB nodes are able to save power by sending their traffic at high data rates. Otherwise a stream is set up between the AP and the Originator node, which is effectively throttled to the data rate of the slowest link in the relay chain and can prevent the full benefits being realized. This is yet another example of how the context in which a Node finds itself determines how close to the maximum energy savings it can get.

8.3 Energy Efficient Handovers in Heterogeneous Radio Networks Showcase

Within this book, we seek to provide energy efficiency for heterogeneous radio access networks with the main focus on the techniques that minimize energy consumption of multi-standard mobile terminals. In this subsection, we concentrate our efforts on showcasing energy efficient smart cognitive handover algorithms. Two possible types of handovers are considered that may eventually lead to energy savings on the mobile terminal side: Vertical Handover and Macrocell-Femtocell Handover.

Vertical Handover approach. The wireless ubiquitous all-IP networking and exponential growth in mobile traffic demand [7] enables us to envision a truly mobile environment where users can freely and globally roam between different points of attachment, with continuous service reception. Such an ability to handover between different heterogeneous networks is typically defined as the Vertical Handover (VHO) [8]. In order to give us a better understanding of how to address the problem, one can find Vertical Handover process as a three phase mechanism, where the consecutive phases are: (1) Network discovery and context evaluation, (2) VHO decision, (3) Handover execution. The first phase "Network discovery" enables the creation of a candidate set of available access networks, to which the VHO decision shall be limited. In the second phase, based on the input parameter set, a radio access network is selected that maximizes or minimizes the objective value (e.g. maximize throughput and/or minimize energy consumption), constrained by user policies (e.g. preference of cheaper service), service requirements (e.g. min. data rate, max. delay) and operator policies (e.g. restricted networks, inter-operator handover policies and load balancing purposes). In the last phase, the point of attachment is changed and an on-going service is seamlessly transferred between two different access networks, which involves radio link transfer, resource reservation in the new access network and mobility management tasks for service continuity. Herein, we mainly concentrate our efforts on validating the energy savings due to the solutions developed for the second step of the VHO process, i.e. the energy efficient VHO decision algorithms.

Macrocell-Femtocell Handover approach (M-FHO). The energy efficient handover algorithms investigated here are not limited to heterogeneous networks (vertical domain), as energy savings can be realized also in homogeneous networks, especially with high rate deployment of Femtocells. This case shall be denoted as the Macrocell-Femtocell Handover (M-FHO). In Macrocell-Femtocell Handover the energy saving gain arises from the proximity of a femto base station. The Macro-Femtocell Handover requires additional procedural enhancements to realize the connectivity with the Core Network along third-party home subscriber lines and to enable efficient femtocell discovery (macrocells do not include femtocells in the neighbor lists).

The demonstrative showcase highlights both the VHO and M-FHO to provide one goal: energy efficiency of the multi-standard terminals.

8.3.1 Testbed Architecture

In order to evaluate the energy efficient Handover strategies (Vertical Handover and Macro to Femto Handover decision) proposed and developed within Chap. 5, a testbed has been put together which integrates together various modules including context aware modules which provide location-specific information about the user (received signal strengths from the available access points, the location of the user, power consumption of each adapter on the user terminal, remaining battery etc.), a centralized database which, together with the Handover decision algorithms, an IP mobility implementation which enables the execution of the handover to ensure session continuity and an energy evaluation module which evaluates the energy consumption of the mobile handsets, demonstrating the energy efficiency achieved by the proposed cognitive handover algorithms. In this showcase, mobile handsets possess two different radio interfaces. The technologies represented in this showcase are WiMAX [9, 10] and WiFi [1, 2]. The WiMAX technology includes both Macro and Femto cells.

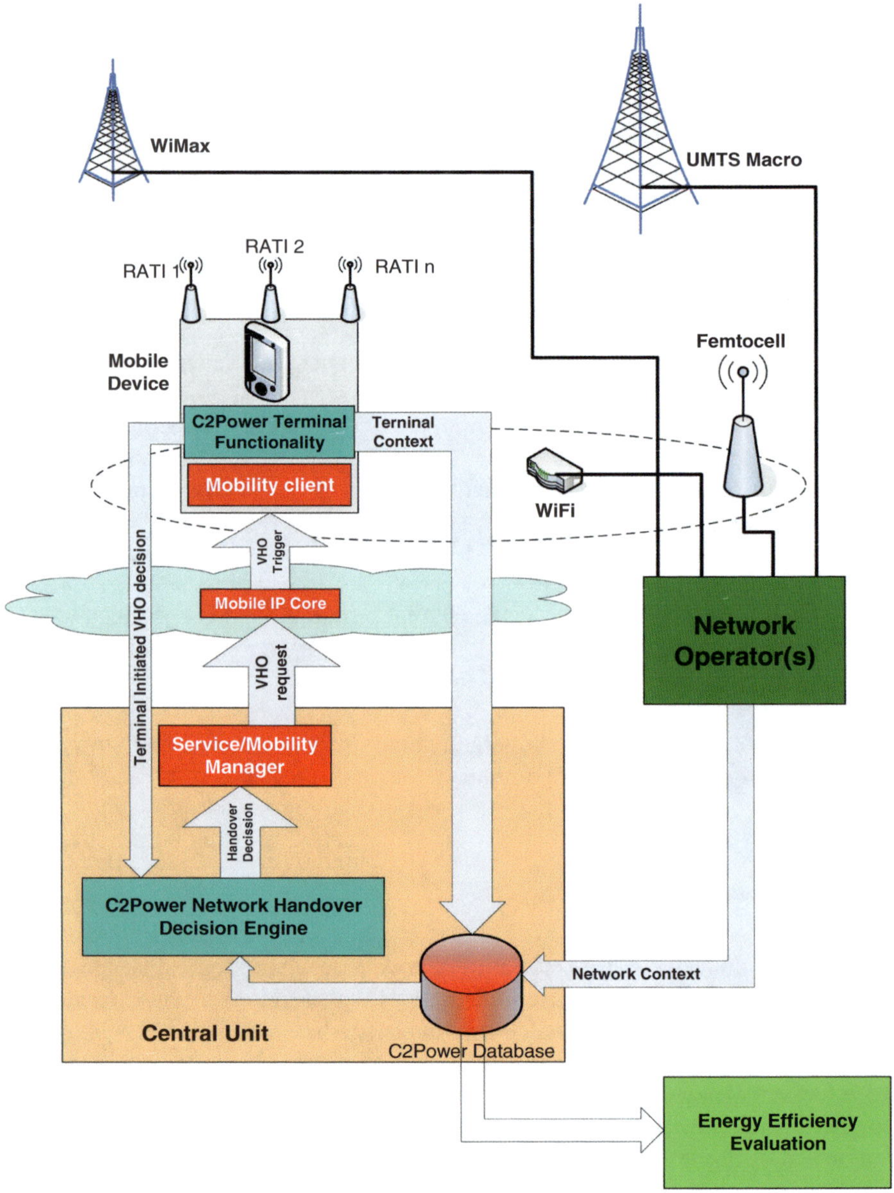

Fig. 8.13 The complete handover architecture framework [5]

The complete architecture of the testbed setup is shown in Fig. 8.13. The testbed integrates together the various entities that are required for the proof of concept of energy efficiency in heterogeneous networking environments through the use of energy efficient handover strategies; this includes:

- The heterogeneous networking infrastructure (the user terminal, the Base stations/access points and their respective radio adapters installed on the terminal)
- The platform which is responsible for executing the handovers (based on the mobile-IP protocol)
- A central processing unit which contains a context-database and the handover decision algorithms to be tested
- The context modules which collect/generate and provide real-time context in the database to be used by the Handover algorithms
- An energy evaluation module to evaluate the proposed architecture in terms of energy efficiency and prove the concept.

The testbed architecture consists of four main modules, which are the mobile IP server, the mobility manager, the terminal client and the database. A brief description of each entity is presented as follows:

Mobile IP Server implements the Mobile IP functionality through the provision of a Home Agent (HA) and a Foreign Agent (FA). The server receives a HO request and sends a trigger to the terminal. The server was implemented so that it can be located anywhere on the Internet and can be accessible through TCP, through a static IP.

Mobility Manager is responsible for generating and forwarding the appropriate commands to the server when a ho decision is generated.

Terminal MIP Client implements the MIP protocol on the terminal side. The client receives the HO decision from the MIP Server and executes it by transferring the IP traffic from one technology adapter to another.

Context database hosts the necessary context needed by the MIP functionality in addition to the context required by the HO algorithms. The database represents a subset of the ANDSF specification and provides the necessary context required by the handover algorithms to generate decisions in terms of energy efficiency. This context involves network-based context as seen by the terminal (e.g. available network links, received signal strength, available connection rates), energy context (e.g. power consumption of the adapters of each user and its remaining battery level), user preferences context (e.g. preferred or restricted networks, required cost etc.), user dynamic information (e.g. user position, velocity, etc.).

It can be seen in Fig. 8.13 that the architecture is generalized to support multiple technologies, which can be more than two, although in the showcase it is restricted to two types of technologies: WiMAX and WLAN (IEEE 802.11 g).

8.3.2 Platform Physical Implementation

For the demonstrative results, all the modules that constitute the demonstration architecture have been integrated in a local network (to allow portability) around a D-Link Router, which serves as a Domain Name Service (DNS) server and also as a WiFi Access Point. For more options, the platform is also designed with the option to be integrated in a distributed network on the internet where every entity could be located in any physical geo-location and would be accessible over the Internet via TCP/IP. This DNS router constitutes the heart of the networking architecture and provides connectivity to any base stations and/or access points that form the heterogeneous radio network. In this demonstrator, the router itself is a WiFi Access Point, and additionally a WiMAX Base station is connected to it. For our demonstrator and the proof of concept, these two Radio Access Technologies (RATs) will be used. A schematic diagram of the practical implementation is shown in Fig. 8.14.

The DNS router also provides direct connectivity to the Mobile IP (MIP) core-Service Manager suite and also to the central processing unit which contains the context database, the Handover Decision algorithms and the energy efficiency Evaluation Module. The MIP-core and the Service manager are implemented on a Linux-Based Machine, which is sitting on a dedicated static IP on the network. The MIP core implements the Mobile IP functionality through the provision of a Home Agent (HA) and a Foreign Agent (FA) in the context of the Mobile IP protocol. It receives the HO request from the Service manager, which is implemented on the same machine and executes the handover (in cooperation with MIP functionality implemented on the client) and if the HO is executed successfully, it updates the database accordingly (updates the Care of Address—CoA which reflects the user

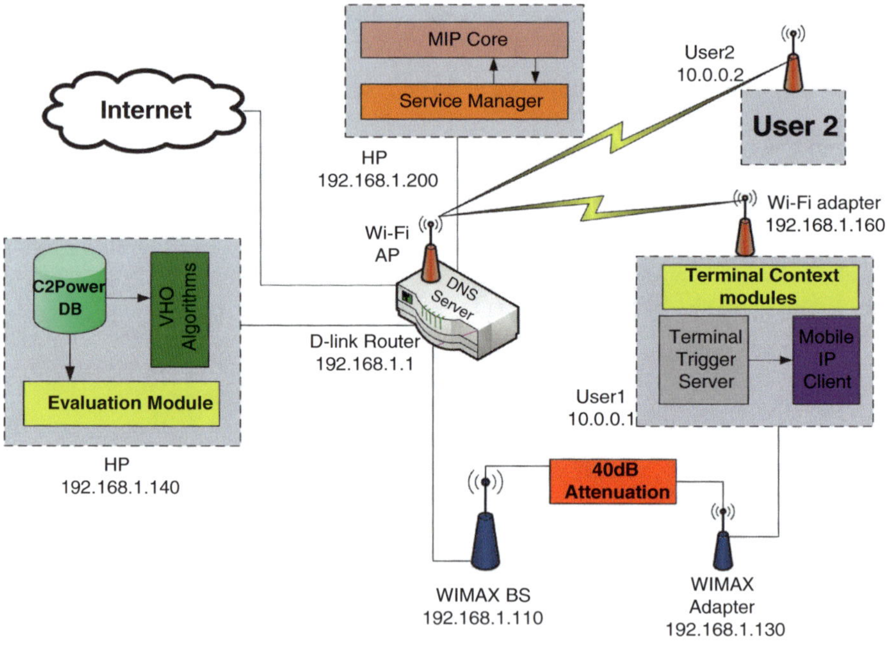

Fig. 8.14 Practical implementation of the handover evaluation architecture [5]

point of attachment to the network). The service connection manager (SCM) is basically a server which listens to the network and detects whether a handover decision has been generated from the HO algorithms, which are sitting on a different machine. It wraps up the HO trigger and sends it to the Mobile Terminal which executes it and replies back with a confirmation of execution. The central unit (database, HO algorithms, evaluation module) is integrated into a Windows-based laptop, which is also attached directly to the DNS router through a static IP. The context database is the central point of context exchange within the whole architecture since it receives terminal and network related context from the terminal and network context aware modules and makes this information available to the Handover Decision algorithms. The database has been implemented in Microsoft SQL together with its communication interfaces to each module in the architecture. The different handover algorithms to be tested use various combinations of this context to make energy efficient decisions.

All algorithms periodically (in predefined interval e.g. 1 s) communicate with the database to pull the required context. They perform the necessary processing based on their algorithm and they generate (if needed) a handover trigger which is forwarded to the Service Manager. All the communications are achieved with dedicated client-server communication applications (either with the database to retrieve information or with the Service connection manager to forward the handover decision).

An energy evaluation module is also implemented on the central unit to monitor a specific user in the heterogeneous environment (through a graphical interface) and evaluate the performance of the integrated algorithms. This module monitors the user while the user moves in the environment, indicates and monitors its available networks and links along with their respective signal strengths and QoS metrics (connection rate, throughput), displays the position of the user and his/her currently connected technology, shows the locations that handover decisions have generated but most importantly it evaluates the performance of the handovers in terms of:

- Energy (in Joules) consumed over a specific route
- The number of handovers generated
- The average connection rate
- The number of network discoveries carried out to discover new links

The terminal is a conventional laptop which employs 2 USB Wireless adapters; one WiFi and one WiMAX adapter. The USB connection is intercepted by two ammeters connected in series to measure the real-time power consumption of the adapters connected to the terminal. These ammeters provide an RS232 connection that allows data passing to the computer. The specific ammeter model RS232 communication protocol has therefore been implemented in order to pass this data to the computer and from there to the context database through the power consumption sub-module which exists in the terminal context module. The power consumption module implementation periodically pulls the measured current consumption of the two adapters and fills in the necessary tables in the database. Practically, this sub-module co-exists in the overall implementation of the Terminal context aware module with the network discovery module, the positioning module, the connectivity module and the user-preferences module.

As mentioned above, the developed architecture is used to demonstrate the benefit in energy efficiency of the developed Handover strategies. This practically means to compare the developed energy efficient handover algorithm to a base-line handover algorithm, which is based on strongest received signal. To do this and obtain comparable results, the radio propagation environment needs to be fixed for all trials, which is difficult to achieve in practice due to the dynamics of the propagation channels. For this reason the propagation channel is modeled using a channel simulator (a 3D Ray Tracing Tool), the channel predictions are saved into a file and are loaded upon the trial.

Finally, in order to generate TCP traffic to the client a second Windows laptop serves as a second user. The purpose of this second user is to serve as a media server for streaming a real-time video to the primary user, while the latter is handing over between heterogeneous networks in an energy efficient way. VLC media server has been setup on user 2 and the primary user remotely views the video through the network.

8.3.3 Experiment Setup

The main scenario revolves around a mobile user who is roaming in heterogeneous wireless environment which employs WiMAX and WiFi Radio Access Technologies. A general presentation of the scenario can be seen in Fig. 8.15.

The more specific scenario that has been implemented in the showcase is shown in Figure, which defines a multi-RAT (WiFi and WiMAX) user which roams in an outdoor/indoor environment where there is WiMAX and WiFi coverage. The WiMAX Base station is located outdoors in a central position so that it provides coverage in the entire area. The WiMAX transmitter is placed 20 m above the ground and its transmit power is 15 dBm. There are also 6 WiFi Access Points (1 in his house, 2 in his office and 3 in the shopping mall) which for the purpose of investigation their transmitting powers have been set equally to 0 dBm. The user is assumed to be on foot with constant speed at 1.83 km/h. The whole trip of the user across the scenario under investigation is 1,306 m and the user requires 43 min to travel across the whole route. In order to generate traffic for the user, a second static user is connected to the platform that transmits a real-time video or TCP traffic (through iperf) to the primary user. Figure 8.16 show the coverage of all the WiFi access points.

In the second test, the WiFi access points are replaced by WiMAX Femto cells, to test the energy efficient Macro to Femto handover algorithm.

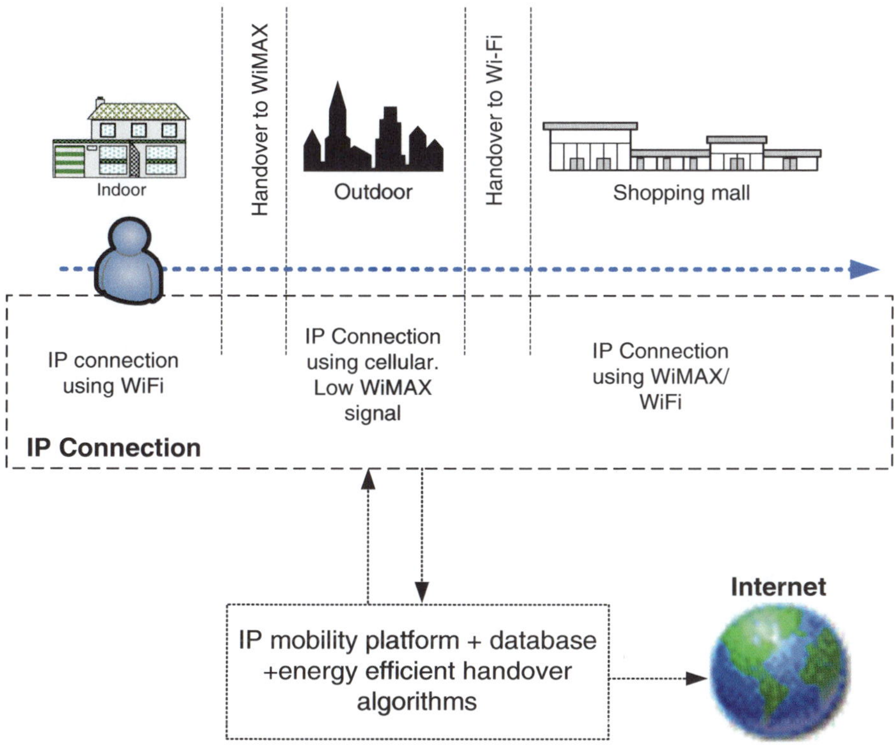

Fig. 8.15 Typical handover scenario [5]

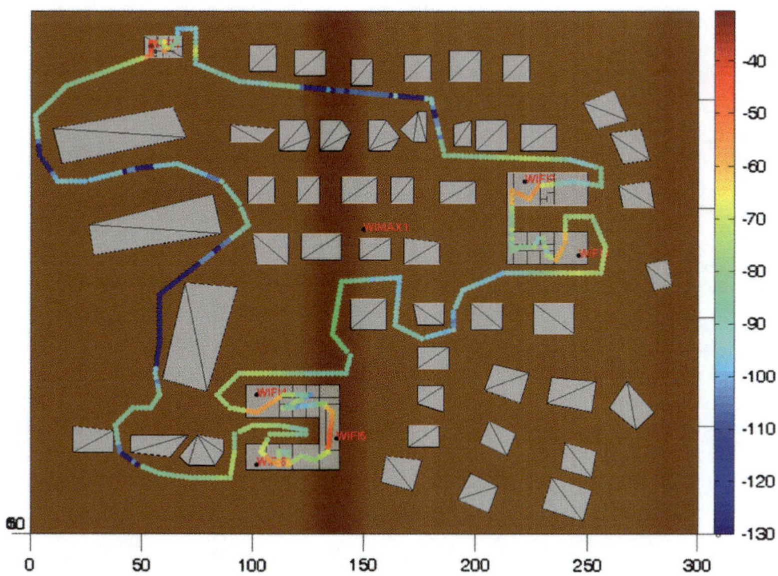

Fig. 8.16 The coverage of the implemented 6 WiFi APs [5]

8.3.4 Measurements and Results

8.3.4.1 Measurement Results for Energy Efficient Vertical Handover Algorithms

For the test-setup, environment layout and network topology described above, the proposed energy efficient vertical handover algorithm has been compared against the case where the user terminal remains connected to the WiMAX Macro cell (base case) which offers coverage in the entire area. The algorithm has been evaluated in terms of the total energy consumed over a complete run (in Joules), the number of performed Handovers and the average connection Rate (QoS) which is offered to the user. The total energy is calculated as the summation of the energy consumed by the connected adapter plus the energy consumed of the not connected adapter which is in idle state.

The Energy Efficient vertical handover algorithm (EE-VHO) has been compared against the baseline scenario where the user remains connected always to the WiMAX Macro cell. In the baseline scenario, no handovers occurs and the total energy consumed is 8.31 Joules and the average connection rate is 7.23 Mbps. For the EE-VHO algorithm, the locations of the vertical handovers are shown in Fig. 8.17 as red circles. In the figure, the black trace indicates that the terminal is connected using the WiMAX adapter whereas the pink trace indicates that the terminal is connected to WiFi (Fig. 8.18).

The energy efficient vertical handover has been compared against typical signal-based VHO algorithm. The signal-based algorithm searches for new links when the signal strength received from the currently connected cell or access point drops below −70 dBm. If the new preferred link is of the same technology then a horizontal handover occurs whereas when the preferred link is of different access

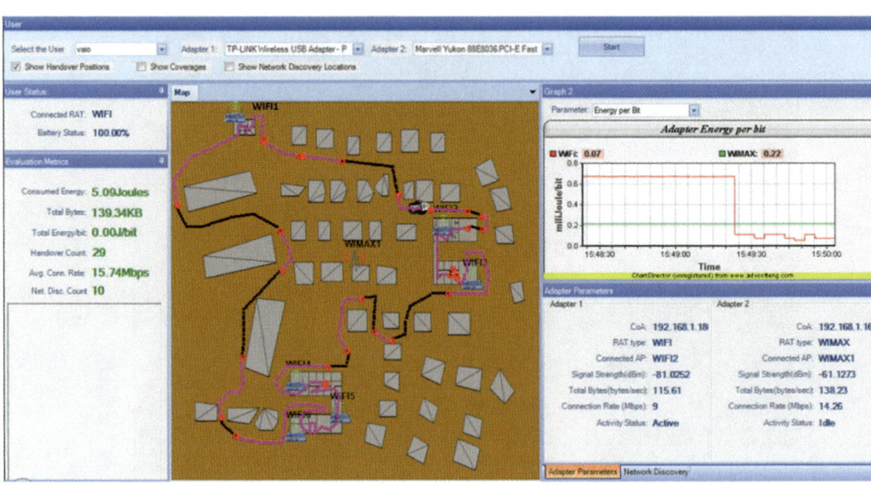

Fig. 8.17 Trace of handovers using energy efficient vertical handover algorithm [5]

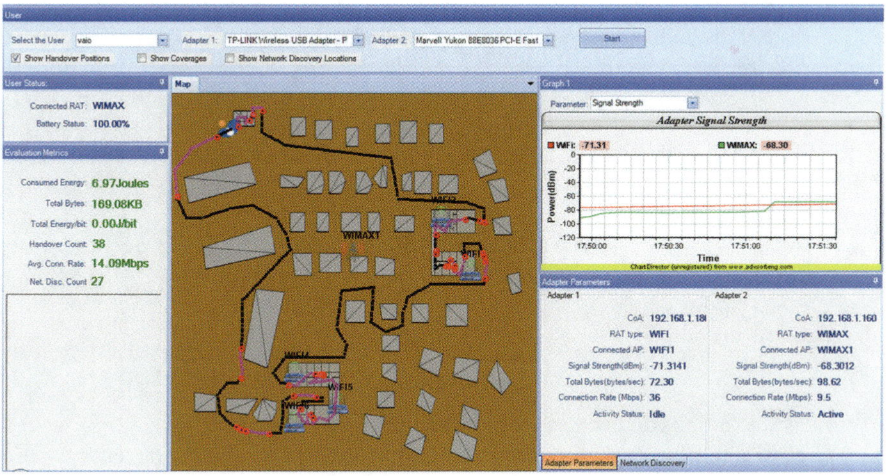

Fig. 8.18 Trace of handovers using signal-based vertical handover algorithm [5]

Table 8.7 Vertical Handover results summary

	Algorithms		
	Always attached to WiMAX	Signal based VHO	Energy efficient VHO
Consumed energy (Joules)	8.31	7.01	5.11
Number of VHOs	0	40	29
Average connection Rate (Mbps)	7.23	14.05	15.72
Energy saving percentage against the baseline case	–	15.6 %	38.5 %
Energy saving against the signal strength-based algorithm	–	–	27.1 %

technology then a VHO occurs. The handover results from the signal-based algorithm are shown in Figure.

Table 8.7 summarizes the results of the energy efficient VHO algorithm and compares it to the baseline case (the terminal is always connected to WiMAX) and the case where the basic signal-strength based algorithm is used. It can be observed from the results that the energy efficient VHO algorithm offers significant improvement compared to the baseline scenario (energy efficiency is improved by 38.5 %. Also compared to the signal-strength based VHO algorithm the algorithm offers an improvement of 27 % in energy efficiency. The algorithm also offers better connection rates since the terminal remains more time connected to the WiFi. The energy benefit is correlated with the time that the user spends in the WiFi areas. This is demonstrated by a second set of results, where the user spends less time in indoor areas and also the transmitting power of the WiFi access points is less. However, relatively the algorithms perform similarly compared to each other. The results of this second scenario are demonstrated in Table 8.8.

Table 8.8 Vertical Handover results summary for scenario 2 (user spends less time indoor)

	Algorithms		
	Always attached to WiMAX	Signa based VHO	Energy efficient VHO
Consumed energy (Joules)	2.64	2.39	1.51
Number of VHOs	0	17	25
Average connection rate (Mbps)	11.11	15.8	13.69
Energy saving percentage against the baseline case	–	9.46 %	42.8 %
Energy saving against the signal strength-based algorithm	–	–	36.8 %

From Table 8.8, it can be seen that, despite the user being less time indoor closer to the WiFi AP, the energy efficient VHO algorithm still reduces the energy consumption even more compared to the baseline scenario (42.8 %) and compared to the signal-based VHO algorithm (36.8 %). It can be noticed that the energy saving ratios are increased in the second scenario, since the algorithm switches to WiFi as long as it can be reached despite slightly decreasing the average achieved data rates. The difference in the energy savings can be related to the comparison between the number of handovers in both scenarios. In the first scenario, where the user spends more time indoors, the energy efficient VHO remains connected to WiFi all the time while indoors, in contrast the signal-based algorithm always keeps switching between WiFi and WiMAX due to the change in the power of the received signal; hence the signal-based algorithm misses on opportunities to decrease the energy consumption by being connected to the WiFi all the time while indoors. On the other hand, in the second scenario, the opposite occurs. Although the user spends a lot of time outdoors, the energy efficient VHO algorithm takes advantage of the availability of WiFi coverage outdoors and hence handovers to WiFi more often whenever possible to save energy by being connected to the WiFi APs. These results show how adaptive the proposed algorithm is and confirms that the algorithm tries to exploit every opportunity to take advantage of all available low energy consuming access points.

8.3.4.2 Measurements Results for Energy Efficient Macro to Femto Handover Algorithms

As mentioned in the experiment setup, the same environment has been used to demonstrate the benefit of macro to femto handovers. The same network topology has been used with the only difference that in the positions of the WiFi access points, it was assumed that Femto WiMAX base stations exist. The algorithm, although originally developed for Macro—Femto HO in LTE, has been adapted to practical implementation on WiMAX standard. The average consumption of the adapter when connected to the Macro cell is around 600 mA and the average consumption when connected to the femto is around 200 mA. As in the case of energy efficient VHO, the developed Macro—Femto HO algorithm has been compared against the baseline case where the user remains connected to the Macro WiMAX cell for the whole duration of the experiment and also against the case where received signal strength-based algorithm is used. In the case that the user remains connected always to the WiMAX Macro cell no handovers occur and the total energy consumed is 7.93 Joules and the average connection rate is 7.23 Mbps.

The trace of the locations where HO stake place using the Macro-Femto HO algorithm is shown in Fig. 8.19, and likewise for the signal strength based algorithm in Fig. 8.20. The results are summarized in Table 8.9. It can be observed from the results that the Macro to Femto Handover algorithm improves the energy consumption by

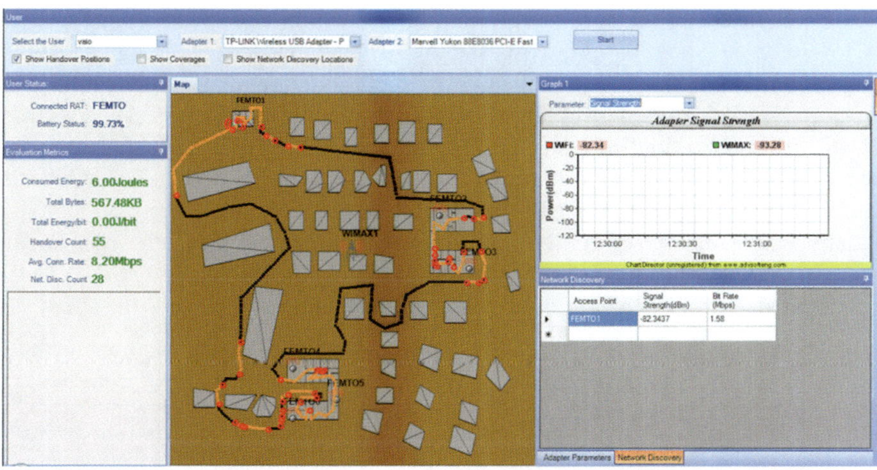

Fig. 8.19 Trace of HOs for the Macro-Femto HO algorithm [5]

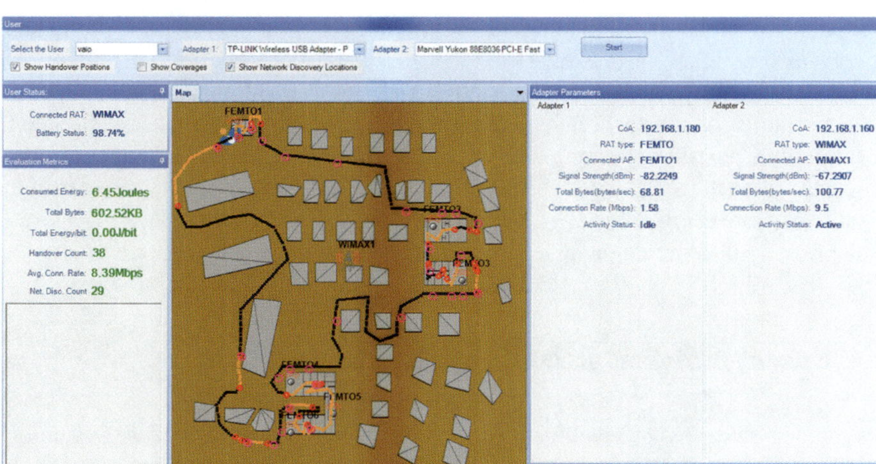

Fig. 8.20 Trace of HOs for the Signal based algorithm [5]

Table 8.9 Macro-Femto HO results summary [5]

	Algorithms		
	Always attached to WiMAX Macro	Signal based HO	Energy efficient Macro-Femto HO
Consumed energy (Joules)	7.93	6.48	6.00
Number of VHOs	0	41	37
Average connection rate (Mbps)	7.23	8.36	8.20
Energy saving percentage against the baseline case	–	18.2 %	24.3 %
Energy saving against the signal strength-based algorithm	–	–	7.4 %

24.3 % compared to the baseline case, and is also better than the simple signal strength based algorithm by 7.4 %, while the QoS is slightly improved.

As in the case of the energy efficient VHO, the energy benefit is correlated with the time the user spends in the Femto areas. The proposed Macro-Femto algorithms, like the energy efficient VHO algorithm, tries to keep the mobile handset connected to the Femto (lower energy requirements), as long as possible to save energy compared to signal strength based algorithms. This can be clearly seen from the comparison between Figs. 8.19 and 8.20, where using energy efficient Macro-Femto HO the mobile handset stays longer periods connected to the Femto cell (Orange trace).

8.3.5 Showcase Wrap up

Summing up the handover showcase, it is clear that the proposed energy efficient HO algorithms (VHO and Macro-Femto) outperform staying connected to one technology (or only to Macro cells); additionally, they outperform signal based HO algorithms. The practical implementation of the showcase acts as a proof of concept of performing energy aware HOs (Vertical or Horizontal) to reduce the energy consumption of mobile handsets, by increasing the periods when mobile handsets are connected to the access point (base station) offering the required quality (i.e. data rates, delays, etc.) at the lowest possible energy requirements. It is worth mentioning that the energy efficiency percentage is highly correlated with the time that the user terminal is in the vicinity of more energy efficient network links; hence those ratios will vary according to the density of more efficient access points (i.e. WiFi or Femto cells).

8.4 Energy Efficient Reconfigurable Radios

The final showcase was developed to verify the benefits of the hardware technology blocks developed in Chaps. 2 and 3. The showcase demonstrates an energy-efficient reconfigurable radio transceiver. The demonstrator comprises an envelope-tracking

Power Amplifier, Doherty Power Amplifier, Tunable Matching Network integrated circuit and a miniature multi-band antenna. The objectives of the proposed set-up were to demonstrate experimentally the performances of the hardware prototypes and the energy efficiency improvement obtained with respect to a standard architecture (i.e. standard PA and no TMN).

8.4.1 Testbed Architecture

In order to demonstrate the benefits of the hardware technology blocks developed in Chaps. 2 and 3, an energy-efficient reconfigurable radio demonstrator has been designed and implemented with the following blocks:

- 1-W envelope-tracking Power Amplifier (PA) at 1,920–1,980 MHz operating with LTE up-link signals,
- 1-W Doherty Power Amplifier (PA) at 3.4–3.6 GHz operating with WiMAX up-link signals,
- Tunable-Matching-Network (TMN) integrated circuit,
- Multiband miniature antenna covering the GSM-900 frequency band, as well as all the bands from 1,900 MHz to 3.6 GHz (i.e. including the two PAs operating bands).

A conceptual description of the demonstrator is presented in Fig. 8.21, representing the hardware blocks developed in previous Chaps. 2 and 3 in green and the laboratory instruments in other colors. Two signal generators are used to generate the LTE and WiMAX up-link modulated signals; these signals are amplified and sent to the TMN antenna module, which integrates a SPDT switch (for LTE/WiMAX selection), the TMN circuit, and the antenna. A computer monitors the PAs' output power through two power sensors, measures the reflection coefficient at the antenna input and controls the TMN circuit. The radiated signal is measured with a signal analyzer through an ad hoc antenna.

The objectives of this set-up are to demonstrate experimentally the performances of these hardware prototypes and the energy efficiency improvement obtained with respect to a standard architecture (i.e. standard PA and no TMN). As presented in Chaps. 2 and 3, a figure of merit, named Power Added Efficiency, was defined to characterize the global power budget of the demonstrator from the input RF power to the radiated RF power taking into account the power consumption of each block [11]. For detailed description of the modules integrated in the showcase, readers are referred to the description, design and implementation of the proposed modules in Chaps. 2 and 3.

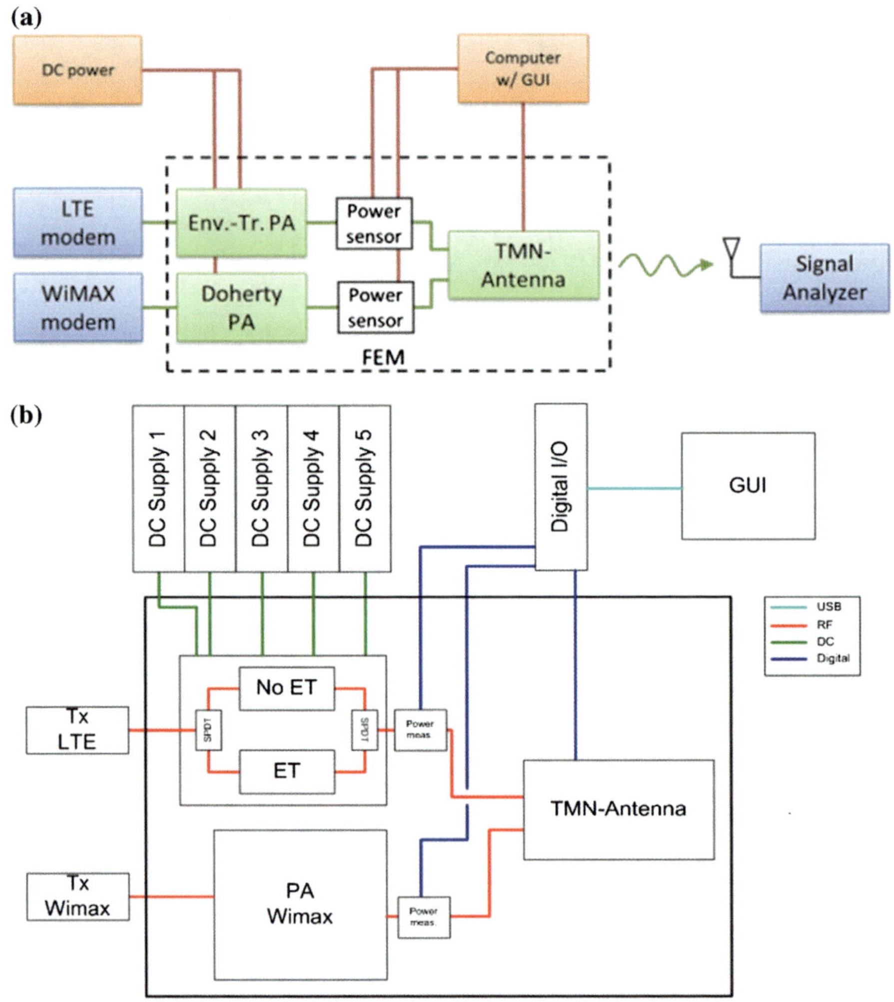

Fig. 8.21 Overview (a) and detailed description (b) of the energy-efficient reconfigurable radio demonstrator [5]

8.4.2 Platform Final Integration

The integrated showcase comprises modules that have been designed, prototyped and implemented in previous Chaps. 2 and 3. In order to ease the integration of such prototypes into the overall demonstration showcase, some modifications have been applied to such modules. In this section, we describe the modifications applied as well as the final overall integration of the full showcase.

8.4.2.1 Envelope-Tracking Power Amplifier

Compared to Chap. 3 prototype, some modifications have been taken into account to simplify the integration with the other components of the TX chain. Different boards and components designed have been assembled in one box, which are:

- 1 board with Envelope Tracking System (2 driver stages + final stage with ETS)
- 1 board with class AB Power Amplifier (2 driver stages + final stage without ETS)
- 2 RF switches
- DC cables

The complete demonstrator allows evaluating the enhancements added by the Envelope Tracking System. The RF switches give the possibility to commute easily between the 2 boards without cable disconnection. A synoptic describing the inside of the box is given in Fig. 8.22.

The demonstrator is composed of two RF paths. The first path is going through a board integrating a class AB Power Amplifier with an Envelope-Tracking System, and the second one is with the same PA topology without ETS. Only one path works at once.

The selection of the RF path is done by two SPDT RF switches (1 input to 2 outputs at the RF input and 2 inputs to 1 output at the RF output). The input of the chain is connected to the RF SMA connector "IN-RF" through one switch and the output is connected to the SMA connector "OUT-RF" through the second one.

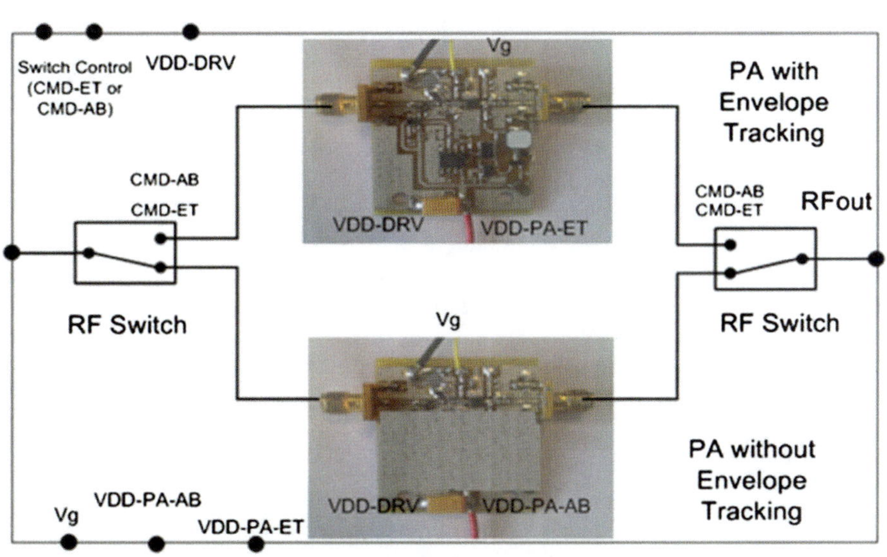

Fig. 8.22 Synoptic of the Envelope-Tracking demonstrator [5]

Only one DC power supply is needed to control the switches (CMD-ET for the ETS board and CMD-AB for the board without ETS).

On each board, the power amplifier is composed of three stages:

- Two driver stages
- One final power amplifier stage.

The two driver stages were not presented in Chap. 3. These components have been added to increase the gain of the TX chain and to be directly driven with the RF signal generator or LTE modem.

Moreover, the ETS board integrates the voltage modulator and the RF coupler with the drivers and final PAs on the same Printed Circuit Board. All drivers are self-biased and are supplied by a common DC voltage named "VDD-DRV". Each final stage is supplied by its own supply voltage named "VDD-PA-ET" for the Envelope tracking board and "VDD-PA-AB" for the class AB drain voltage.

The gates of each board are connected together in order to use only one DC Power supply.

8.4.2.2 Doherty Power Amplifier

A 3.5-GHz, 30-dBm Mobile WiMAX handset load modulation RF power amplifier has been designed using the TOM3 large signal model and FPD1500 transistor, and measured with RF measuring equipment, which include spectrum analyzer, network analyzer, power supplies, power meter, and a signal generator. The details of the Power Amplifier design have been previously introduced in Chap. 3. This design comprised several design steps for which the optimization is applied to each in order to obtain global high performances of the entire load modulation RF power amplifier. Initially, the design of carrier and peak amplifiers, input 3 dB 90° hybrid coupler designs, Output 90° offset line and impedance transformer designs were performed.

First of all, the 3-dB quadrature splitter was tested separately to see its performance. A 3-dB quadrature splitter is part of load modulation and if designed properly can contribute a lot to the total efficiency of the system. Our investigation shows that the operation of this technique is strongly influenced by the coupling factor of the input splitter. In fact, in this research 3-dB quadrature splitter have been designed (Fig. 8.23a) and tested in terms of the operating frequency and bandwidth, and this showed good results as shown in Fig. 8.23b, c. It should be noted that this splitter is at the input of amplifier and divides the input signal equally between the carrier and peaking amplifiers. The splitter, the Carrier Class B, the peaking Class C, and impedance transformer at the output are combined to form a load modulation RF power amplifier.

Figure 8.24a shows the prototype diagram of the proposed load modulation RF power amplifier with offset transmission line at both output and input circuit which maximize the overall system efficiency with the configuration of Class B amplifier.

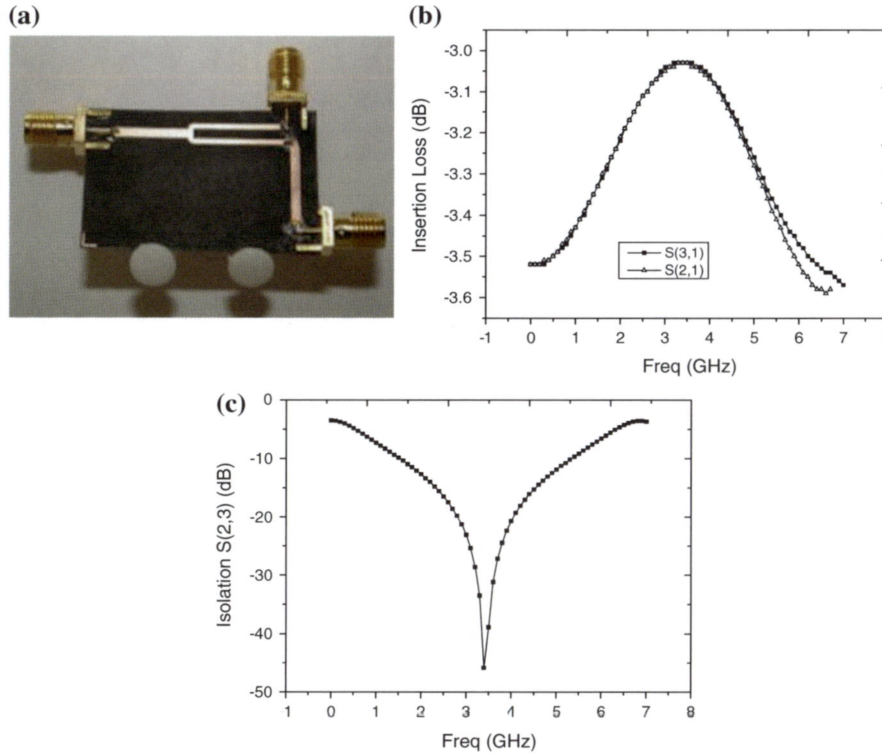

Fig. 8.23 Power splitter photograph (**a**), insertion loss (**b**) and isolation (**c**) [5]

The load modulation was initially characterized for AM-AM and AM-PM responses, as well as output power and efficiency. The performance comparisons between the load-modulation amplifier and Class-B amplifier are performed and the output power increased to 30 dBm at 1-dB compression point, while the efficiency increased to 53 %. Figure 8.24b presents the variation of the input power versus output power of the load modulation. It clearly shows that the 30 dBm output power level is in the linear region of the amplifier and this was achieved due to the characteristic of gain compression and expansion of the load modulation. The peaking amplifier Class C late gain expansion can compensate the carrier Class-B amplifier gain compression. Figure 8.24c presents the gain characteristic versus output power; the graph shows the power gain of load modulation amplifier is degraded drastically compared to Class-B due to the arrangement of lower biasing.

Figure 8.24d shows the power added efficiency (PAE) versus output power. The load modulation amplifier has a higher efficiency over the range of wide output power levels compared to Class B amplifier (Table 8.10).

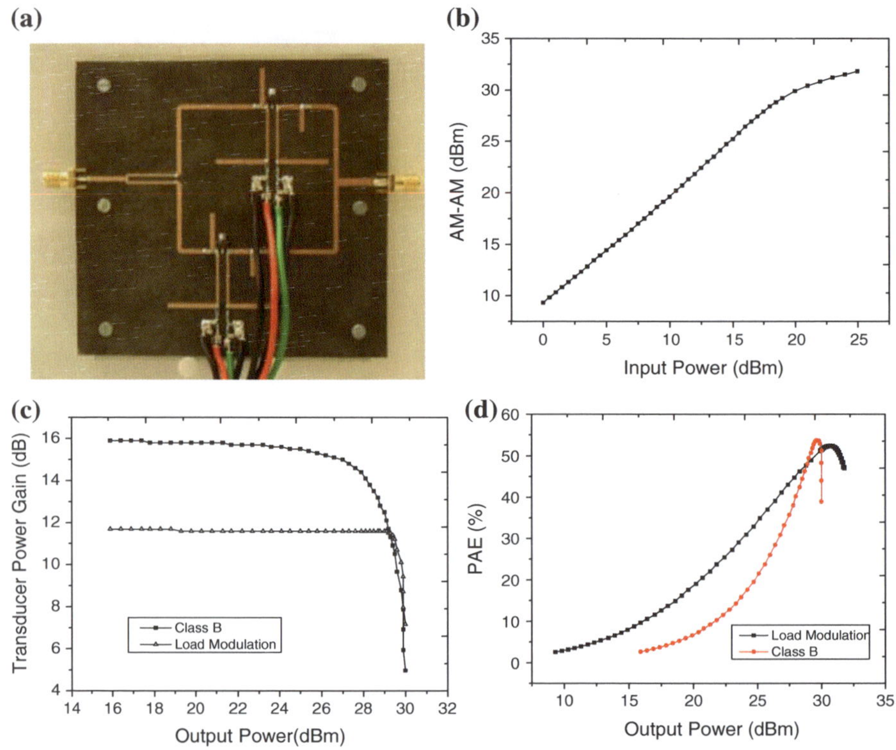

Fig. 8.24 Implemented prototype of energy-efficient load-modulation power amplifier (**a**), AM-AM characteristics of load modulation amplifier (**b**), gain characteristics (**c**), and Power-Added Efficiency (**d**) [5]

Table 8.10 Performance comparison of class-B and load-modulation power amplifiers at Pout 1 dB compression point

Amplifier	Gain (dB)	Pout (dBm) at 1 dB	PAE (%) at P1 dB
Class B amplifier	15.4	27.5	37
Load modulation	11.8	30	53

8.4.2.3 Tunable Matching Network and Antenna

The third main block of the platform is the TMN-Antenna module (Fig. 8.25), which includes:

- Dual-band antenna
- Tunable matching network (TMN)
- A coupler and two detectors to measure the incident and reflected power at the input of the TMN

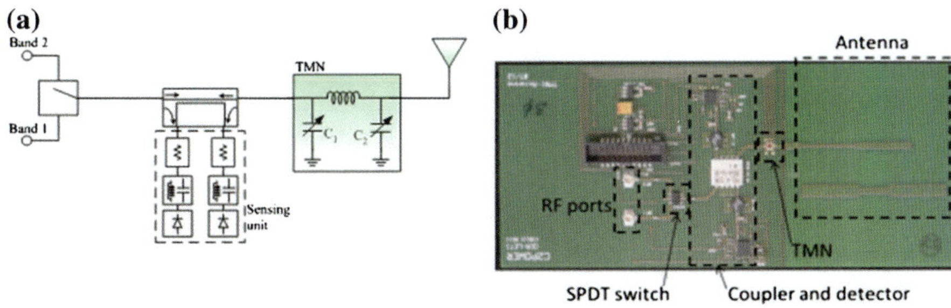

Fig. 8.25 Description of the TMN-Antenna module (size 110 × 50 mm^2) [5]

- A SPDT switch to select one of the two RF input ports (UFL connectors) associated to each PA of the platform.

A. ***Dual-band antenna***

The antenna performance was presented in Chap. 2. The miniature notch slot antenna is shown in Fig. 8.25. It exhibits a dual-band response with a lower band at 815–1,010 MHz (VSWR < 3:1) covering the GSM standard and a higher band at 1,800–3,600 MHz (VSWR < 3:1) covering, among others, the 1,920–1,980 MHz and 3,400–3,600 MHz bands chosen for this demonstration. Its efficiency is higher than 70 %.

B. ***Tunable matching network***

The Tunable Matching Network (TMN) is based on a Pi-network composed of a fixed external inductor and two integrated SOI CMOS Digitally Tunable Capacitors (DTC). In Chap. 2, the measured results of a first TMN circuit (TMN1) including two identical DTCs were presented. The obtained smith chart coverage was relatively good in the 1.7–2.7 GHz frequency range and was in line with simulations but insertion loss was high, partly due to DTC connections (bondwires). In this work, a new TMN circuit (TMN2) has been mounted and tested with/without the antenna. TMN2 includes two different DTC which have been designed to cover the range 0.7–2.8 pF (C1) and 1.5–5.0 pF (C2) respectively, with a 5 bits resolution (32 states for each capacitor).

The TMN has been characterized independently on a dedicated board to evaluate its insertion loss. As can be seen in Fig. 8.26, the TMN is centered on 50 Ohm and provides good impedance coverage at 1.95 GHz. The corresponding Insertion Loss (IL) has also been measured and is given for the DTC states that provide less than 1.5:1 VSWR at the input of the TMN-Antenna module. In the considered capacitance range, IL remains below 2 dB, with a minimum of 1.1 dB obtained for C1 = 17 and C2 = 6 which corresponds to a S11 of −14 dB. For this state, the simulated IL is lower than 1 dB.

First, the TMN-Antenna module has been tested in small signal as shown in Fig. 8.27, where the antenna is not perturbed. The S11 of the module is given for

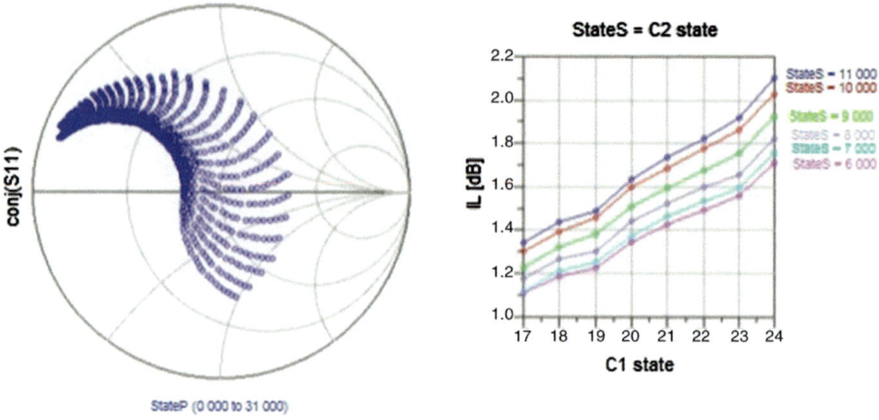

Fig. 8.26 Measured smith chart coverage (*left*) and insertion loss (*right*) of the TMN at 1.95 GHz [5]

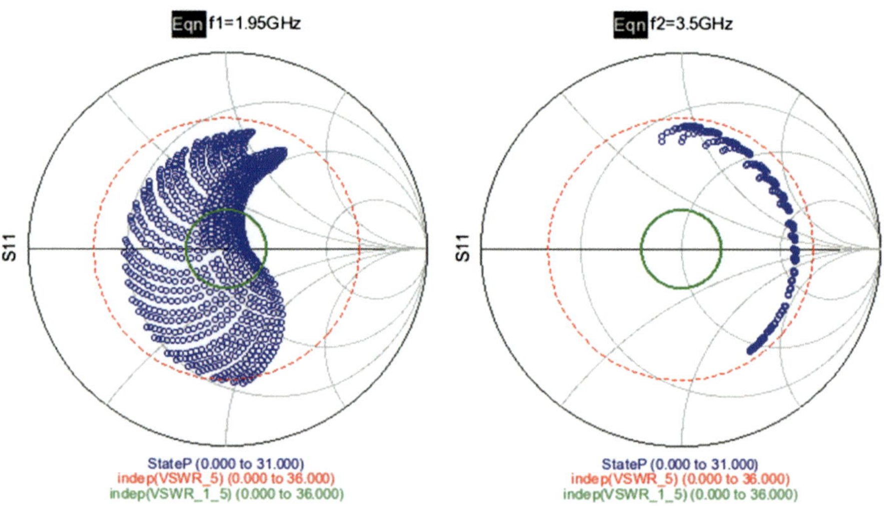

Fig. 8.27 Measured S11 of the TMN-Antenna module at 1.95 GHz (*left*) and 3.5 GHz (*right*); case with no antenna perturbation [5]

the 1,024 states of the TMN at two different frequencies (1.95 and 3.5 GHz). The TMN can provide a very good matching (VSWR < 1.5:1—green circle) at 1.95 GHz for the indicated DTC states (C1 = [17::24] and C2 = [6::11]). At 3.5 GHz, the smith chart coverage is very limited and the best matching (VSWR = 1.67:1) is obtained for state 0 (C1 = C2 = 0).

Figure 8.28 presents several cases where the antenna is perturbed and shows all the TMN impedance states at 1.95 and 3.5 GHz. In every cases, a very good

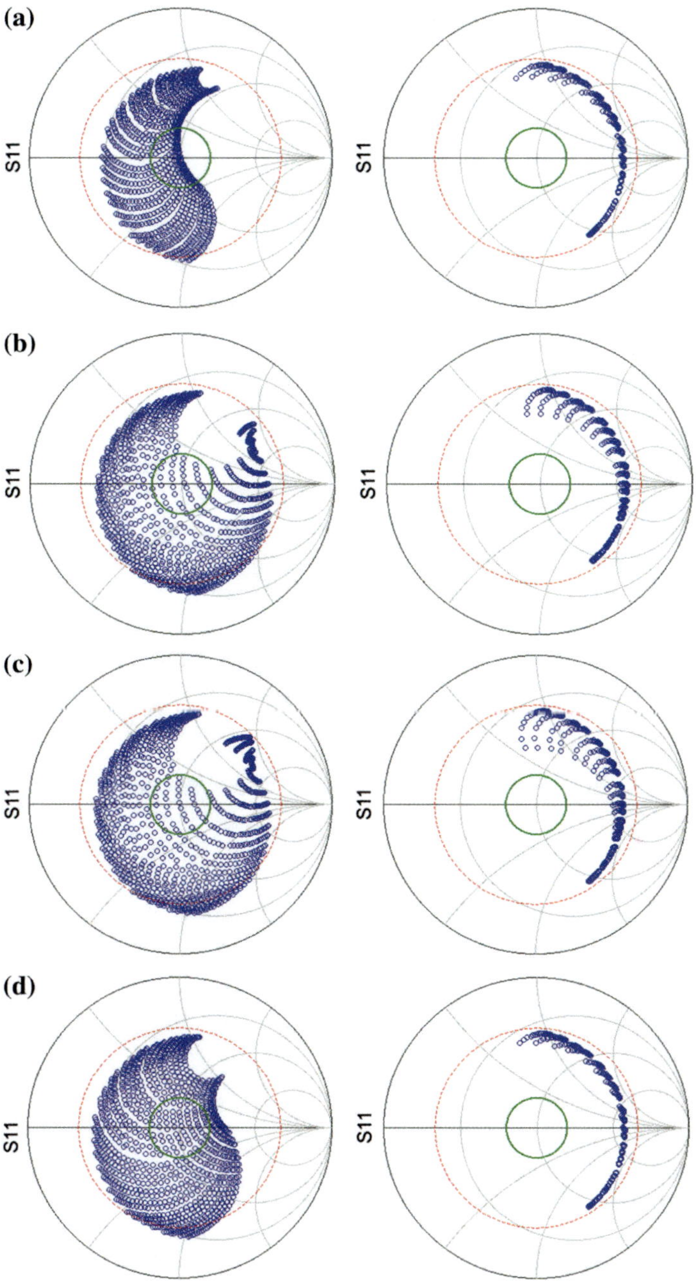

Fig. 8.28 Measured S11 of the TMN-Antenna module at 1.95 GHz (*left*) and 3.5 GHz (*right*) for several perturbation cases; ground plane at 15 mm (**a**), 7 mm (**b**), and 3 mm (**c**); Perturbation by a dielectric block of permittivity equal to 15 (**d**). The green and red circles correspond to VSWR = 1.5 and 5, respectively [5]

Table 8.11 Summary of the tests of the TMN-antenna module

Freq. (MHz)	Power (dBm)	State 0 (ref.)	Optimal state			Configuration
		S11 (dB)	C1	C2	S11 (dB)	
1950	−10	−19.2	0	2	−28	No perturbation
1950	−10	−17	6	0	−37	Ground plane at 15 mm
1950	−10	−14.3	15	6	−38.5	Ground plane at 10 mm
1950	−10	−5.9	27	8	−34.4	Ground plane at 7 mm
1950	−10	−5.5	27	9	−32	Ground plane at 5 mm
1950	−10	−5.9	27	9	−40.6	Ground plane at 3 mm

performance is obtained at 1.95 GHz, i.e. numerous impedance states are obtained in the VSWR = 1.5 circle. These performances are summarized in Table 8.11, where the reflection coefficient in the reference state and the optimal state are reported. At 3.5 GHz, the performances of the TMN are significantly degraded because the TMN was initially designed for operation at 1.7–2.7 GHz. More optimization for the band 3.5 GHz is required, which is the topic of future work.

Additionally, a validation test using a DSP development board based on Cyclone II FPGA from Altera was designed. A search and control algorithm has been implemented into the DSP to automatically find the optimal TMN state based on the incident and reflected power measured at the input of the TMN by the detectors of the TMN-Antenna board. The current search algorithm based on an exhaustive sweep of all the TMN states allows an automatic antenna tuning in approximately 1 s for a clock frequency of 1 MHz.

8.4.2.4 Full Platform

The full platform is presented in Fig. 8.29; it is composed of the three main blocks (two PAs and TMN-Antenna board), two power sensing boards, and a Digital Input/Output (DIO) USB interface. This interface allows the control of the TMN-Antenna board and the power sensors from the computer (through Matlab) in order to read the power levels and control the TMN circuit.

Through this interface, the TMN can be controlled without the DSP board previously presented and in a simpler way. The inconvenient is the rather low running speed of communication between the computer and the USB interface; the optimization of the TMN state runs at about 1 s per tested state, which makes testing of the 1,024 possible states impractical. In order to perform the measurements presented further, we have limited the TMN control to the three most significant bits of each capacitor, which corresponds to 64 states.

In order to test the functionality of the TMN, a controlled perturbation of the antenna is needed. To this purpose, we used the system presented in Fig. 8.30,

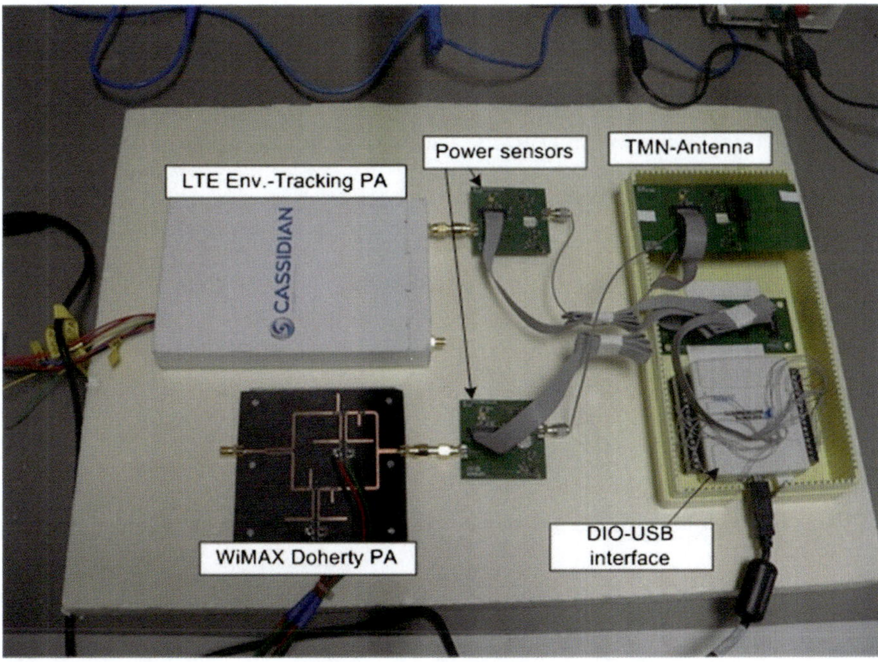

Fig. 8.29 Photograph of the platform (58 × 35 cm) [5]

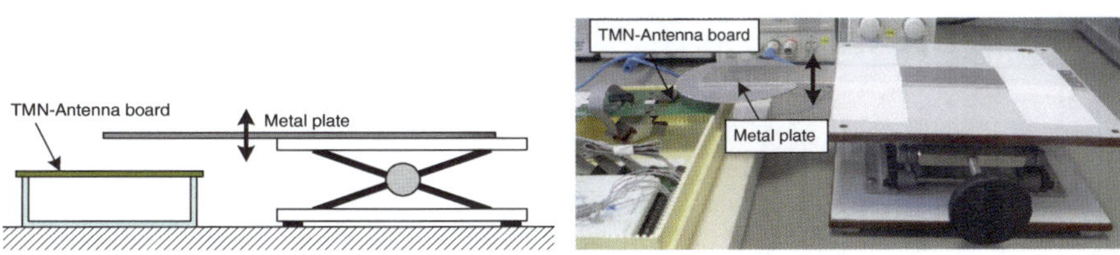

Fig. 8.30 Schematic and photograph of the antenna perturbation system [5]

where the height of a metal plate can be manually controlled above the antenna with a good accuracy. Another set of tests were performed with a block of dielectric material with a permittivity equal to 15. A main reason for choosing such a system as compared to an actual user hand or head is the convergence time of the TMN algorithm which requires the perturbation to be constant in order to have meaningful and reproducible results.

The output power of the two PAs is monitored with two power sensing boards using the same power detectors as the TMN-Antenna module. These boards are connected to the computer with the DIO interface so that the output power of the PAs is displayed in real time.

8.4.2.5 Computer Interface (GUI)

The graphical user interface (GUI) was implemented in Matlab and is presented in Fig. 8.31. The two PAs and the TMN-Antenna are represented graphically.

Three buttons at the top right of the window control the application (Init, Start, Stop). Two buttons (RF1, RF2) control the RF switch of the TMN-Antenna board to select one of the RF input ports. Two buttons (Manual, Auto) control the operation of the TMN:

- In manual mode, it is possible to select the states of the 10 bits of control of the TMN using the 10 small circular buttons under the TMN block;
- In Auto mode, the application goes through all the possible combinations and finally selects the optimal one.

For each PA, the input RF power and the biasing values are given by the user. The output power as well as the incident power are measured and displayed by the application. The incident power is useful to detect an anomalous high reflection at the input of the TMN-Antenna board, such as if the SPDT switch is in the wrong state for instance.

At the input of the TMN, the application measures and displays the input and reflected power. The reflection coefficient (Γ) and Voltage Standing Wave Ratio (VSWR) are calculated and displayed as well. Next to the antenna, a text field allows the user to indicate the efficiency of the antenna. At the bottom right of the

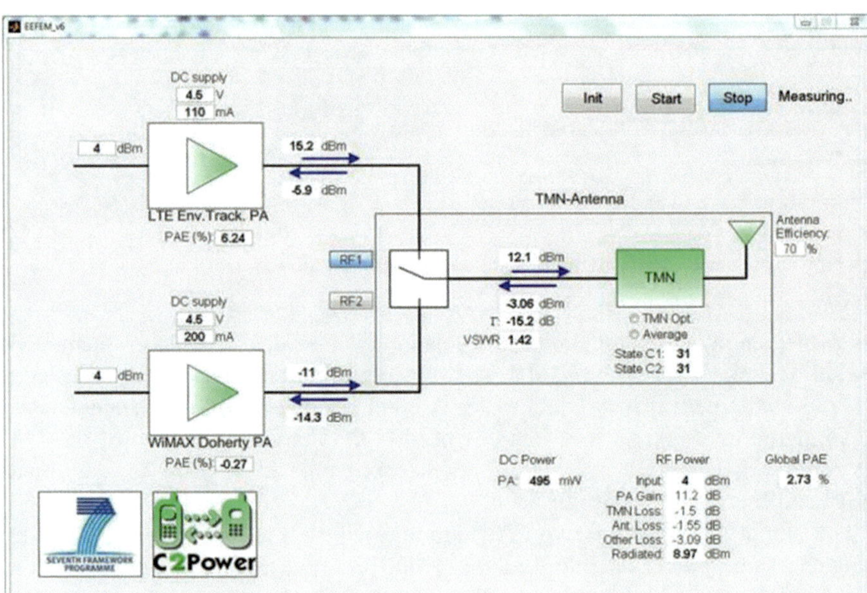

Fig. 8.31 Screenshot of the user interface (GUI) [5]

window, a summary of the power budget is presented. The first column corresponds to the DC power consumption which consists only of the PA consumption. The power consumption of the main other components are given below and are neglected in the power budget:

- TMN: 2.5 V, <1 mA.
- Power detectors (2 circuits on the TMN-Antenna board): 3.3 V, <3 mA
- SPDT switch: 5 V, 2 μA typ.

The second column displays the RF power budget along the transmission chain:

- Input power level (PA input);
- Gain of the PA (calculated from the measured output power and the input power);
- Insertion losses of the TMN: known from characterizations of the TMN circuit;
- Antenna losses: calculated from the efficiency
- Other losses: this term represents the losses of the cables, SPDT switch, and transmission lines between the PA output and the TMN input. It is calculated as the difference between the PA output power and the TMN input power. This term is related to the implementation of the demonstrator and would be reduced to less than 0.1 dB in an actual terminal, thanks to a much higher integration of all the circuits on a single board. Therefore, it is not taken into account in the calculation of the global power added efficiency.
- Radiated power level: the calculation of the radiated power (P_{Rad}) is given by the equation below, taking into account the measured incident power (P_{inc}) at the input of the TMN, the insertion losses (IL_{TMN}) of the TMN (which depends on its state and the frequency), and the efficiency of the antenna (η_{rad}).

$$P_{Rad} = P_{inc} - IL_{TMN} + 10\log(\eta_{rad}) \ \ in\, \mathrm{dBm}$$

The third column presents the calculated global Power Added Efficiency (PAE) which is the representation of the figure of merit of the platform; the PAE is defined as the ratio between the added RF power (difference between the radiated power (P_{Rad}) and the input power of the PA (P_{in})) and the DC power consumption (P_{DC}). The "Other Losses" mentioned above are taken into account in this calculation in order to extract the losses due to the implementation of the platform:

$$PAE = (P_{Rad} - P_{in})/P_{DC} \, in\, \%$$

8.4.3 Experimental Setup

Only the LTE mode of the platform is presented here, as an exemplary results of the designed modules and the full platform. Nevertheless, the experimental results presented hereafter are representative of the functions investigated in Chaps. 2 and 3, and can be easily reproduced for other frequency bands.

Measurements have been performed on the platform at 1,950 MHz in different configurations:

- Power amplifier with envelope-tracking disabled ("no ET") or enabled ("ET"),
- Tunable matching network disabled ("OFF") or enabled ("ON"),
- Antenna unperturbed ("Free space") or several cases of antenna perturbation: metal plate at 3, 4, 6, 8, 12 mm, dielectric block with a permittivity equal to 15 ("Eps = 15").

8.4.4 Measurements and Results

This section presents the experimental results obtained during the testing of this platform. The measured values presented hereafter were taken directly from the GUI interface presented in previous section and are reported in Table 8.12.

As previously explained, the power amplifier is composed of two stages: a driver amplifier and an output stage. The output stage is a regular class-AB amplifier in the "no ET" mode and an envelope-tracking amplifier in the "ET" mode. As the driver is common to the two operating modes and is a standard design, its power consumption is not taken into account in the global power budget of the platform.

The power consumption of the PA without envelope-tracking is 855 mW for 23 dBm of output power and about 30 dB of gain; this corresponds to a power-added efficiency of 23 %.

In the envelope-tracking mode, the power consumption is 450 mW for 23 dBm of output power and about 28.5 dB of gain; this corresponds to a PAE of 44 %. Therefore, the difference between the two modes represents a reduction of the power consumption by about 400 mW or a division by two, which is quite significant.

Taking into account the losses in the TMN and the antenna, the global PAE of the platform in nominal state, i.e. no antenna perturbation, ends up around 9.5 % without ET and 18.7 % with ET.

Figure 8.32a reports the reflection losses calculated as $\mathrm{RL} = 10\log\left(1 - \Gamma^2\right)$. (in dB), where Γ is the reflection coefficient measured by the TMN-Antenna module and reported in Table 8.12. In the "TMN OFF" mode, significant reflection losses up to 6.7 dB are observed when the metal plate is very close to the antenna. Consequently, the radiated power (Fig. 8.32b) is reduced by the same amount in these cases and the global PAE of the platform is degraded down to 3–4 % (Fig. 8.32c).

In all cases of antenna perturbation, the TMN ("TMN ON" mode) succeeds to reduce the reflection losses down to less than 0.5 dB. As a result, the radiated power is quite constant around 16 dBm and the global PAE is maintained in the range 19–21 %.

Table 8.12 Summary of the measurements performed at 1950 MHz (LTE mode)

PA	TMN	Configuration	P_{DC} (mW)	P_{OUT} (dBm)	Ref. Coeff. (dB)	P_{RAD} (dBm)	Global PAE (%)
no ET	OFF	MP @ 3 mm	855	24.1	−1.05	8.62	2.90
no ET	OFF	MP @ 4 mm	855	24	−1.85	11.2	4.67
no ET	OFF	MP @ 6 mm	855	23.8	−3.1	12.9	6.55
no ET	OFF	MP @ 8 mm	855	23.6	−5.08	14.2	8.20
no ET	OFF	MP @ 12 mm	855	23.2	−7.3	14.8	9.06
no ET	OFF	Free space	855	22.9	−11.3	15.2	9.51
no ET	OFF	Eps = 15	855	24	−7.31	15.4	10.97
no ET	ON	MP @ 3 mm	855	23.7	−14.7	16.5	11.94
no ET	ON	MP @ 4 mm	855	24.1	−20.6	17.3	13.58
no ET	ON	MP @ 6 mm	855	23.8	−15.6	16.7	12.11
no ET	ON	MP @ 8 mm	855	23.5	−17	16.5	11.56
no ET	ON	MP @ 12 mm	855	23.6	−15.1	16.3	11.75
no ET	ON	Free space	855	24	−19.3	16.6	13.09
no ET	ON	Eps = 15	855	23.4	−15.9	16.3	11.30
ET	OFF	MP @ 3 mm	450	22.9	−1.05	7.74	4.21
ET	OFF	MP @ 4 mm	450	23.5	−1.2	8.5	5.47
ET	OFF	MP @ 6 mm	450	23.6	−2.9	11.9	11.06
ET	OFF	MP @ 8 mm	450	23.5	−5.15	13.8	14.35
ET	OFF	MP @ 12 mm	450	23.3	−8.73	14.9	18.48
ET	OFF	Free space	450	22.8	−16.8	15.6	18.70
ET	OFF	Eps = 15	450	23.2	−8.16	15.1	17.77
ET	ON	MP @ 3 mm	450	23.4	−11.8	15.9	20.45
ET	ON	MP @ 4 mm	450	23.3	−14.6	16.1	20.64
ET	ON	MP @ 6 mm	450	23.3	−11.9	15.8	20.45
ET	ON	MP @ 8 mm	450	23.3	−14.2	15.9	20.69
ET	ON	MP @ 12 mm	450	23.2	−10.4	15.3	18.87
ET	ON	Free Space	450	23.1	−30.2	16	20.74
ET	ON	Eps = 15	450	23	−15.6	15.7	19.53

8.4.5 Showcase Wrap up

A demonstration platform based on the hardware technology blocks developed in Chaps. 2 and 3 was implemented and characterized. This platform operates as a front-end power module and is composed of two power amplifiers for LTE

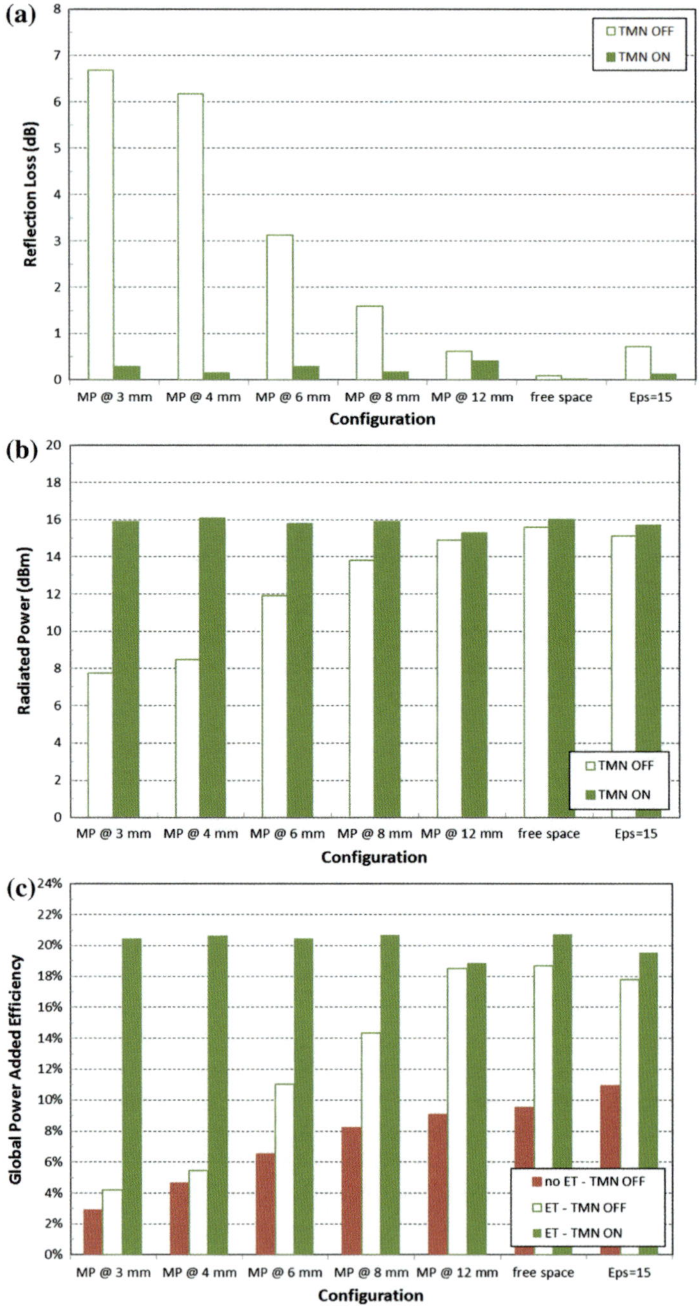

Fig. 8.32 Measurements performed at 1,950 MHz (LTE mode); reflection losses (**a**), radiated power (**b**) and global power-added efficiency (**c**)

(1,920–1,980 MHz band) and WiMAX (3.4–3.6 GHz band), a tunable matching network and a miniature dual-band antenna.

The first PA implements an envelope-tracking architecture and demonstrated a PAE up to 45 % at 1.95 GHz for 23 dBm of average output power. Its linearity performances showed a slight degradation of the EVM and a more significant increase of the Adjacent Channel Leakage Ratio (ACLR) with respect to a standard class-AB amplifier; solutions to these issues are the subject of a future design. The envelope-tracking leads to a reduction of power consumption by about a factor of two (400 mW in this case) as compared to a classical class-AB amplifier.

The second PA implements a Doherty architecture and demonstrated a PAE up to 53 % and 11–12 dB of gain at 3.5 GHz for 30 dBm peak output power.

The tunable matching network (TMN) and the antenna were jointly designed and more specifically optimized for the 2 GHz band. The TMN was fabricated in CMOS-SOI technology; it exhibits 1,024 states and insertion losses in the range 1.1–2 dB. The circuit can still be improved and it is believed that insertion losses lower than 1 dB can be expected in future optimization. In the current experimental results, the TMN succeeds in maintaining a fairly constant radiated power even in cases of strong perturbations created by a metallic plane close to the antenna. It is worth noting that the antenna developed in this work is quite well matched and robust to perturbations. Many antenna types used in actual phones are smaller and more sensitive to their environment (folded monopole or PIFA types); therefore, the future work may require investigating these other antenna types in combination with the TMN. In addition, the integration of the TMN Antenna prototype into a realistic phone case is also envisioned is order to enable real-case experiments.

The full platform was demonstrated in LTE mode at 1.95 GHz. Considering a baseline configuration (no ET, TMN OFF) where the global PAE is in the range 2.9–11 %, the combination of the envelope-tracking and the TMN allows a very significant increase of the global PAE to the range 18–21 % (Table 8.12).

8.5 Conclusion

This chapter presents two showcases to validate the two main envisioned scenarios tackled in Chap. 1, namely energy saving through short-range cooperation and smart energy-efficient handover within heterogeneous environment. Additionally, the chapter presents an energy efficient reconfigurable radio showcase, which strictly speaking should be considered as a horizontal showcase that provides a proof of concept of the hardware components designed and prototyped in Chaps. 2 and 3; horizontal in the sense that proposed RF architecture is applicable to both of the aforementioned scenarios and can be representative layout for a candidate 5G handset.

The short-range cooperation showcase addresses the use of short-range clustering, represented by Ultra-Wideband technology to save energy when connecting to an infrastructure-based long-range technology (WiFi is used in the showcase). The showcase covers different use-cases including a wired 2-node and 4-node setups, as well as a wireless 2-node setup. The results show that significant energy savings are achievable when using a mix of high-speed short-range and low speed long range technologies. Clearly, the energy savings vary depending on user locations and channel conditions. The savings are higher when the handset is experiencing bad channel conditions on the path to the access point of the infrastructure technology representing a shadowed region, whilst at the same time experiencing good channel conditions on the path between the handset and the relaying device within the near proximity, and eventually towards the access point. The showcase highlights high energy savings up to 68–76 %, in the 2-node and 4-node use-case respectively.

The energy efficient handover showcase studies the achieved energy savings using cognitive energy-aware handover algorithms to stay connected to the technology (VHO) or access point (Macro-Femto HO), that provides the best energy efficiency. First, the showcase integrates WiMAX technology with the WiFi access points. Second, the showcase analyzes the case of a mix between WiMAX Macro and Femto cells. The experiment setup covers a scenario, where a user with a smart mobile handset roams between the office, house and the shopping mall, in a networking environment that has total WiMAX coverage and installed WiFi APs (Femto cells) in certain locations providing partial coverage. The proposed energy aware HO algorithms highly outperform the benchmark that involves staying connected to one technology (or only Macro cells), with energy savings up to 42.8–24.3 %, in case of VHO and Macro-Femto HO, respectively. Additionally, the proposed algorithms are compared to signal strength based HO algorithms and still show performance enhancements up to 36.8 %.

Finally, a showcase covering an energy efficient front-end reconfigurable radio is presented. The showcase integrates components proposed earlier in Chaps. 2 and 3. The showcase integrates two designed power amplifiers (envelope tracking and Doherty base PAs), the matching network and the miniature antenna. The showcase is designed to transmit LTE and WiMAX uplink modulated signals. The full platform was demonstrated in LTE mode at 1.95 GHz. The platform can operate, with or without Envelope Tracking, with or without Matching Network, and with or without perturbation to test the effect of all components and their gains. Considering a baseline configuration (no ET, TMN OFF) where the global PAE is in the range 2. 9–11 %, the combination of the envelope-tracking and the TMN allows a very significant increase of the global PAE to the range 18–21 %.

References

1. IEEE Std 802.11-2007 (Revision of IEEE Std 802.11-1999): IEEE Standard for Information Technology-Telecommunications and Information Exchange between Systems-Local and Metropolitan Area Networks-Specific Requirements—Part 11: Wireless LAN Medium Access Control (MAC) and Physical Layer (PHY) Specifications, Jun. 2007
2. IEEE Std 802.11-2012 (Revision of IEEE Std 802.11-2007): IEEE Standard for Information Technology-Telecommunications and Information Exchange between Systems-Local and

Metropolitan Area Networks-Specific Requirements—Part 11: Wireless LAN Medium Access Control (MAC) and Physical Layer (PHY) Specifications, Mar. 2012
3. ECMA-368: High rate ultra wideband PHY and MAC standard, 3rd edn, Dec. 2008
4. Hittite: 0.5 dB LSB GaSs 6-Bit Digital Positive Control Attenuator, 2.2–8.0 GHz http://www.hittite.com/content/documents/data_sheet/hmc425lp3.pdf
5. ICT-248577 C2POWER Consortium: D7.3: Proof of Concept Test Report, Measurements and Global Perspective on C2POWER Power Saving Strategies. Deliverable (D7.3) of C2POWER, 2012
6. Netgear Datasheet: http://www.netgear.com/images/wndap350_ds_11nov091-5813.pdf
7. AdMob: Admob mobile metrics: Metrics highlights. Technical report, AdMob, May 2010
8. Yan, X., Ahmet Şekercioğlu, Y., Narayanan, S.: A survey of vertical handover decision algorithms in fourth generation heterogeneous wireless networks. Comput. Netw. **54**(11), 1848–1863 (2010)
9. UM230 WiMAX USB Modem Data Sheet: http://www.datasheetarchive.com/UM232-datasheet.html
10. WiMAX Product guide: http://www.wimaxforum.org/files/wimax_product_ brochure_12-2011.pdf, Dec. 2011
11. Niamien, M.A.C., Dussopt, L., Delaveaud, C.: A compact dual-band notch antenna for wireless multi-standard terminals. IEEE Antennas Wirel. Propag. Lett. **11**, 877–890 (2012)

Chapter 9
Business Models for Cooperation

Tinku Rasheed, Ayman Radwan, Jonathan Rodriguez, Jacek Kibilda, Radoslaw Piesiewicz, Christos Verikoukis, Lorenzo di Gregorio, Álvaro Gomes and Tiago Moreira

Abstract Identifying attractive business models for the network/service provider and users based on cooperation is essential in order to secure the adoption of this technology. In this chapter, we identify new business models showing how we exploit cooperation between user terminals and heterogeneous networks and operators. However, enabling technology based on cooperation raises some hurdles that need to be addressed such as motivation, selfish behaviours, among others; all if which are challenges that can be best addressed at the business level rather than technology wise; for example, without tailored incentive mechanisms, operators will face difficulties to implement cooperation among users. This chapter focuses mainly on the business models, addressing the importance of reducing the global power consumption. This chapter provides a survey of existing business models, which can be tailored to fit the approaches proposed within the book. Based on our survey, motivations and challenges are identified. Different business solutions are proposed using use cases to emphasize the selection of cooperation scenarios and network sharing.

T. Rasheed
CREATE-NET Research Center, 38123 Trento, Italy

A. Radwan (✉) · J. Rodriguez
Instituto de Telecomunicações, Campus Universitário de Santiago, 3810-193 Aveiro, Portugal
e-mail: aradwan@av.it.pt

J. Kibilda · R. Piesiewicz
Information and Communications Technology Research Center, Wroclawskie Centrum Badan EIT+, Wrocław, Poland

C. Verikoukis
Telecommunications Technological Centre of Catalonia, Barcelona, Spain

L. di Gregorio
Intel Mobile Communications GmbH, Munich, Germany

Á. Gomes · T. Moreira
Portugal Telecom Inovação, Aveiro, Portugal

A. Radwan and J. Rodriguez (eds.), *Energy Efficient Smart Phones for 5G Networks*, Signals and Communication Technology,
DOI 10.1007/978-3-319-10314-3_9

9.1 Introduction

The ability to use mobile terminals at any time and any place without being weary of battery supply seems to be a futuristic approach. However by exploiting smart cooperative networking concepts, this vision can take a step closer to reality. Exploiting these technology paradigm in tandem, can lead to new radio topologies that are able to provide energy efficient connectivity and thus battery lifetime in mobile phones enabling the use of services that require greater bandwidth than legacy services currently provided by UMTS technology. Most of the research in cooperative networking technology has addressed the technical and engineering ways to save energy.

In present days, to support cooperative networking in the cellular network market, it is absolutely necessary to have good business models that show the main benefits of this new architecture and the revenues associated. To build a successful business model, it is important to follow a strategy that defines how the entity attracts customers to pay for a service; how those services are delivered to the final customers; how the entity converts customers' payments into profits and finally how the entity is organized to be able to meet the customer needs. With these goals in mind, it is also important to understand that the term business model is also closely related to innovation which means that this concept is related to a whole new range of business design opportunities. The most obvious is the innovation in the value proposition. The value proposition is a statement of the tangible results a customer attains from using a certain product or service. Meaning that, the value proposition has to be clear and concise to achieve good results in the mobile market. For example, when mobile phones appeared in the market, they offered a different value proposition than the fixed line phones, which brought new ways to obtain profits for network operators [1].

Cooperative techniques have received substantial attention recently since they offer potential improvements in link and network performance, including capacity, coverage and QoS. Moreover, cooperation can enhance greatly the efficiency in the utilization of network resources, offering new services for end users, namely cooperative services. On the other hand, there are the non-cooperative behaviours in communication networks that can adversely affect the entire network. This analysis is important at a time where wireless systems are emerging that integrate conventional cellular network paradigm with the emerging wireless relay network paradigm. Wireless relay networks leverage the respective advantages of cellular network and WLAN relay networks to achieve better system performance. In order to encourage nodes to share their resources, it is necessary to have some attractive features and incentives that bring clear benefits to mobile users when they participate in this cooperation action. There are several approaches to elicit cooperation between network nodes. Some of these benefits encourage the voluntarily cooperative behaviour (e.g. by offering better services for the customers who cooperate or even enhancing the e-reputation of a cooperative customer). A rational entity chooses to cooperate as the result of an incentive mechanism that aims to bring benefits for the cooperating user and eventually for all collaborating users.

For a large expansion in the market, network operators must invest in aggressive business models with ambitious features that enable a rapid uptake by subscribers. The main objective of this chapter is to demonstrate some mechanisms that encourage various forms of cooperative behaviour, showing the main advantages of this cooperative system for consumers. In addition, we illustrate new solutions to overcome the difficulties faced in cooperative networks. Moreover, it is interesting to observe the efficient cooperation rules that exist in nature, and transport such rules to cooperative networking context.

9.2 Motivation and Incentives for Cooperative Networks

Cooperation is a strategy of a group of entities working or acting together towards a common or individual goal. Correspondingly, the connotation of cooperative networking is described as devices working together to achieve a goal, within a network. Wireless devices, controlled ultimately by human, can be considered as selfish, without any incentive to cooperate by nature. The reason, that this phenomenon occurs, is that a wireless device is always interested in maximizing its own benefits. So cooperation adds costs to mobile devices, but bringing new services that can increase the performance of a mobile device. However, in any cooperation network, the communication ultimately depends on the willingness of the nodes to cooperate. Such cooperation can only be established and maintained if fairness and profitability are guaranteed among cooperating nodes. Therefore, to avoid the collapse of cooperation, robust cooperation rules and good incentives are required.

To understand the incentives and basic rules for cooperation, an observation from cooperation used in nature is needed. The cooperative network is often realized in the form of a composite access network, which is composed of heterogeneous networks. With the rapid development of technologies and mobile networks, especially with the arrival of LTE, certain characteristics have not been able to compete or keep up with the technology growth, mainly the battery autonomy of mobile devices and other features such as quality of service, higher throughput or spectral efficiency. Dealing with this paradigm provided the impetus for new ideas and possible innovations in the area of cooperation, aimed at combating the negative features resulting from the progress in mobile networks. This cooperation could create, through network operators, mechanisms to encourage customers to cooperate using their mobile devices.

As shown in Fig. 9.1, one of the main methods of cooperation is the cooperation among mobile devices that can be done through short-range communication technologies such as Bluetooth, UWB or WLAN. These technologies promote low energy consumption, while providing high quality of service. The main advantages of cooperative communications are [3]:

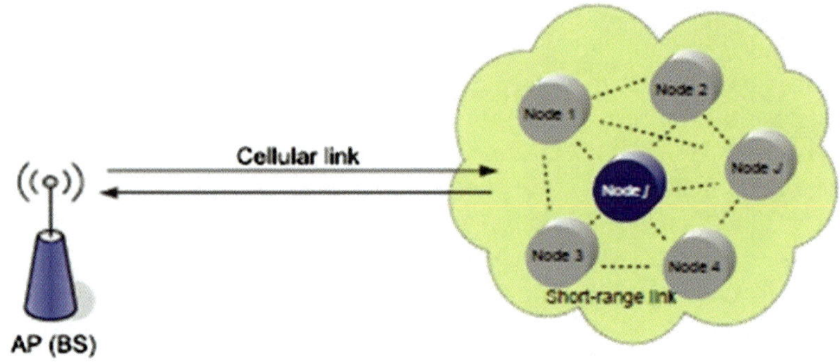

Fig. 9.1 Cooperative nodes [2]

- Reducing the power consumption of mobile devices and decreasing the transmission power of base stations;
- Increasing the quality of service, since with strong signal, customers can achieve better services without failure, e.g. higher data rates;
- Lower delay between cooperative users;
- Possibility to create more services like context exchanges between nodes or parallel processing;
- Decreasing the carbon footprint. Base stations and mobile devices consume much electricity in their use phase, which makes energy saving a valuable contribution for these facilities;
- Better use of wireless spectrum;
- Overcoming the limited cellular capacity. For multicast services such as video streaming or file downloading the benefit is obvious. In this case the cooperative devices receive partially the original stream and collaboratively merge it over the short-range communication links.

9.2.1 Challenges in Cooperative Networks

Nowadays, infrastructure networks (i.e. cellular networks) do not expect or anyhow permit cooperation among connected mobile devices. However, as some wireless environments enable the realization of ad hoc cooperative behaviours, it seems natural that infrastructure networks could also benefit from such cooperation. From a general perspective, the most interesting business cases are related to battery power savings at mobile devices. Due to its locality and gain from statistical multiplexing (traffic aggregation), such energy saving gain can be observed with ad hoc cooperative networks. Still, wireless ad hoc cooperative networks are to be successful only if they are able to align the current individual and selfish assumptions of the mobile devices into a cooperative paradigm that succeeds to benefit all the entities involved in the cooperation.

Current assumptions for wireless ad hoc networking are:

- Mobile devices are expected to achieve selfish goals, enforced by users or running services;
- Mobile devices belong to different users;
- Mobile devices are served by different PLMNs (different network operators).

The enlisted assumptions, which are obstacles to cooperation, can be overcome with the introduction of motivational and incentives systems, which encourage various forms of cooperative behaviour. Examples of such cooperative behaviour are altruistic cooperation, non-altruistic cooperation or reciprocal cooperation. Each of these forms of behaviour has its stability and thus, it has to be resilient to the number of challenges derived from the transient nature of the wireless communication system. In classical cooperation theoretic approach, cooperation is modelled and presented using Prisoner's Dilemma game. The simplest model of Prisoner's Dilemma is a two player game[1] of mutual cooperation, where players choose from a pair of two possible actions either to *cooperate* or *defect*. The payoff is given to players based on the effect of the actions of both players. The principle rule of the payoffs is that regardless from the other player's choice, defection always provides higher expected payoff than cooperation. However, it shall be noted that in the case that both players defect the payoff is lower, than if they would both cooperate.

In practice however, two individuals can meet more than once and therefore they can use knowledge about previous interaction(s) to influence the strategy. Such a game then becomes an iterated Prisoner's Dilemma and strategy takes a form of decision policy. Thorough analysis of iterated Prisoner's Dilemma with a priori unknown meeting number results in the extraction of optimal strategy that outperforms ALL D—TIT FOR TAT (TFT). TFT is a simple strategy, where in the first game individual always cooperates and in the subsequent games it does whatever the opponent did in the preceding game. Hence, the strategy is based on reciprocity of interactions. It is worth noting that TFT not only provides optimal payoff result but is also evolutionary stable and is invulnerable to invasion from any other strategy, provided that the probability of the next meeting between two individuals w is sufficiently high:

$$w \geq \frac{T-R}{T-P} \cap w \geq \frac{T-S}{R-S} \tag{9.1}$$

where T is temptation, R is Reward; P is punishment, while S represents Sucker.

One more important observation that can be helpful to the introduction of applicable incentives mechanism treats the evolutionary stability of ALL D strategy. ALL D strategy being also evolutionary stable cannot be invaded by any single strategy. However one can find a particular case, where it can be overwhelmed by cooperative clustering. Herein by cluster one shall understand representation of

[1] Prisoner's Dilemma is derived as a model of two-player game, hence for larger number of players (>2), it is presented as a sum of individual two-player (bilateral) games.

individuals using the same strategy, which have high w and high rate of interactions with other cluster member's p (in proportion to the whole number of interactions). Using clustering with TFT, individuals can achieve higher payoff than the defective environment (adopting ALL D). Thus, TFT could also be adopted by the rest of the population as the strategy leading to higher payoff. Cluster adopting TFT is able to achieve higher payoff than defective environment, only if:

$$p\left(\frac{R}{1-w}\right) + (1-p)\left(S + \frac{wP}{1-w}\right) > \frac{P}{1-w} \tag{9.2}$$

Equation (9.2) suggests also how to construct the clustering algorithms, so as to encourage cooperative behaviour in selfish or even hostile environment. In general, the basic analysis of the problem from cooperation theory provides an indication on how to design payoff (incentives) and reputation mechanisms, to encourage users towards cooperation. In practical cases, the problem of cooperation receives additional complications, due to existence of altruism, contractual obligations and availability of central authority, which provides additional means for authentication, promotion and punishment. Considering the problem network-wise, if the mobile devices would mutually relay traffic for each other, they would obtain higher payoff than in the case when they do not cooperate at all (defect all interactions). Such a simple model provides an attractive starting point for further studies on strategies and challenges for cooperation; thus as indicated in the introduction of the current section, cooperation in ad hoc wireless networks was and still is topic of much interest for wireless research community, especially since, the cooperation decisions have a richer strategy space than just cooperate/defect.

Based on the state of the art study for cooperative ad hoc networks a number of cooperation challenges were identified:

- *Motivation and fairness*—the two concepts need to be addressed by the cooperative network to encourage cooperation among users, to provide fair access to the resources and to punish the adversaries (to discourage malicious and greedy behaviours). Fairness problems affect also the design of the motivational mechanisms. Furthermore, another known fairness problem is related to location, as mobile devices with more favourable location receive more incentives, leading to even higher income.
- *Malicious behaviours*—although authentication and access control can reinforce cooperation through prevention against external attacker, there are still possibilities for users to exploit the cooperation even in the presence of effective authentication and access control mechanisms.
- *Privacy protection*: the key to the success of any reciprocity based cooperation strategy is the ability to identify (and possibly also punish) the defective nodes [4], and thus, mobile devices cannot stay anonymous. Furthermore, the reputation ratings have to be assigned to correct mobile device and payments or reciprocal behaviour returned to correct initiator.

- *Cooperation maintenance*: as a consequence of mobility or in general, transient changes in channel quality, the node which realizes a cooperative scenario may experience reduction of the quality in provided relaying service or decrease in the reliability of provided information [5]. This may lead to a situation where users are unfairly treated due to decrease in incentives or reputation. Thus, reputation system has to recognize faulty behaviour and distinguish it from malicious behaviour.
- *Observability*—the results of the cooperation are highly dependent also on the ability to identify and distinguish cooperative behaviour from selfish. In wireless networks, a typical situation is that nodes have non-equal information, leading to information asymmetry [6].

9.2.2 A Business Model Based on Incentives and Reputation

A widely accepted way to improve the coverage of mobile networks is to organize a cooperation scheme among network terminals: if terminals allow each other to "bridge" communication, access points can be reached by terminals which are at larger distance than what can be covered by one "hop", i.e. the distance under which a network node can be reached. A mobile terminal operates as a bridge and transports data between an access point and another remote mobile terminal. Consequently, the remote mobile terminal can reach the access point in two hops, with a better quality and consequently consuming less energy. This sort of scheme requires cooperation between nodes, because intermediate nodes must sacrifice resources in order to enable the participation of other remote nodes. They do so because they will benefit from the same cooperation scheme when they move to a remote location.

Taking for granted that cooperative schemes present overall benefits and require individual nodes to make contributions to the community, a question which immediately arises for an individual node is whether its own benefit from the overall cooperative scheme is greater than its own contribution to such a scheme. A closely related question is how to avoid faulty, selfish or malicious behaviours, which break the cooperation scheme at an individual node's sole advantage.

9.2.2.1 Cooperation Model

In this section, we define a cooperation model for a network of nodes which represent, without loss of generality, generic mobile terminals. This includes also the case that some terminals are fixed access points and the model can be extended to the case where different types of terminals, which present different characteristics. The cooperation model describes the equilibrium states under which the nodes cooperate by roaming their data through each other at reciprocal advantage in saving power by reversing their roles periodically. These equilibriums are characterized by the fact that the nodes know the overall likelihood that other nodes will participate to the collaborative scheme, but they do not have any information about individual participation of specific nodes in the collaborative scheme.

Every node can decide whether to participate or not in the cooperation scheme. If a node decides to cooperate, it allocates a certain power budget for sustaining communication routes defined by other nodes, but it does not post any information about its individual participation to the scheme. In return, this node is allowed to save power by routing its own data over other nodes participating in the same scheme. Every individual node will cooperate only if it can expect to save power individually and regardless of the decisions of the other nodes, based only on the overall likelihood that other nodes will participate. On the other hand, every node will contribute in a measure which matches its individual benefit, so the scheme must ensure that the highest power saving can be achieved by nodes who do not claim any lower benefit than the one they actually experiences.

In general, this model demands that every node contributes a certain amount of power for sustaining communication routes and excludes from the scheme those nodes which are not contributing enough power. Among the contributors, each individual node is guaranteed to individually save power, on average, out of the overall balance (Fig. 9.2).

To begin with the model, we first define θ as an individual utilization vector for the cooperative scheme, defined as

$$\theta \doteq [\theta_1, \ldots, \theta_n] \tag{9.3}$$

where θ_i is a function of the individual preference parameters of node i, for example:

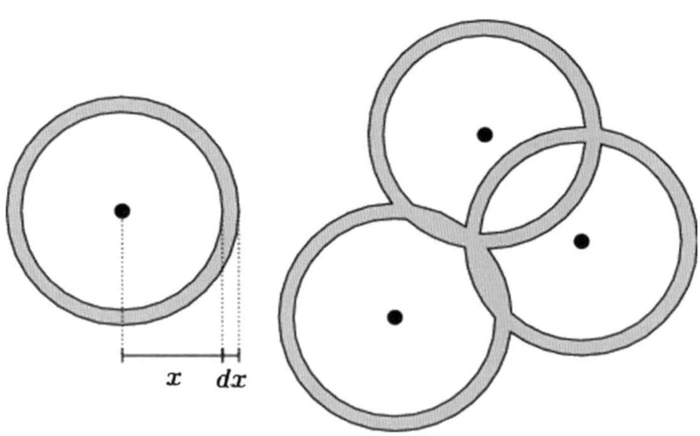

Fig. 9.2 Differential area at distance x from any node [7]

$$\theta_i = \frac{\text{average active time}}{\text{total active time}} \tag{9.4}$$

In general, θ_i is the fraction of average saved power on a nominal communication, which gets exploited by node i due to individual characteristics.

We define a differential operator with the purpose of computing averages as

$$dF(\theta) \doteq \prod_{i=1}^{n} f_i(\theta_i)\, d\theta_i \tag{9.5}$$

$\pi_i(\theta)$ is the probability that node i participates in the scheme, given the chosen utilization, $Q(\theta)$ is the coverage as access probability to a node within a range depending on the node's power, $u(\cdot)$ is the power efficiency curve given a coverage. This curve returns the average saved power on a nominal communication, given the available coverage. $p_i(\theta)$ is the power which a node is willing to provide to the cooperative scheme for participation, ultimately depending on the preference parameters of all participants.

Given this interpretation, now we can define the equilibrium which maximizes the efficiency of the network. We define $\Theta \doteq \mathrm{X}_1^n[0,\ 1]$ as the n-dimensional space of the vector θ and $\overline{\Theta}_i = \mathrm{X}_1^{i-1}[0,\ 1]\, \mathrm{X}_{i+1}^{n}\, [0, 1]$ as the n − 1 dimensional subspace projected along dimension i, i.e. "without" dimension i.

The social welfare is a measure of the global efficiency of the scheme and consists of the average power saved by every individual node $i(\theta_i u(Q(\theta)))$ minus the average power required by that individual node to sustain the communication routes defined by other nodes $(C_i(\theta))$. The contribution of every individual node to the scheme is determined by its probability of participation $\pi_i(\theta)$ and the resulting average social welfare must be maximized according to the non-linear optimization problem:

$$\underset{\pi_i(\cdot),\ldots,\pi_n(\cdot),\,Q(\cdot)}{\text{maximize}} \int_{\Theta} \sum_{i=1}^{n} \pi_i(\theta)\,[\theta_i\, u(Q(\theta)) \;-\; c_i(\theta)]dF(\theta) \tag{9.6}$$

The maximization of the social welfare presented in (9.3) is subject to three constraints. The first constraint is known as the weak feasibility constraint and states that the sum of powers $p_i(\theta)$, which every node is willing to sacrifice for the cooperation scheme, is higher than the sum of costs $c_i(\theta)$ to sustain the same scheme. The first constraint can be expressed as

$$\int_{\Theta} \sum_{i=1}^{n} \pi_i(\theta)\,[p_i(\theta) - c_i(\theta)]dF(\theta) \geq 0 \tag{9.7}$$

The second constraint is known as individual rationality constraint and states that every individual node i will only participate in the scheme if it can individually expect that the saved power $\theta_i u(Q(\theta))$ is higher than the one it has to budget for participating to the scheme $p_i(\theta)$. This difference must be at least K, because this power gain is the gained benefit which can be charged in the business model presented in the following subsection. This fixed benefit is applied only to participants; hence it must be multiplied by the probability of participation $\pi_i(\theta)$. Inequality (9.8) presents the second constraint in average:

$$\int_{\overline{\Theta}_i} \pi_i(\theta)\,[\theta_i u(Q(\theta)) - p_i(\theta)] d\bar{F}_i(\overline{\theta}_i) \geq \pi_i(\theta)K,\ \forall\theta_i \tag{9.8}$$

The last constraint is known as incentive compatibility and states that the power $\theta_i u(Q(\theta))$ saved by node i over the power budget $p_i(\theta)$ is higher than the one saved over a power budget $p_i(\theta^*)$ obtained by posting any $\theta_i^* \neq \theta_i$. Defining the vector θ^* as θ with the ith element replaced by θ_i^*. The incentive compatibility constraint is represented by

$$\begin{aligned} &\int_{\overline{\Theta}_i} \pi_i(\theta)[\theta_i u(Q(\theta)) - p_i(\theta)] d\overline{F}_i(\overline{\theta}_i) \geq \\ &\int_{\overline{\Theta}_i} \pi_i(\theta^*)[\theta_i u(Q(\theta^*)) - p_i(\theta^*)] d\overline{F}_i(\overline{\theta}_i),\ \forall_i, \theta_i^* \end{aligned} \tag{9.9}$$

These constraints can be shown to be equivalent to the constraint presented by inequality as shown below.

$$\begin{aligned} &\int_{\Theta} \sum_{i=1}^{n} \pi_i(\theta)[g(\theta_i)u(Q(\theta)) - c_i(\theta) - K] dF(\theta) \geq 0 \\ &\text{where} \qquad g(\theta_i) \doteq \theta_i - \frac{1 - F_i(\theta_i)}{f_i(\theta_i)} \end{aligned} \tag{9.10}$$

The problem can be shown to be convex and solved by Lagrangian methods, which consist of finding $\pi_1(\cdot), \ldots, \pi_n(\cdot)$ and $Q(\cdot)$ which maximize a Lagrangian of:

$$\frac{\sum_{i=1}^{n} \pi_i(\theta)[\theta_i + \lambda g(\theta_i)]u(Q(\theta))}{1 + \lambda} - \sum_{i=1}^{n} \pi_i(\theta)[c_i(\theta) + K] \tag{9.11}$$

This problem is complex, but there are simple asymptotic characteristics which makes it attractive for devising a business model, described below.

9.2.2.2 Business Model

The previous section presented the model of a cooperative scheme for power saving. The constraints ensure that:

- every node saves power of at least a value K in its individual average,
- every node must individually contribute a power $p_i(\theta)$.

From the analysis, if the power efficiency $u(\cdot)$ and power cost $c_i(\cdot)$ are bounded by exponential laws of the coverage $Q(\theta)$, increasing the number of participants in the scheme drives the individual contributions $p_i(\theta)$ to converge to a fixed value [8].

This observation implies that, with some approximation, an attractive flat rate pricing policy can be introduced, given that every participating node agrees to forward enough traffic to guarantee the contribution of power $p_i(\theta)$ by sustaining roamed communication routes. Each node can then enjoy a power win of at the least K on average, which translates into a benefit in terms of battery duration and can be priced with a flat rate as the market size grows. A very similar business model can be setup for the extended coverage $Q(\theta)$, so that a participating node can benefit from both coverage and power savings.

In particular, the fact that participants bear different types and can be priced at different weights of a nominal flat rate is attractive, because a mobile phone user could rely on the contribution provided by his/her WLAN access point (or Femtocell) at home in order to participate with his/her UMTS mobile phone. The presented cooperation model does not maximize profit for a given participation, but rather maximizes participation for a given fixed revenue per participant, because large participation provides a more stable revenue, and in addition a complex paying scheme is likely to find bad acceptance among participants.

Consequently, the introduction of this scheme demands some investment. As long as the number of participants is small, the coverage $Q(\theta)$ is small and the individual contributions $p_i(\theta)$ must be relatively high to present an individual convenience. In fact, the equilibrium, described above, excludes from participation those nodes, which provide a too low $p_i(\theta)$.

The scheme can be shown to stabilize with the increasing number of participants, i.e. the coverage $Q(\theta)$ grows while demanding smaller individual contributions $p_i(\theta)$ and providing a linear growth in revenues due to the fixed return per participant (K).

Figure 9.3 provides a qualitative representation of this business development. After an initial and non-recurring cost, the infrastructure setup is made available to a base of high-end customers, which reach some social welfare equilibrium with high $p_i(\theta)$ values. This group could be composed, for example, of enterprises willing to adopt the scheme within their office spaces. In order to drive the market toward low-end customers, some special offers are necessary to cross the gaps to subsequent customer groups. For example, this could be the case for extending the collaboration to airports and subsequently to shopping malls, travelling trains, highways, etc. Such offers cause additional costs, but as the customer base gets larger, the gaps in this down-market trend are expected to become always smaller and since the costs for the infrastructure scale less-than-linearly with the increase of participation, a break-even point must be reachable.

Lowering $p_i(\theta)$ is key in accelerating the down-market trend of the scheme and consequently reducing the investment to profitability. This can be achieved by increasing the efficiency of the routed communication, thereby reducing $c_i(\theta)$, which is the main driver for $p_i(\theta)$. A way to do so is by introducing some reputation model.

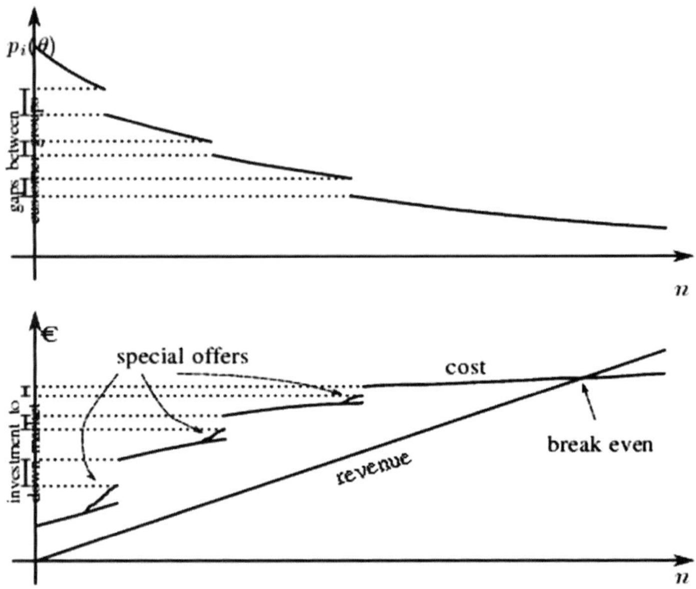

Fig. 9.3 Qualitative business development for down-market. *Note* Horizontal axes represent number of participants n, not time [7]

The adoption of the scheme outlined above demands that all participating nodes contribute a certain amount of power. This can be achieved by demanding traffic forwarding on roamed routes from every node within the scheme until every power contribution has been delivered. If a node exploits the scheme without providing traffic forwarding on roamed routes, either due to malicious or faulty behaviour, it breaks the equilibrium to its own advantage, i.e. it uses the network without contributing power. Such nodes are called 'free riders', because they exploit the scheme 'for free'. A free riding behaviour is highly likely to be an optimal strategy in a one-shot game, i.e. a single use of the scheme. If the exploit of the scheme is repeatedly used, participants know about past behaviours and a tit-for-tat strategy, i.e. a strategy which penalizes past free riders, is successful at eliciting cooperation. Consequently, correctly identifying and penalizing free riders is paramount in supporting the cooperative scheme.

Since every node can expect a win of at least K, a corresponding amount of credit can be assigned to each node. This credit is transferred between nodes according to the securely reported traffic and validated by trust authorities. The overall goal of the reciprocal credits is that all nodes must settle down at an average credit corresponding to a power contribution of K. If after sufficient traffic has been processed, some nodes still fail at reaching this level, they are not providing the minimum power contribution; hence they must consequently fall off the cooperative scheme and get penalized on re-entering, in order to avoid free riding.

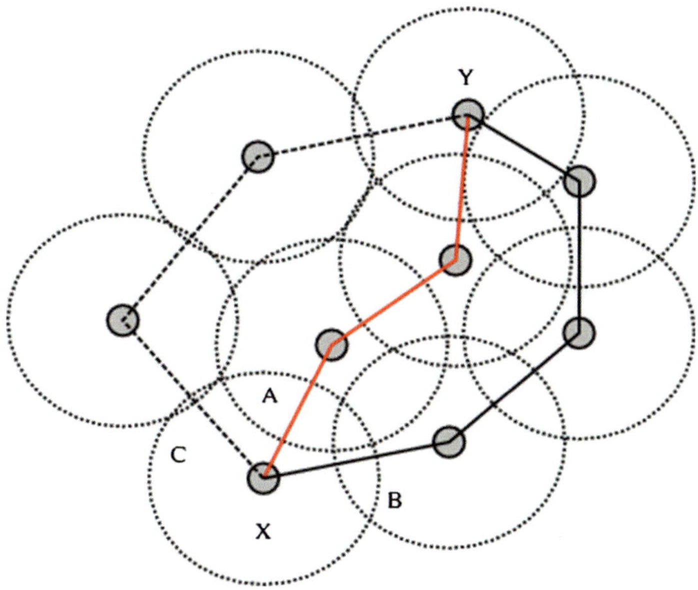

Fig. 9.4 Node X must find the most efficient route (route A) to node Y Route C has less hops but is less power efficient because of longer distance [7]

A way to increase the efficiency of the scheme, hence lowering $c_i(\theta)$ and consequently the demand for $p_i(\theta)$, is to choose the most convenient node to route communication. Reputation systems assign some qualification to nodes based on past experiences, in order to provide users of a cooperative scheme with information about the effectiveness of their peers. Figure 9.4 presents a route search problem. Route A is the most efficient between node X and node Y because it has two hops and less distance than route C. Route B could be the second best efficient route because despite its three hops, nodes have shorter distances between each other. A reputation system could specify how reliable is route A. If the nodes on route A are likely to move away or fail, the second best route could be a better alternative.

In general, a node might present several different patterns of behaviour depending on circumstances and, apart from other technical considerations, a centralized reputation system which classifies nodes in terms of "good" and "bad" has discouraging social and legal implications. Search theory offers an approach to a distributed reputation system, which does not merely classify the nodes, but rather characterizes them. A Markov reward model is a Markov model in which each state is assigned an additive reward to be collected during the state evolution of the Markov chain. In our context, this reward can specify a quality parameter of the communication. For example, the transferred quantity of data while a node is in a certain state. Such Markov reward models are characterized by a terminal state, which is an absorbing state of the Markov chain at zero reward and models the fact that the node is out-of-range and no further communication is possible. A user of multiple nodes may not search for a node which merely offers a high bandwidth. If this node is a mobile phone in a car on the highway, it will quickly become out of range and a new route will need to be found. Consequently, a user must prefer nodes which are likely to provide roaming over time, despite a lower immediate reward, also known as myopic reward.

9.2.3 A Brokerage Model for Cooperative Networking

We present an auction based cooperation mechanism including the quality of service constraints to an applicable business model. Particularly, we consider the brokerage based business model as explained subsequently. Considering a source node S in need of help to connect to the access point AP, we define the following procedures assuming that S is intending to get help from a neighbouring short range radio device that can connect to its own AP.

- The source node broadcasts a cooperation request with its quality of service requirements hoping to get response from a neighbouring device.
- The neighbouring devices after receiving the request from the source node reply back with a price for the cooperation in terms of Euros/bit by taking into account the quality of service requirements and the associated local cost for them.

In the case when no neighbouring device exists or in the case of no device that could provide the requested quality of service exists, there will be no way for the source node to communicate, and hence there exists no business case in such situation. When a response is received from any neighbouring node, the source node would then select a node with the minimum cost for cooperation and send back its agreement to its offered price. In the case where the offered price (bids) from neighbouring nodes is too expensive and/or not cost effective and/or not affordable, then the source node does not respond back.

Once the source node selects the cost effective neighbouring node, hereafter known as the relay node R, for its cooperation, the relay node would act as a 'broker' between the access point (network service provider) and the source node (client) for business purposes. Note that for communicational (technical) purposes the relay node would simply receive the data from the source node and send it to the access point by means of various cooperation protocols (such as decode and forward, amplify and forward etc.).

The relay node acts as a broker by activating its 'brokerage-agent' (an application protocol as described later) internally in order to setup the cooperative communication link between the source and access point. By activating the brokerage-agent the network identifies that the corresponding node (relay node) is acting as a relay on behalf of the source node such that several technical/logistical things are taken care of by the network. Once the cooperation session is complete, the 'brokerage-agent' is deactivated and the relay node returns to its normal mode

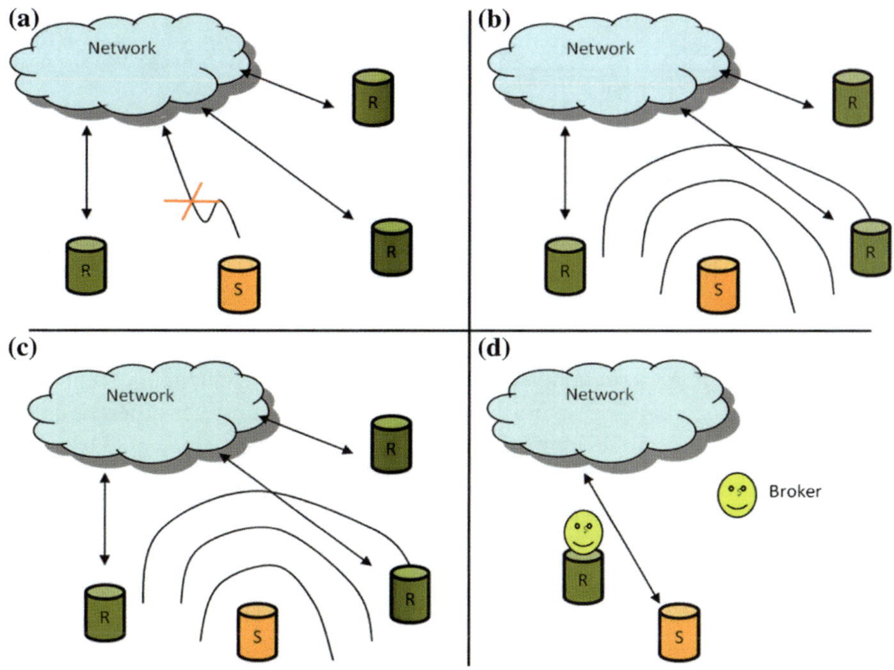

Fig. 9.5 Description of the process for the auction based business model with a brokerage agent for short range cooperation scenarios [7]. **a** Source node unable to reach the network. **b** Source node broadcasting cooperation request. **c** Relays cooresponding to the source node's request with their bids. **d** The selected relay then acts as a broker by connecting the source and the network

of operation. The above mentioned cooperation process is depicted in Fig. 9.5, where initially a Source node cannot reach the network (Fig. 9.5a), hence it broadcasts a cooperation request (Fig. 9.5b). All Relay nodes receiving the broadcast request, reply with their offered bid (Fig. 9.5c). Finally, one Relay is selected and acts as a broker to connect the Source node to the network (Fig. 9.5d). The brokerage agent is activated and deactivated every time a cooperation agreement is made between the source and the relay nodes.

The role of the brokerage agent in such a cooperative communication application is quite eminent. The relay node acts as a broker between the network service provider and the source node by basically connecting the source node and the network through its link, which is a classical operation of a broker in business models.

The broker here provides the following services:

- It brings in revenue to the network service provider by linking the source node (customer) to the service, and
- It provides relay service to the source node.

The relay node therefore can expect to be paid off from both the source node (as per the auction based agreement prior to the cooperation session) and the network service provider. The payoff can be in terms of either cash or credit towards future communications related activities of the relay node. The billing model can vary depending on the service providers.

In case the relay network is the same as the source node network, the billing could be simply to credit the relay node with the prescribed amount and send the invoice to the source node to be paid during the next billing period.

9.3 Cooperation Use Cases

Cooperative ad hoc networking and cooperative relaying, in particular between users of infrastructure networks, is not only a potentially challenging technology and a driver for new services, but also an attractive opportunity for different parties of the mobile value chain. Within this section we provide a three potential use cases for cooperative networks. The use cases are based on the chosen real life examples. In each of the use cases, one can identify three main parties that belong to mobile value chain:

- **Client Consumer**—require data/voice service with minimum energy consumption and possibly also decreased cost of service. The consumer objectives are realized through cooperative connectivity using short-range interfaces.
- **Relay consumer**—seek additional incentive in return for increased power consumption. The incentive is received by sharing the connection through short-range interface cooperation with other devices. The sharing occurs only if legitimate measures are met and/or radio conditions enable effective cooperation (e.g. the device is connected to a power supply).
- **Mobile Network Operator**—provide voice and data services for Client Consumer and Relay Consumer, and enables tethering connectivity between their mobile devices.

The parties and their relations are described through the revenue model. However, the main driver for the introduction and success of cooperative networking/relaying will be related to the potential energy saving gain received in a scenario when a mobile device handovers to a short-range relaying connection through other mobile device. To rationalize this thought, Table 9.1, we compare the power and energy consumption for two wireless interfaces that are envisioned to be available in

Table 9.1 Energy per bit comparison for different wireless access

Technology	Power consumption for voice transmission [mW]	Energy per bit metric for data transmission [μJ/bit]	Average transmission rate [Mbps]
3G—HSPA	1265.7	600	2
UWB	250	2.5	100

The presented figures in the table are gathered and calculated based on the numbers in [10] and [11]

smartphones: 3G-HSPA communication and UWB short-range communication, as shown in Table 9.1. As can be seen from the table, a significant gain can be expected in terms of energy saving for the mobile user. The gain is calculated for iPhone 3G, which has 3.7 V operational voltage and 1150 mAh of battery capacity. In case the power consumption figures from Table 9.1 would be applied to this particular type of a device, the data transfer of 1 Gb using 3G interface would cause 389.21 mAh of energy consumption, which applies to roughly 33 % of battery capacity [9]. However, if the cooperative relaying is applied for the whole transmission time, then energy consumption for the same transfer would deplete to approximately 6 % of the battery lifetime, which gives 6.5 times better energy expenditure.

The obtained results are clearly a motivation for users to seek cooperative transmission. However, proportional incentives need to be rewarded also to the relaying user and Mobile Network Operator (MNO), which needs to be rewarded for spent energy and for enabling the service.

9.3.1 Home Short-Range Tethering Access

"Joe arrives home in the middle of a voice call. When he enters his house, his mobile device discovers using the short-range (WiMedia) interface his docked laptop, which has Mobile Internet dongle and short range interface (WiMedia). The laptop indicates that it is main-powered and is tethering enabled. The devices start cooperation procedure and the network handovers the voice call session (VoIP) to tethering connection. After the voice call session ends, the mobile device stays connected to the network through the laptop, since Joe is using data connection to

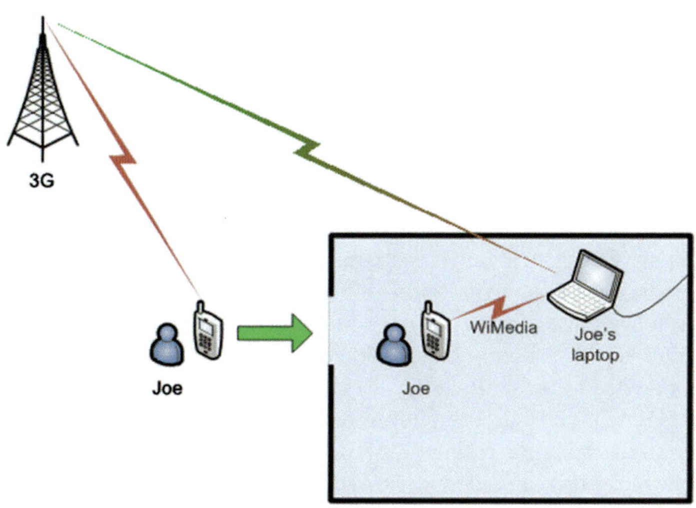

Fig. 9.6 Home short-range tethering access scenario [7]

check his email and browse the Internet". The provided use case is depicted in Fig. 9.6, where a mobile device having voice call connectivity moves to indoor location with laptop that has cooperative relaying service enabled. The connection is handed over to short-range interface and the incoming/outgoing traffic is relayed through the laptop.

The customer owns two devices, which can in favourable conditions (e.g. the Relaying device is main-connected) help each other, as the Relaying device is able to provide energy saving gains to the Client device. From the perspective of the Consumer owning the devices, this is beneficial not only due to energy savings of one device but also because of decreased price per bit of transferred data. The Operator enables cost reduction because this type of connectivity decreases the number of utilized radio links and provides a gain of statistical multiplexing, thus providing better resource utilization at the cell level (more users per cell). Increased number of available resources means either increased revenue during peak hours or less optimization cost.

While the consumer receives decreased price per bit, the MNO benefits from better link utilization and an increased number of available radio resources. The cooperative relaying service enables the consumer to distribute energy consumption between his two devices to enable longer battery lifetime for the mobile device (the Client). However, cooperative relaying can be implemented in the real networks, only if there is a business case for the MNO. In this scenario, the gain of the operator not only comes from the freed radio resources (radio links and channel elements in the baseband). Additionally operator experiences higher link utilization due to statistical multiplexing of traffic, and this enables better network planning and optimization, possibly also in case of dynamic spectrum access systems depletes the chance for secondary user activity in the particular channel. Effectively operator can optimize the network to include cooperative relaying, by e.g. employing lower number of carriers (in case of multicarrier systems), more efficient resource planning, possibly resulting also in the decrease of energy consumption in the network.

9.3.2 Short-Range Tethering Hot Spot

"Dave is in his garden listening to his music phone and decides to purchase and download some new tunes from an on-line music store. The phone's battery is well-charged but it has discovered the relaying-enabled docked laptop, which belongs to Dave's neighbour. To preserve battery power, the phone relays its IP session via the neighbour's device to the cellular network". The considered use case is depicted in Fig. 9.7, where a mobile device having data connectivity decides to utilize short-range cooperative relaying connectivity offered by another device. The incoming traffic is jointly sent to a cellular radio link of laptop device, which then forwards packets destined to the mobile device using short-range interface (WiMedia).

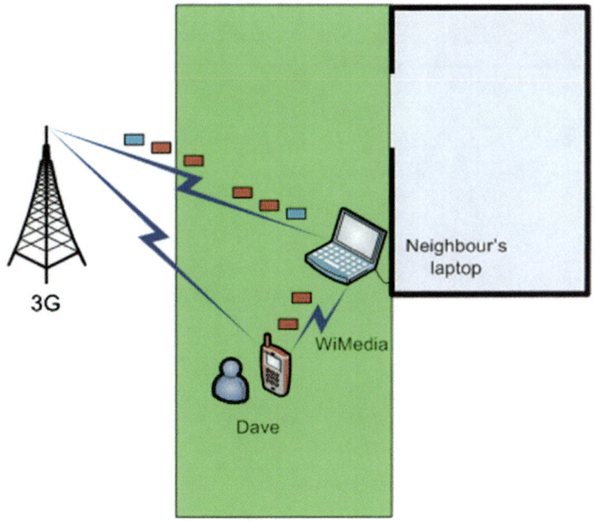

Fig. 9.7 Short-range tethering hot spot scenario [7]

Dave uses data connectivity on his mobile device very sparsely. However, each time he uses data connectivity he downloads large files (even up to 100 MB per session). Thus, he uses "Pay as you go" tariff as his activity is very irregular and he does not want to have monthly plan. Nevertheless, he is aware of the possible energy and money savings (extra Mbits of data transmission) available from exploitation of cooperative relaying; therefore he enables cooperative networking application over short-range interface of his mobile device. His neighbour has Internet Dongle attached to his docked laptop, which also has tethering service enabled (it enables to provide cooperative relaying). The relationships between the parties in this scenario are described in Fig. 9.8. In cooperative relaying, both the Client and the Relay Consumer receive decreased price per bit. At the same time, the MNO benefits from better link utilization and stable income coming from monthly contracts. The cooperation itself enables energy saving for the Client, while the Relay receives additional income (which is observed as a decrease in monthly plan cost).

9.3.3 Short-Range Contract Resellers

"Susan is walking down the street, while making 3G voice call. Her phone battery is critically low. As she comes into range of another phone with short-range capability, the devices set up to relay the call via short-range to the 3G interface of a passer-by with full battery. Optionally due to mobility, the network can also perform handovers to different pedestrians as they move in and out of range until she hangs up the call". The perspective use case is depicted in Fig. 9.9, where a mobile device having an ongoing voice connection moves along a certain pathway. During

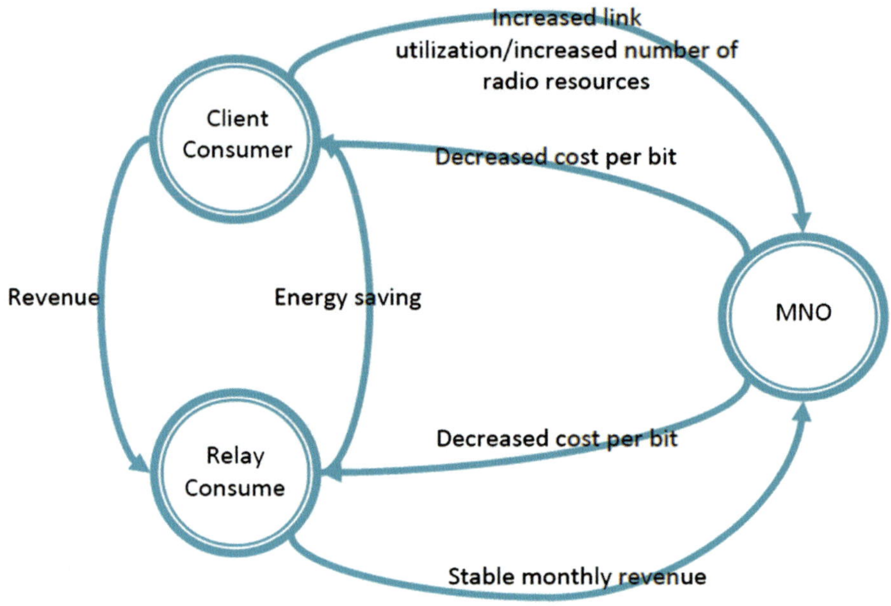

Fig. 9.8 Relationship diagram for tethering hot spot [7]

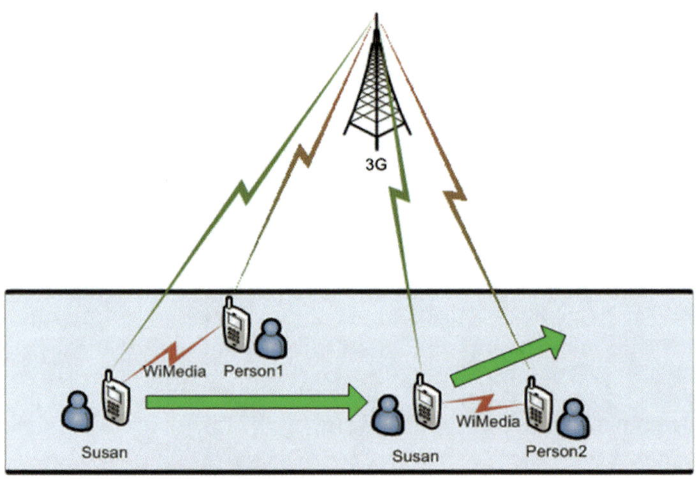

Fig. 9.9 Short-range contract reseller scenario [7]

the movement, the device decides to handover the connection to a short-range relay which then transfers the traffic via its cellular radio link. The handover to cooperative relaying connection can occur multiple times during the connection lifetime.

Each of the Consumers in this use case has the same operator price plan. The operator enables cooperative relaying by including the cooperative application as pre-installed software on the mobile devices supplied by the operator. Thus, utilization of cooperation capabilities is only dependant on the user policies. In case the user would like to exploit such capabilities, he/she would have a choice between two policies available:

- seek cooperation to save battery power or
- offer cooperation in order to receive incentives.

In this approach, Consumers exchange goods using a bartering mechanism based on exchange of price plan minutes. The Client User while being relayed is charged by the MNO according to his price plan (1 min from the price plan per 1 min of connection). However as the connection is relayed, the user, who provides the relaying service, receives an incentive from the operator in the form of a certain percentage p of the relaying time.[2] In such a scenario, the operator is eager to enable incentives for the relaying user as by letting the two users use the same radio links, some cell resources are freed. Furthermore, the revenue of the operator from monthly plans remains untouched (as Consumers are exchanging minutes from their already paid tariffs). The relationships between the parties in this scenario are described in Fig. 9.10. Herein, the parties provide incentives to each other based on indirect reciprocity. Once the Relay enables the Client energy saving, the Operator

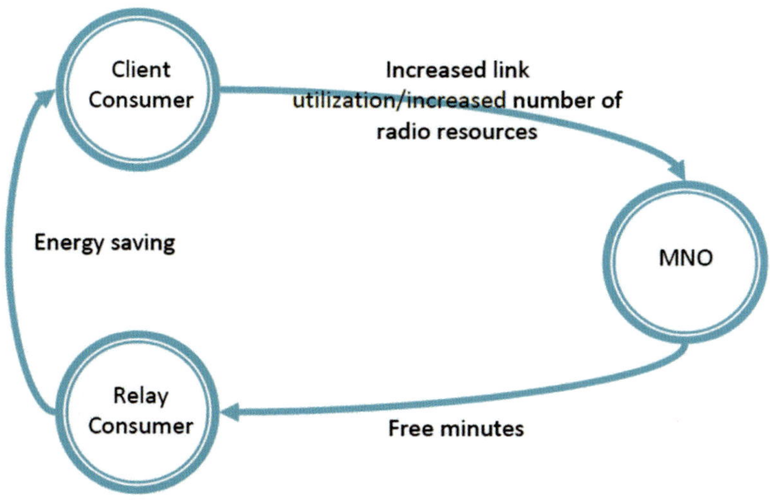

Fig. 9.10 Relationship diagram for contract reseller scenario [7]

[2] If the Relay Consumer has relayed the connection for t minutes he/she will receive a payoff from the operator of p times t free minutes.

provides extra free minutes to the Relay, receiving in return radio resources freed by the Client.

In the presented use cases, cooperation is encouraged not only by potential energy saving but also by a reduction in the cost of connectivity. The possible payoff and decreased cost of connectivity is assumed to be additional motivation for both client and relaying users, where the latter enable statistical multiplexing of traffic at the cellular radio link, at the same time enabling the operators to expand a pool of available radio resources.

9.4 Networks Cooperation Use Cases

Recently a vastly spreading paradigm appeared in mobile networks: the Femto-cells (or Small Cells). The Femto-cells are small base stations, which are expected to revolutionize the concept of indoor network coverage and further intensify in the near future. This is due to increased indoor traffic, such as home, office or schools. According to recent surveys, around 90 % of data services and 2/3 of the phone calls take place in indoor environments. Beyond the higher bit rates, QoS and coverage, Femto-cell can also decrease significantly the energy consumption of mobile terminals when compared with macro cells. For this reason, Femto-cells or small-cells are expected to be more broadly adopted in the future.

9.4.1 Femtocell Cooperation Use-Case

"John is a police officer and has to be reachable all the time for emergency purposes. When John arrives home, he figures out that his mobile terminal has weak signal strength. Moving around inside his house, John finds that the signal strength is getting even weaker in other rooms, where signal is completely lost. So, after verifying the best location at home to get the strongest signal strength, John concludes that the best place to leave the mobile terminal is on a windowsill. However, this meant that he cannot always hear the phone ring, or the receipt of a text message. To overcome this problem, John decides to install a Femto-cell at home. In order to share the costs of the broadband and the Femto-cell access, John decides to implement a hybrid access instead of a close access mode. Consequently after installing this simple device, John verifies that his mobile always has high signal strength anywhere in the house, and some meters outside too".

The use case shows the main reason that motivates users to choose the Femto-cell solution. From the three types of femto cell access, John decides to install the hybrid mode due to the fact that this solutions leads to best outside coverage for sharing its connection, hence bringing him some extra benefits. In fact, the limiting factor of the Femto-cell connection will be the broadband connection. In this particular case, the network operator could update the broadband connection, at no

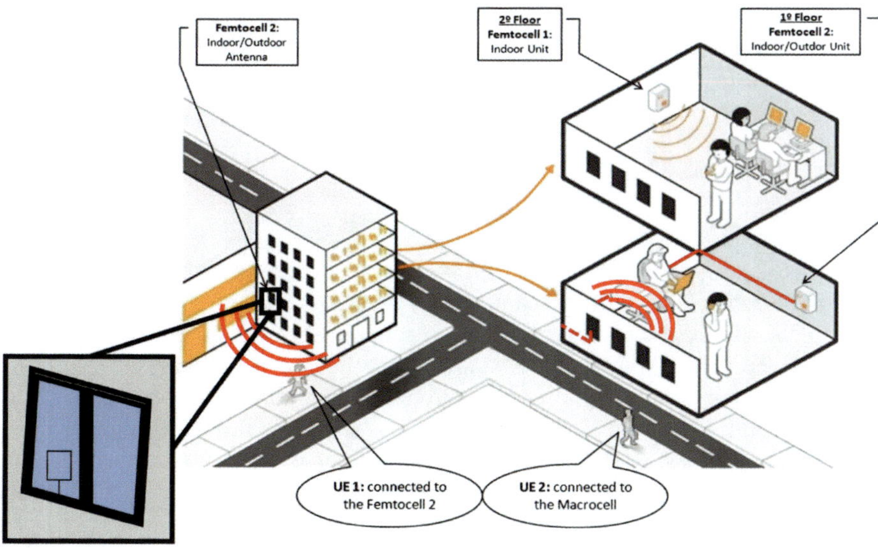

Fig. 9.11 Indoor and outdoor Femto cell coverage [7]

extra cost for John, or could also offer some extra incentives, such as cost reduction in monthly payments of the mobile access by allowing other customer's access to John's Femto-cell. The simple fact of allowing data traffic by outside customers will be converted into revenues for John. The presented use case shows the main benefits that could motivate users to choose hybrid Femto-cells. John could also choose the appropriate pricing option to buy his own Femto-cell.

Femto-cells can be freely deployed at customer homes. However, in certain countries where operators are required, by law, to know the precise location of Femto-cells, it may be necessary to place the Femto-cell near a window, where it can receive a GPS signal. This important information could be used for the business model of hybrid Femto-cell represented in Fig. 9.11. With the requirement to install the hybrid Femto-cell near windows, it is possible to guarantee the outdoor coverage for users in the vicinity of the Femto-cell, assuring the support for users who perform VHO for power saving.

Figure 9.11 depicts the use case after John installation of his own hybrid femto cell (Femto-cell 2) on the first floor of his apartment. The innovative view of outdoor coverage of John's LTE hybrid Femto-cell brings benefits by giving free access to users in the neighbourhood. For example, UE1 is moving in 3G network and discover John's Femto-cell in its vicinity. Even though not belonging to the with list of John's Femto-cell, with the cooperation application enabled, the UE1 automatically evaluates the decision to perform vertical handover to John's Femto-cell for energy saving purposes. In this particular case, UE1 cooperation application decides that the terminal will connect to the Femto cell 2 based on context information. With the available context information, UE1 can tune more efficiently the decision to perform VHO from the actual network connection (3G) and the future Femto-cell connection (LTE) by making some calculations based on power consumption.

9.4.2 Incentive for Subscribers

In this section, some possible business models, that network operators could propose for Femto-cell subscribers to encourage them to open their Femto-cells to non-subscribers, are described. According to recent surveys [12], closed access is the preferred access method of the customers for the home Femto-cells. The main reason is that most of the customers would only accept having a Femto-cell, if they have full control over the list of authorized users. After this, it is logically concluded that business models should follow an aggressive line to convince subscribers. This line may be composed of marketing and/or financial/service offers. All business models have common advantages and disadvantages for operators and subscribers. Apart of the advantages and disadvantages of hybrid Femto-cells, it is interesting to show how this kind of network access affects the Femto-cell users, the client users and the mobile operators.

Figure 9.12 describes the relationship between Femto-cell users (owners of Femto-cell) and the client users who connect to the Femto-cell. With this cooperation, there are two levels of benefits, one for the Femto-cell user and another for the client user. The first level of cooperation is mainly triggered by the client user, when performing handover to the Femto-cell for power saving purposes. To proceed with this transition, the client user must pay an extra bill to the MNO to have access to the target Femto-cell. On the other hand, the Femto-cell user receives some benefits for providing the Femto-cell connection to client users. Finally the MNO will be an intermediate regulator between the client user and the Femto-cell user, by receiving the revenues from the client user and reducing the cost per bit of

Fig. 9.12 Relationship diagram for Femto-cell users [7]

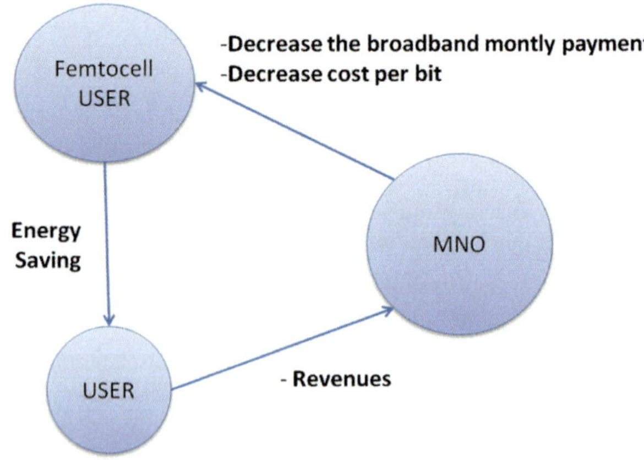

the Femto-cell connection. Below there are some network business model examples that could be applied to introduce the hybrid Femto-cells in the network market.

Free services for subscribers—This model is probably the most aggressive of all, because subscribers have free services. The subscribers accept to share the Femto-cell resources with non-subscribers. In exchange, subscribers have a free service. In this model, as in all others, the important issue to determine is who will pay for the Femto-cell. Depending on the business model used, it can be the subscriber or the operator, but in past experiences such as Wi-Fi, it is usually the client who pays the equipment. Even though the customer pays, there may be a contribution, to lower the price, from the operator side, whose purpose is to encourage the purchase of Femto-cell equipment by customers.

Some free services for subscribers—In this model, subscribers, who accept to share the Femto-cell resources with non-subscribers, are awarded with some benefits. For example, the network operator can reduce the fees for the subscriber broadband access or can offer Femto-cell free phone calls, if they share the Femto-cell connection with more than one non-subscriber at the same time.

Points for connections—In this model, subscribers receive points for each non-subscriber connection. When the subscriber collects a certain number of points, the subscriber is entitled to some free services or free equipments. In this case, subscribers are rewarded by the number of connections made by non-subscribers. The information is controlled by the network operator.

Flat rate—In this model, the subscribers pay a fixed bill, much smaller than the closed model. This bill may vary, depending on whether or not subscribers require a minimum quality of service. In this case, subscribers may accept lower quality of service due to the number of non-subscribers connections or may require a minimum quality of service by paying a little more, compared to no minimum quality of service guarantee situation.

Free broadband—In this model, the network operator offers subscribers the broadband connection that is necessary to connect the Femto-cell, for free. The broadband supplier could be the same as the Femto-cell operator. In this case, it is easier for the network operator to integrate all services and offer the broadband connection. If the broadband supplier belongs to a different company, then the network operator is obliged to negotiate in order to get the best deal for both companies.

9.4.3 Deployment Impact

For each particular mobile system (UMTS, LTE), the average throughput and QoS per user can be increased deploying more small cells (or sectors, cell sectorization) and optimizing the radio network, particularly minimizing the interference. However small cells imply a high cost for the network operators (CAPEX and OPEX) that can jeopardize any expectation of future profits. For example, the cost of site rent can be as high as 20 % of the operator permanent costs (OPEX). One of the most attractive characteristics of the Hybrid access Femto-cell for the mobile operator is that it can provide a high quality service at low expense (no site rent). In fact, if the operator is also a fixed network operator, then it can increase its revenue providing broadband connection that supports Femto-cells. Despite the fact that Femto-cell deployment is typically indoors to overcome the poor indoor radio coverage, other new design should be investigated that allows also outdoor transmission. Femto-cells, that provide indoor and outdoor coverage, could be an interesting hybrid approach to address at the same time the private and the public service. The main drawback of using open access (public) instead of closed access (private) is the high number of HOs that is required to provide continuity of service to a user with high mobility.

9.5 Conclusion

Throughout this book, we presented multiple innovative solutions to tackle the issue of energy efficiency at the mobile handset. Previous chapters have addressed the technical aspect of such solutions, with detailed analysis and evaluation, showing the expected gains in terms of energy savings. On the other hand, in order to implement such solutions in the actual cellular market, it is very important to have solid business models that motivate operators as well as customers to adopt them. The business model must show the costs and benefits for all players involved in the equation; i.e. operators, clients, and maybe service providers. For instance, for network operators, the cost to implement the technology and the revenues after the implementation have to be well defined and measured; additionally, the revenues have to be more than the cost for operators to be interested in implementing such techniques.

This chapter focused on the aspect of business models, which would motivate the players in the value chain of the mobile environment to adopt the innovative solutions for energy savings proposed within the book. The chapter starts with an overview of certain existing business models in the Internet world, which can be easily adopted for the proposed cooperative solutions. The chapter hence identifies the difficulties facing cooperation in the mobile world. It then discusses two business models which can be adopted to encourage energy efficient cooperation among mobile devices. The first model is build around incentive and reputation, while the second model is centralized around a brokerage. The two models are explained in details, with some analytical modelling to highlight the benefits gained from such models.

Moreover, the chapter goes in more details about use-cases for cooperation. Such use-cases are built on actual pricing models of existing mobile operators. The models take advantage of the tethering feature of mobile devices, to save energy while reducing the load on cellular networks and decreasing the cost for operators as well as clients. Three different use-cases are presented, with detailed description of each, in addition to the relationship chain, between the different players, highlighting the cost paid by each player as well as the profits gained.

Finally, the chapter concludes with use-case for network cooperation. The use-case discusses the sharing of Femto-cell among users. The model presents different incentives for users to keep their Femto-cells open for other customers, including free broadband, flat rate, and some other free services. This model benefits the three involved players, namely operator, user (owner of Femto-cell) and other customers (who will use the Femto-cell). The benefits are shown in the model and elaborated with a diagram showing the interconnection among the players.

Without appealing business models, innovative solutions may never make it to actual market; hence the chapter has presented different business models, elaborating how the innovative energy saving solutions proposed throughout the book can be adopted in the actual mobile wireless market.

References

1. Osterwalder, A.: What is a business model? Business Model Alchemist. http://www.businessmodelalchemist.com/2005/11/what-is-business-model.html(Nov 2005)
2. Zhang, Q.: Novel Concepts and Applications of Cooperative Wireless Networking, pp. 25–34. Technical University of Denmark, Denmark (2008)
3. Fitzek, F.H.P., Katz, M., Zhang, Q.: Cellular controlled short-range communication for cooperative P2P networking. Springer Wirel. Pers. Commun. **48**(1), 141–155 (2009)
4. Wrona, K., Mähönen, P.: Cooperative and cognitive networks with reputation and trust. Wirel. Netw., pp. 64–74 (2004)
5. Yu, W., Liu, K.J.: Secure cooperation in autonomous mobile ad-hoc networks under noise and imperfect monitoring: a game theory approach. IEEE Trans. Inf. Forensics Secur. **3**(2), 317–330 (2008)
6. Buchegger, S., Chuang, J.: Encouraging cooperative interaction among network entities. In: Fitzek, Katz (ed.) Cognitive Wireless Networks: Concepts, Methodologies and Visions Inspiring the Age of Enlightenment of Wireless Communications. Springer, Berlin (2007)
7. ICT-248577 C2POWER consortium. D2.3: Business Models and Incentive Mechanisms for Cooperation, Deliverable (D2.3) of C2POWER (2011)
8. Courcoubetis C., Weber, R.: Incentives for large peer-to-peer systems. IEEE J. Sel. Areas Commun. **24**(5), 1034–1050 (2006) (available online)
9. http://www.ehow.com/list_7640171_iphone-3g-charge-specifications.html
10. Perrucci, G.P., Fitzek, F.H.P., Sasso, G., Kellerer, W., Widmer, J.: On the impact of 2G and 3G network usage for mobile phones' battery life. In: Proceedings of European wireless conference, pp. 255–259 (2009)
11. Shor, G.: How bluetooth, UWB, and 802.11 stack up on power consumption. Wisair, http://www.wirelessnetdesignline.com/howto/20720048 (Mar 2008)
12. de la Roche, G., Valcarce, A., López-Pérez, D., Zhang, J.: Access control mechanisms for femtocells. IEEE Commun. Mag. **48**(1), 33–39 (2010)

Printed by Printforce, the Netherlands